P9-CJR-816

International Review of Cytology

A Survey of Cell Biology

VOLUME 139

International Review of Cytology

A Survey of Cell Biology

Edited by

Kwang W. Jeon
Department of Zoology
The University of Tennessee
Knoxville, Tennessee

Martin Friedlander
Howard Hughes Medical Institute
Jules Stein Eye Institute and
Department of Physiology
UCLA School of Medicine
Los Angeles, California

VOLUME 139

Academic Press, Inc.
Harcourt Brace Jovanovich, Publishers
San Diego New York Boston London Sydney Tokyo Toronto

This book is printed on acid-free paper. ∞

Academic Press, Inc.
1250 Sixth Avenue, San Diego, California 92101-4311

United Kingdom Edition published by
Academic Press Limited
24–28 Oval Road, London NW1 7DX

Library of Congress Catalog Number: 52-5203

International Standard Book Number: 0-12-364542-5

PRINTED IN THE UNITED STATES OF AMERICA

92 93 94 95 96 97 EB 9 8 7 6 5 4 3 2 1

CONTENTS

The Cell Biology of Pattern Formation during *Drosophila* Development

Teresa V. Orenic and Sean B. Carroll

Assays of Random Motility of Polymorphonuclear Leukocytes *in Vitro*

Leon P. Bignold

Molecular Origins of Cellular Differentiation in *Volvox* and Its Relatives

Rüdiger Schmitt, Stefan Fabry, and David L. Kirk

Actin Matrix of Dendritic Spines, Synaptic Plasticity, and Long-Term Potentiation

Eva Fifková and Marisela Morales

Role of Signal Transduction Systems in Cell Proliferation in Yeast

Isao Uno

CONTRIBUTORS

Numbers in parentheses indicate the pages on which the authors' contributions begin.

Leon P. Bignold (157), *Department of Pathology, University of Adelaide, Adelaide, South Australia, Australia*

Sean B. Carroll (121), *Howard Hughes Medical Institute and Laboratory of Molecular Biology, University of Wisconsin, Madison, Wisconsin 53706*

Stefan Fabry (189), *Lehrstuhl für Genetik, Universität Regensburg, D-8400 Regensburg, Germany*

Eva Fifková (267), *Department of Psychology, Center for Neuroscience, University of Colorado, Boulder, Colorado 80309*

Isabelle Howald (59), *Institute of Molecular Biology, University of Oregon, Eugene, Oregon 97403*

Taisen Iguchi (1), *Department of Biology, Yokohama City University, Yokohama 236, Japan*

David L. Kirk (189), *Department of Biology, Washington University, St. Louis, Missouri 63130*

Karen E. Moore (59), *Institute of Molecular Biology, University of Oregon, Eugene, Oregon 97403*

Marisela Morales (267), *Department of Psychology, Center for Neuroscience, University of Colorado, Boulder, Colorado 80309*

Teresa V. Orenic (121), *Howard Hughes Medical Institute and Laboratory of Molecular Biology, University of Wisconsin, Madison, Wisconsin 53706*

Christopher K. Raymond (59), *Institute of Molecular Biology, University of Oregon, Eugene, Oregon 97403*

Christopher J. Roberts (59), *Institute of Molecular Biology, University of Oregon, Eugene, Oregon 97403*

Rüdiger Schmitt (189), *Lehrstuhl für Genetik, Universität Regensburg, D-8400 Regensburg, Germany*

Tom H. Stevens (59), *Institute of Molecular Biology, University of Oregon, Eugene, Oregon 97403*

Isao Uno (309), *Life Science Research Center, Nippon Steel Corporation, Kawasaki, Japan*

Cellular Effects of Early Exposure to Sex Hormones and Antihormones

Taisen Iguchi
Department of Biology, Yokohama City University, Yokohama, Japan

I. Introduction

The discovery that the simple hydrocarbons, stilbene and triphenylethylene, are weakly active estrogens in the mouse (Robson and Schonberg, 1937) led to the synthesis of a potent nonsteroidal estrogen, diethylstilbestrol (DES) (Dodds *et al.*, 1938a,b) and a nonsteroidal antiestrogen, ethamoxytriphetol (MER-25) (Lerner *et al.*, 1958). MER-25 is antiestrogenic in all species tested and has antifertility activity in rats (Lerner *et al.*,1958; Segal and Nelson, 1958; Chang, 1959). However, because of its low potency and its side effects in humans (Kistner and Smith, 1959, 1961; Smith and Kistner, 1963), other derivatives of triphenylethylene, namely, clomiphene (Holtkamp *et al.*, 1960) and tamoxifen (Harper and Walpole, 1966), have been synthesized, which are more potent than MER-25 but have some estrogenic properties in laboratory animals. Clomiphene has antifertility properties in animals but stimulates ovulation in women (Greenblatt *et al.*, 1961, 1962; Huppert, 1979; Clark and Markaverich, 1982). Tamoxifen is a complete estrogen antagonist in the chick oviduct (Sutherland *et al.*, 1977), a partial agonist with antiestrogenic activity in the immature and ovariectomized rat (Harper and Walpole, 1967; Jordan and Koerner, 1976), and a full estrogen in the immature and ovariectomized mouse (Harper and Walpole, 1966; Terenius, 1971; Jordan *et al.*, 1978). Antiestrogens can inhibit the estradiol-stimulated increase in uterine wet weight (Harper and Walpole, 1967); however, they simultaneously induce hypertrophy of uterine luminal epithelial cells (Kang *et al.*, 1975) and progesterone receptor synthesis (Dix and Jordan, 1980; Jordan and Gosden, 1982). In the 1960s, with the belief that some human breast tumors are directly dependent on estrogen for growth, several antiestrogens, including clomiphene (Herbst *et al.*, 1964), nafoxidine (EORTC, 1972), and tamoxifen (Cole *et al.*, 1971; Baum *et al.*, 1985), were tested clinically as agents for breast cancer therapy; only tamoxifen treatment proved to be at least as effective as other endocrine therapies.

DES is an extremely potent nonsteroidal estrogen (Dodds *et al.*, 1938a,b). The compound was recommended for preventing miscarriages as well as some of the later complications of pregnancy (Smith, 1948). Some investigators reported beneficial effects of DES in preventing pregnancy complications while others did not (Dieckmann *et al.*, 1953). In 1970 Herbst and Scully reported on vaginal tumors found in eight adolescent women, seven of whom were born from mothers who had received DES for prevention of threatened abortion during pregnancy (Herbst and Scully, 1970). Clear cell adenocarcinomas of the vagina and cervix frequently accompanied by adenosis are found in young women exposed to DES *in utero* (Herbst *et al.*, 1971, 1975; Johnson *et al.*, 1979; for reviews see Herbst and Bern, 1981; Mori and Nagasawa, 1988).

The study of permanent changes in target organs induced by sex hormones administered during a critical period of development began with experiments on the neonatal mouse treated with estrogen by Takasugi *et al.* (1962). The vaginal epithelium of mice treated neonatally with 17β-estradiol showed persistent proliferation and cornification, which were not abolished by adrenalectomy and hypophysectomy following ovariectomy (Takasugi, 1963). This persistent vaginal cornification was thus estrogen-independent and could be induced by estrogen, or by aromatizable or nonaromatizable androgens only when the treatment started within 3 days after birth, indicating the presence of a critical period (Kimura *et al.*, 1967; Takasugi *et al.*, 1970; Iguchi and Takasugi, 1976; Tanaka and Takasugi, 1982). The vaginal epithelium showing estrogen-independent persistent proliferation and cornification frequently resulted in precancerous and cancerous lesions (Dunn and Green, 1963; Takasugi and Bern, 1964; Kimura and Nandi, 1967). Neonatal treatment of male mice with estrogen or androgen induces persistent suppression of spermatogenesis and atrophy of seminal vesicles and prostates (Ohta and Takasugi, 1974). Effects of perinatal exposure to sex hormones on the reproductive system of female and male animals have been reviewed by Bern *et al.* (1976), Bern (1979, 1991), Takasugi (1976, 1979), Forsberg (1979), Bern and Talamantes (1981), Arai *et al.* (1983), Newbold and McLachlan (1988), Mori and Iguchi (1988), and Kincle (1990). In addition, various abnormalities have been demonstrated in rats and mice treated neonatally with antihormones (Iguchi *et al.*, 1986a).

This review presents an update on the genital effects of perinatal exposure to sex hormones, including DES, and antihormones and also the induction of nongenital abnormalities, with special consideration of the cellular basis for the changes observed.

II. Effects of Perinatal Exposure to Sex Hormones on Genital Organs

A. Genital Abnormalities in Mice, Guinea Pigs, and Rats

Short-term administration of large doses of sex hormones to rodents during a critical period results in persistent changes in the hypothalamo-hypophysio-gonadal system and in reproductive organs that develop precancerous or cancerous lesions with age (for reviews see Takasugi, 1976, 1979; Bern *et al.*, 1976; Bern, 1979, 1992; Bern and Talamantes, 1981; Forsberg, 1979; Herbst and Bern, 1981; Walker, 1984; Johnson, 1987; Mori and Iguchi, 1988; Takasugi and Bern, 1988). Early exposure of mice to estrogens also induces hyperplastic and dysplastic changes, as well as neoplastic changes, in the mammary glands that show increased sensitivity to hormones and to carcinogens in the adult (for reviews see Mori *et al.*, 1980; Bern and Talamantes, 1981; Bern *et al.*, 1985). Levay-Young and Bern (1989) showed that neonatally estrogen-exposed mouse mammary epithelial cells are generally less sensitive to prolactin and growth hormone in casein production *in vitro*. Rothschild *et al.* (1987) showed that prenatal DES exposure alone (0.8–8 μg on days 15 and 18 of gestation), or postnatal DES treatment alone, or the combination of the two yielded significantly high mammary tumor incidence and decreased tumor latency in ACI rats.

The neonatal mouse is a useful animal model for analyzing the possible reproductive consequences of transplacental exposure of the human fetus to sex hormones (Takasugi, 1976; Bern *et al.*, 1976). Bern *et al.* (1987) examined the relation of the dosage of DES (10^{-5} to 5×10^{-1} μg/day for 5 days from the day of birth) administered neonatally to the incidence and the severity of genital tract and mammary gland lesions and to the levels of sex hormone receptors in BALB/cCrgl mice. Neonatal doses of $5 \times 10^{-2} \mu$g DES resulted in vaginal lesions and uterine metaplasia at 2 months of age. With age this threshold level decreased, implying interaction with an altered hormonal milieu. The ovary (absence of corpora lutea) and mammary glands (hyperplastic alveolar nodules and abnormal secretory state) were 10- to 100-fold more sensitive to neonatal DES exposure than vagina and uterus. Iguchi *et al.* (1988b) demonstrated that subepithelial nodules of polygonal cells in the Müllerian vagina during postnatal life (Takasugi, 1976; Iguchi *et al.*, 1976; Mori *et al.*, 1983) were associated with the later occurrence of ovary-independent persistent stratification in BALB/cCrgl mice treated neonatally with 0.1–10 μg DES/day. The thresholds for the induction of ovary-independent pegs or downgrowths and of adenosis (gland formation) in neonatally DES-exposed mice were 0.1 μg/day and 0.5 μg/day DES, respectively. Daily injections of female BALB/cCrgl mice with 3–10 μg of DES for 5 days from the day of birth resulted in squamous metaplasia in the uterine epithelium (Iguchi *et al.*, 1987b). Three major cervicovaginal abnormalities, namely, adenosis, aberrantly simplified cervical lumen, and twin fornices, were induced by neonatal injections of DES or androgens (Bern *et al.*, 1984b). Adenosis-like lesions occurred in vaginae of newborn mice transplanted into C57BL and BALB/c mice for 30 days. These lesions were located in Müllerian-derived genital tract regions, cervix and/or fornix, and in the middle vagina but never in the urogenital sinus-derived portion of the vagina, after month-long exposure to endogenous ovarian hormones or exogenous estradiol from the day of transplantation (Iguchi *et al.*, 1985b). Vaginal grafts in ovariectomized hosts did not exhibit adenosis, suggesting its dependence on estrogen. Iguchi *et al.* (1986e) also demonstrated that an ovarian hormone (probably estrogen) participates in the prepubertal induction of adenosis-like lesions in prenatally DES-exposed mice. The incidence of ovary-independent adenosis-like lesions was low in neonatally DES-treated BALB/c mice (Iguchi *et al.*, 1988b). Estrogen-independent vaginal proliferation and cornification induced by neonatal treatment with estradiol, DES, or androgen could be suppressed by neonatal injections of retinoids (Takasugi, 1976; Iguchi and Takasugi, 1979; Tachibana *et al.*, 1984a,b; Iguchi *et al.*, 1985a). Jones *et al.* (1984) also showed that instillation of progesterone into the vaginal lumen of mice treated neonatally with 17β-estradiol resulted in a significant decrease in vaginal concretions, cornification, and lesions. On the contrary, long-term intravaginal implantation of a vaginal concretion significantly increased the incidence of cervicovaginal abnormalities in mice treated neonatally with estrogen (Bern *et al.*, 1984a). Uterine adenocarcinoma in CD-1 mice exposed neonatally to DES for 5 days occurred in a time- and dose-related manner; at 18 months neoplastic lesions were seen in 90% of the mice exposed neonatally to 2 μg DES. These DES-induced uterine tumors were estrogen-dependent (Newbold *et al.*, 1990). Mice exposed neonatally to DES, or to the DES analogs, hexestrol and tetrafluoro-DES, developed uterine adenocarcinoma by 8–12 months of age. The estrogenic potency of the compounds were ranked in the order of hexestrol > trifluoro-DES > DES > estradiol (Newbold *et al.*, 1990).

Transplacental exposure to 0.01–100 μg DES/kg on days 9–16 of gestation resulted in various abnormalities in genital organs, including vaginal adenosis, adenocarcinomas, uterine cancers, oviducal malformation, and ovarian cysts in female mice, and epididymal cysts, inflammation, cryptorchidism, and lesions in the epididymis, seminal vesicles and/or prostate glands, Müllerian remnants, and testicular adenocarcinomas in male mice (McLachlan, 1979; Newbold and McLachlan, 1985, 1988; Newbold *et al.*, 1985a,b,1987a,b; Haney *et al.*, 1986; Bullock *et al.*, 1988). Ozawa *et al.*(1991a) demonstrated that nodules of polygonal cells reacting to epidermal growth factor (EGF) antibody in the Müllerian vagina during postnatal life were associated with the later occurrence of ovary-independent persistent stratification in ICR/JCL mice treated prenatally with 2–2000 μg DES/day from days 15 to 18 of gestation. The threshold for the induction of ovary-independent stratification in mice was 2 μg DES/day. Suzuki and Arai (1986) demonstrated that prenatal exposure to DES (4 μg, from days 10 to 17 of gestation) caused an atrophy of the gubernaculum testis and suggested that the atrophic gubernaculum was associated with the intraabdominal position of the testes. Stumpf *et al.* (1980) showed that radiolabeled DES injected prenatally to males was concentrated in the gubernaculum testis rather than in the testis, thus acting directly on the developing gubernaculum testis to cause its regression. In mice high level expression of estrogen receptor was found in the undifferentiated cells that comprise the mesenchymal core of gubernaculum in early development (T. Sato and T. Iguchi, unpublished observations).

Continuous intravenous infusion of human chorionic gonadotropin (hCG) into pregnant mice from days 15 to 19 of gestation induced disorganization of the germinal epithelium and impaired spermatogenesis in male offspring and some persistent vaginal cornification in female offspring (Takasugi *et al.*, 1985). Persistent anovulation was also induced in female mice given long-term administration of hCG from the day of birth (Iguchi *et al.*, 1986c).

Female guinea pigs exposed transplacentally to 2.5 μg DES/day for 32–37 days, but not to 50 μg/day estradiol, from the 28th day of pregnancy to term resulted in acceleration of genital tract differentiation and cystic glandular hyperplasia of the endometrial glands and squamous metaplasia. However, DES exposure for less than 15 days before term or after birth for 3 days failed to induce abnormal changes in the adults; therefore, the critical period for induction of genital tract abnormalities in the guinea pig lies from the 28th to about the 45th day of gestation (Davies *et al.*, 1985; Davies and Lefkowitz, 1987). Prenatal DES injections (0.8–8 μg on days 15 and 18 of gestation) induced atypical epithelia, cystically dilated uterine glands, and thickened vaginal epithelium in ACI rats (Rothschild *et al.*, 1988). In these prenatally DES-exposed rats, later implantation of DES pellet increased the incidence of squamous metaplasia of the uterine epithelium and of cystic uterine glands. Baggs *et al.* (1991) demonstrated that *in utero* DES exposure (0.1–50 mg/kg body weight on days 18–20 of gestation) induced vaginal tumors (adenocarcinoma, squamous cell carcinoma, and mixed carcinoma) in a dose-related manner in Wistar rat. Tumors of other reproductive tissues showed no discernible DES dose–response relationship. They also showed that oral contraceptive treatment did not increase the risk of neoplasia in prenatally DES-exposed rats.

B. Changes in Receptor Systems

In 1- to 12-month-old BALB/cCrgl mice treated neonatally with 10^{-1} to 1 μg DES/day for 5 days from the day of birth, mammary gland development was inhibited (Tomooka and Bern, 1982) and estrogen receptor levels were decreased (Bern *et al.*, 1985). In the vagina and uterus of mice treated neonatally with estradiol or DES, cytosolic estrogen receptors were consistently decreased; however, cytosolic progestin receptors were increased (Shyamala *et al.*, 1974; Aihara *et al.*, 1980; Bern *et al.*, 1987). The localization of steroid receptor levels changed in the two tissue compartments (epithelium and fibromuscular wall) of the neonatally DES-exposed mouse vagina (Eiger *et al.*, 1990): neonatal DES exposure caused a marked decrease in total estrogen binding in the epithelium, an increase in nuclear estrogen binding in the fibromuscular wall, a decrease in cytosolic estrogen binding, and an increase in cytosolic progestin binding in both the epithelium and the fibromuscular wall. Edery *et al.* (1989) showed that neonatal injections of 5 μg testosterone for 5 days and/or continuous subsequent exposure to implanted Silastic capsules containing testosterone lowered estrogen and progestin receptors in both the uterus and mammary gland in mice, indicating that testosterone is a negative modulator of progestin receptors in these tissues. In the uterus of 6-week-old CFY rats given a single injection of 8.8 μg DES or 17.5 μg allylestrenol, the number of Type II binding sites for estradiol was consistently decreased without altering receptor affinity, suggesting that sex hormones are able to induce hormonal imprinting during the critical stage of receptor maturation (Csaba *et al.*, 1986). In male mice exposed neonatally to 1 μg DES, significant decreases were seen in organ weights and protein levels in all glands. Neonatal DES exposure caused a significant decrease in cytosolic androgen and cytosolic and nuclear estrogen receptor levels in the anterior prostate and in cytosolic estrogen receptor levels in the ventral prostate, with a significant increase in cytosolic estrogen receptor levels in the seminal vesicle of DES-exposed mice (T. Turner *et al.*, 1989). Edery *et al.* (1990) showed that neonatal DES exposure caused significant decreases in prolactin receptor levels in the seminal vesicle, ductus deferens, and anterior and ventral prostates. Nelson *et al.* (1991) demonstrated that EGF has estrogen-like effects in the promotion of cell growth and differentiation in the mouse uterus and vagina and that EGF may serve as an important mediator of estrogen action *in vivo* because an antibody specific for EGF significantly inhibited estrogen-induced uterine and vaginal growth. Iguchi *et al.* (submitted) found that neonatal treatment of BALB/cCrgl female mice with 1 μg DES reduced EGF receptor levels in the vagina but not in other organs (uterus, oviduct, ovary, and liver). On the basis of these findings, it appears that ovary-independent vaginal proliferation is correlated with the reduction of estrogen and EGF receptors.

C. Changes in Cell Growth *in Vitro*

Cell culture is a powerful method for studying the factors involved in the growth and differentiation of various organs under defined conditions. A collagen gel culture system has been developed to grow epithelial cells of mammary gland (for review see Yang and Nandi, 1983; Imagawa *et al.*,1990), vagina (Iguchi *et al.*, 1983, 1987c; Tsai *et al.*, 1991), uterus (Iguchi *et al.*,1985c; Uchima *et al.*, 1991a), and prostate gland (T. Turner *et al.*, 1990). Using serum-free collagen gel culture system, Tomooka *et al.* (1983) demonstrated that mammary epithelial cells from 2- to 3-month-old BALB/cCrgl mice treated neonatally with DES showed a reduced growth response to insulin, EGF, and lithium ion than control cells. Levay-Young and Bern (1989) showed that mammary epithelial cells from DES-exposed ovariectomized mice accumulated less casein content of normal cells when cultured in released collagen gels with a serum-free medium containing prolactin. Combinations of prolactin and growth hormone enhanced the difference between casein accumulation in DES-exposed and control cells, and DES-exposed cells were much less responsive to growth hormone. Uchima *et al.* (1991b) demonstrated that vaginal epithelial cells from DES-exposed ovariectomized 40-day-old mice showed decreased initial rate of proliferation and decreased sensitivity to insulin and EGF than control cells in serum-free collagen gel culture. DES-exposed vagina showed greater sensitivity to collagenase digestion than the control tissue. Estrogen did not stimulate proliferation but rather inhibited vaginal epithelial cells from both control (Iguchi *et al.*, 1983, 1987c) and DES-exposed mice (Uchima *et al.*, 1991b), although estrogen promoted an accumulation in estrogen receptor in the nucleus and induction of progestin receptors in normal cells (Uchima *et al.*, 1987). Vaginal epithelial cells from DES-exposed 21-day-old mice showed a reduced growth response to EGF and an increased growth response to insulin compared with control cells, and transforming growth factor-β inhibited cell growth from both DES-exposed and control mice (Ozawa *et al.*, 1991b). Turner and Bern (1990) demonstrated that DES-exposed ventral prostatic epithelial cells cultured in serum-free collagen gel showed a longer maintenance of the initial plating density (lag in growth) than control cells. Neonatal DES exposure resulted in two colony types: one similar to colonies arising from control cells and one that appears to be nongrowing, suggesting heterogeneous populations of cells. Keratinization was observed in some DES-exposed anterior prostatic epithelial cell colonies. EGF was required for growth of prostatic epithelial cells from both DES-exposed and control mice. T. Magaña and H. A. Bern (personal communication) recently demonstrated that DES-exposed ventral prostatic epithelial cells were more sensitive to various concentrations of EGF than control cells. These *in vitro* studies reveal an alteration in the responsiveness to hormones and growth factors of DES-exposed epithelial cells. Some of these alterations are different from those demonstrated *in vivo* or organ culture (Levay-Young and Bern, 1989), both cases in which the normal stroma is present, and suggest that the presence of stromal cells alters the responsiveness of epithelial cells to hormones and growth factors. Cultures of stromal cells of the respective organs from DES-exposed mice to determine their responsiveness and to use them in a co-culture system might reveal other differences between normal and DES-exposed organs.

D. Effects of Enzyme Inhibitors in Induction of Changes

Perinatal injections of estrogens, including DES and aromatizable or non-aromatizable androgens, induce ovary-independent vaginal and uterine changes in female mice and impaired spermatogenesis in male mice (Iguchi and Takasugi, 1976, 1981, 1983, 1987; Ohta and Iguchi, 1976; Iguchi and Ohta, 1979; Takasugi *et al.*, 1983; Tanaka *et al.*, 1984; Iguchi *et al.*, 1985 a,b, 1986b,e). Exposure of neonatal vaginae to estradiol, DES, or 5α-dihydrotestosterone (DHT) *in vitro* caused estrogen-independent epithelial proliferation (Iguchi, 1984). Using inhibitors of 5α-reductase and aromatase, Iguchi *et al.* (1988c) also demonstrated that development of persistent anovulation is due to the direct action of estrogen converted from testosterone on the neonatal hypothalamohypophyseal system. However, testosterone itself induced ovary-independent vaginal changes, although 5α-reduced androgen and testosterone-derived estrogen seemed to be more effective.

E. Polyovular Follicles

The spontaneous occurrence of polyovular follicles (PFs) in the ovaries has been reported in intact mice (Engle, 1927; Takewaki, 1937; Fekete, 1950; Kent, 1960), rats (Sigemoro, 1947; Sigemoro and Makino, 1947; Davis and Hall, 1950; Dawson, 1951), hamsters (Kent, 1959), dogs (Telfer and Gosden, 1987), and several other species of mammals, including humans (Arnold, 1912; O'Donoghue, 1912; Corner, 1923; Hartman, 1926; Dederer, 1934). PFs were rarely found in most mouse strains (Fekete, 1946, 1950; Kent, 1960). However, neonatal injections of DES or 17β-estradiol were found to cause an increased occurrence of PFs with 2–23 oocytes per follicle in ovaries of immature mice (Forsberg *et al.*, 1985; Iguchi, 1985; Iguchi *et al.*, 1986d) in addition to abnormalities of the oviduct and ovary in rats and mice exposed to DES perinatally (Newbold *et al.*, 1983a,b; Wordinger and Highman, 1984; Wordinger and Morrill, 1985). PF incidence (percentage of PF per ovary) and PF frequency (percentage of mice with PF) were significantly greater in 34-day-old BALB/cCrgl mice receiving neonatal five daily injections of 0.1–2 μg DES, 100 μg progesterone, 137 μg 17α-hydroxyprogesterone caproate, and 20 μg testosterone, but not in mice receiving 20 μg 5α-DHT, than in the controls. In DES-treated mice, PF incidence was 120–340 times higher than in the controls. In 30-day-old C57BL/Tw mice treated neonatally for 5 days with a daily dose of 50 μg testosterone, 20 μg estradiol, or 1 μg DES, PF incidence also increased by 2–50 times. However, 50 μg of 5α- and 5β-DHT failed to increase PF incidence. PF incidence increased gradually from 10 to 30 days of age only when neonatal DES treatment was begun at 0–3 days (Iguchi *et al.*, 1986d), although PF occurrence was reported in squirrel monkeys by prolonged DES implantation in adults (Graham and Bradley, 1971). Natural and synthetic estrogens induced a higher PF incidence than did aromatizable androgen (Iguchi *et al.*, 1986d). Simultaneous injections of an aromatase inhibitor lowered the induction of PF by testosterone, indicating that testosterone enhanced PF formation as a result of its conversion to estrogen (Iguchi *et al.*, 1988c). The minimum daily dose of DES necessary to induce PF was 10^{-2} μg in ICR mice and 10^{-3} μg in C57BL mice (Iguchi, 1985). PF also occurred in the ovary of 30-day-old offspring of ICR mice given four daily injections of 20–2000 μg DES from days 15 to 18 of gestation. PF incidence in prenatally DES-exposed offspring was increased 33–112 times compared with controls. PF appeared as early as at 5 days in DES-exposed mice, and the incidence increased linearly until 30 days of age (Iguchi and Takasugi, 1986). PF was induced in all strains of mice examined (C57BL, C3H, BALB/c, ICR, and SHN) and in T strain rats by neonatal DES exposure. There was a negative correlation between PF incidence in DES-exposed, 30-day-old mice and incidence of onset of luteinization in DES-unexposed control mice at 30 days (Iguchi *et al.*, 1987a). A high incidence of PF was also found in newborn mouse ovaries transplanted for 30 days into ovariectomized adult hosts given DES injections. When neonatal ovaries were cultured in a serum-free medium containing DES for 5 days and then transplanted into ovariectomized hosts, PFs were formed in the grafts. PFs induced by neonatal DES exposure were ovulated by injections of pregnant mare's serum gonadotropin and hCG. These findings indicate that neonatal ovaries exposed to estrogen *in vivo* or *in vitro* are capable of responding to gonadotropins given later, resulting in a reduction of PF incidence (Iguchi *et al.*, 1990a). These ovulated ova from PFs were capable of fertilization after insemination, although the percentage of fertilized ova was smaller than in ova from uniovular follicles of DES-exposed and control mice (Iguchi *et al.*, 1991a). Halling and Forsberg (1990b) showed that ova from neonatally DES-exposed mouse ovaries grafted to control females gave rise to normal living offspring. In DES-exposed mice, a few zygotes developed only to the 4-cell stage. Under *in vitro* condition, zygotes from DES-exposed mice developed into blastocysts and to the implantation stage, but the incidence of these stages was lower than with zygotes from controls (Halling and Forsberg, 1991).

Wordinger and Derrenbacker (1989) demonstrated that a single injection of DES (10 μg/kg on day 15 of gestation) decreased the time between the stages of follicular development, resulting in a greater number of developmentally advanced stage of follicles during neonatal ovarian development. Wordinger *et al.* (1989) also demonstrated that ovaries of mice exposed prenatally to DES are capable of responding to gonadotropins and that the second-generation progeny of prenatally DES-exposed mice have the potential for normal development, suggesting that *in vivo* decline in developmental potential may be attributable to genital tract abnormalities rather than ovum/embryo defects. Menczer *et al.* (1986) reported that 22 of 40 women exposed to DES *in utero* had primary infertility, showing a higher rate of anatomical structural defects. After treatment with ovulation-stimulating drugs, spontaneous abortion and tubal pregnancy were frequent in both fertile and infertile groups of DES-exposed women. About 30% of the infertile women had mild hyperprolactinemia.

F. Changes in Gonadal Steroidogenesis

Ovaries of 3- to 14-month-old mice exposed prenatally to DES (100 μg/kg body weight on days 9–16 of gestation) showed an increased *in vitro* production of estrogen, progesterone, and testosterone, the value of which was calculated per milligram of tissue; however, only testosterone production remained elevated "per ovary" (Haney *et al.*, 1984). Steroid synthesis in ovarian homogenates from 8-week-old NMRI mice given five daily injections of 5 μg DES was examined using [^{3}H]-pregnenolone as precursor (Tenenbaum and Forsberg, 1985; Halling and Forsberg, 1990a). Homogenates from neonatally DES-exposed mouse ovaries showed increased synthesis of progesterone and androstenedione, whereas the synthesis of 17

α-hydroxyprogesterone and testosterone was reduced (Tenenbaum and Forsberg, 1985). Plasma concentrations of testosterone, but not progesterone, were significantly lower in DES-exposed females than in controls (Halling and Forsberg, 1989). Synthesis of progesterone and testosterone in the females was demonstrated at 6 days of age, androstenedione at 12 days, 17α-hydroxyprogesterone at 21 days, and estradiol at 28 days. After 28 days progesterone, androstenedione, and estradiol were higher in the homogenates of the DES-exposed ovaries than in the control homogenates, whereas 17α-hydroxyprogesterone and testosterone were higher in controls (Halling and Forsberg, 1990a). In 60-day-old C57BL/Tw mice, levels of 17α-hydroxyprogesterone, androstenedione, and estradiol were higher in the homogenates of neonatally DES-exposed ovaries than in the controls. Only estradiol was lower in the homogenates of neonatally DES-exposed testes than in the controls. In 60-day-old C57BL female mice treated neonatally with tamoxifen, levels of testosterone and estradiol were higher in the homogenates of ovaries than in the control ones. In the testicular homogenate, only estradiol production was lower in the controls (Y. Uesugi and T. Iguchi, unpublished observations). These findings indicate that steroidogenesis of ovary and testis is impaired directly by neonatal exposure to DES or tamoxifen, as well as indirectly through alterations of the hypothalamohypophyseal system.

G. Molecular Changes

Biochemical and molecular changes after perinatal exposure to DES were reported in reproductive organs of mice (Newbold *et al.*, 1984; Pentecost *et al.*, 1988). Newbold *et al.* (1984) showed that prenatal DES exposure (100 μg/kg body weight) resulted in the disappearance of a protein (70 kDa) in the fetal reproductive tract, providing an early marker for alterations in normal genital tract function. However, it has been demonstrated that lactotransferrin (lactoferrin) is an estrogen-inducible secretory protein in mouse uteri (Teng *et al.*, 1986, 1989; Pentecost and Teng, 1987). Lactotransferrin mRNA is constitutive and estrogen-inducible in seminal vesicles; amounts were negligibly small in control mice (Pentecost *et al.*, 1988). After castration treatment with 17β-estradiol for 3 days induced the lactotransferrin mRNA in the seminal vesicle of both control and prenatally DES-exposed mice; however, the levels in DES-treated tissues were approximately six-fold higher than those in control tissue (Newbold *et al.*, 1989). These findings suggest that hormonal manipulation in development induces alterations in mRNA and protein expression.

Normand *et al.* (1990) showed that neonatal administration of 10 μg estradiol to male CD-1 mice reduced protein concentrations to 39–56% in epididymis, ductus deferens, and seminal vesicle in the adult. The protein profiles were persistently altered. In the epididymis of estrogenized males, three protein bands were differentially increased and one was reduced. In the ductus deferens, four proteins were increased and one was virtually absent. In the seminal vesicle, 20 proteins were increased or decreased. Testosterone replacement at adulthood was unable to reverse these effects. Treatment with estradiol during adult life induced persistent alterations in the protein profiles of the three organs, but these alterations were reversed by androgen therapy.

In neonatally 1 μg DES-exposed BALB/cCrgl female mice, expression of two proteins was increased and that of three proteins was decreased in the separated vaginal epithelial cells relative to the control epithelium. In the DES-exposed vaginal fibromuscular walls, expression of nine proteins was increased and that of 21 proteins was decreased (Uchima *et al.*, 1990). Takamatsu *et al.* (1990, 1992a,b,c) examined protein profiles by two-dimensional gel electrophoresis, and demonstrated that two proteins specifically appeared and one protein specifically disappeared in the vagina showing ovary-independent irreversible proliferation and cornification in neonatally 2 μg/day DES-exposed C57BL/Tw mice compared with vaginae of postpubertally 0.1 μg/day DES-exposed mice and the controls. In the uterus the protein expression pattern of neonatally DES-exposed mice closely resembled that of the controls except for one protein that was specifically increased. These findings indicate that long-term tissue-specific alterations in the synthesis of a select number of cellular proteins occur in neonatally DES-exposed mice.

III. Genital Abnormalities Induced by Perinatal Exposure to Phytoestrogens

The possible effects of dietary "environmental" estrogens in humans and laboratory rodents during critical periods of reproductive development are a matter of considerable concern (Verdeal and Ryan, 1979; Horwitz *et al.*, 1983; Burroughs *et al.*, 1985, 1990a; Williams *et al.*, 1989). The predominant dietary estrogens are naturally occurring phytoestrogens, such as coumestrol isolated from ladino clover (Bickoff *et al.*, 1957, 1958) and zearalenone isolated from *Fusarium* (McNutt *et al.*, 1928; Bennett and Shotwell, 1979). Coumestrol stimulates uterine enlargement (Bickoff *et al.*, 1962), decreases ovulatory rate (Leavitt, 1965), and increases embryo degeneration in pregnant mice (Fredericks *et al.*, 1981). Zearalenone expands the neonatal rat uterus (Sheehan *et al.*, 1984). Coumestrol (Martin *et al.*, 1978) and zearalenone (Boyd and Wittliff, 1978; Kiang *et al.*, 1978;

Greenman *et al.*, 1979; Katzenellenbogen *et al.*, 1979; Powell-Jones *et al.*, 1981; Fitzpatrick *et al.*, 1989) bind to estrogen receptors with less affinity than 17β-estradiol and DES.

The toxic effects of zearalenone vary with the species. Chicken, rat, and pig were relatively insensitive, moderately sensitive, and relatively sensitive to dietary zearalenone, respectively (Allen *et al.*, 1981; Etienne and Jemmali, 1982; Kumagai and Shimizu, 1982). Fitzpatrick *et al.* (1989) determined the relative binding affinity of zearalenone and its metabolites, α-zearalenol and β-zearalenol, for uterine and oviduct estrogen receptors in pig, rat, and chicken. The relative binding affinity of α-zearalenol was 10–20 times greater than that of zearalenone and 100 times greater than that of β-zearalenol; the affinity of α-zearalenol was greater in the pig than in the rat and the chicken, which may partially explain the interspecies differences in sensitivity to dietary zearalenone.

Neonatal treatment of female rats with coumestrol induced persistent vaginal cornification and reduced ovarian weight (Leavitt and Meismer, 1968). Neonatal injections of 8×10^{-2} to 100 μg/day coumestrol for 5 days resulted in precocious vaginal opening, PFs, hemorrhagic follicles, ovary-independent persistent vaginal cornification, cervicovaginal hyperplasia, downgrowths, adenosis, cysts, and uterine squamous metaplasia in C57BL/Crgl mice. In mice receiving more than 5 μg/day coumestrol neonatally, mammary glands showed castration-like morphology, but the glands were similar to those of the controls when exposed to less than 5 μg/day (Burroughs *et al.*, 1985, 1990a). Neonatal exposure to coumestrol has long-term effects: ovary-independent persistent vaginal cornification, cervicovaginal pegs and downgrowths, and uterine squamous metaplasia (Burroughs *et al.*, 1990b).

Neonatal exposure of female C57BL/Crgl mice to 1 μg/day zearalenone caused ovary-dependent genital tract alterations, ovarian dysfunction (74%), dense collagen deposition in the uterine stroma, absence of uterine glands, and squamous metaplasia and dysplastic lesions in the vagina at 8 months of age. No such alterations were found in ovariectomized mice treated neonatally with zearalenone (Williams *et al.*, 1989).

IV. Genital Abnormalities Induced by Perinatal Exposure to Antihormones

A. Female Guinea Pigs

Fetal guinea pig uterus has a high content of estrogen receptors (Pasqualini *et al.*, 1976a,b; Sumida and Pasqualini, 1979) and responds to estrogens

(Pasqualini and Nguyen, 1980) and antiestrogens (Gulino and Pasqualini, 1980; Pasqualini *et al.*, 1986c). Prenatal exposure to 100 μg DES/kg/day on days 8–11 of pregnancy induced uterine cystadenomatous papillomata and hypoplasia in 250-day-old Syrian hamsters (Gilloteaux *et al.*, 1982; Gilloteaux and Steggles, 1985, 1986). In guinea pigs treated prenatally with tamoxifen (5 mg/kg/day) for 12 days from approximately the 50th day of pregnancy and neonatally with 100 μg tamoxifen for 2–12 days, there was substantial alteration of the mitochondria and rough endoplasmic reticulum with the formation of numerous vacuoles and secretory granules, indicating an enhancement of synthetic and secretory activities in the uterine epithelium. These ultrastructural changes were similar to those induced by estrogen (Pasqualini and Lecerf, 1986). In newborn guinea pigs given tamoxifen together with 17β-estradiol, stimulating effects on growth of uterus (thickness of the uterine epithelium, protein, and DNA contents) were more intense than in those given tamoxifen or estradiol alone (Pasqualini *et al.*, 1986a). Tamoxifen stimulated the fetal and newborn guinea pig vagina, resulting in an increase in DNA content and wet weight. This compound also induced progesterone receptor expression in the vagina of fetal and newborn guinea pigs (Nguyen *et al.*, 1986). When a combination of tamoxifen and progesterone was given to neonatal guinea pigs, progesterone blocked the production of progesterone receptors by tamoxifen (Pasqualini *et al.*, 1986b). Lecerf *et al.* (1988) showed that N-desmethyltamoxifen and 4-hydroxytamoxifen are more effective than *cis*-tamoxifen on growth and progesterone receptor induction in the uterus and vagina of newborn guinea pigs. These findings indicate that tamoxifen acted as a real estrogen agonist on the uterus and vagina of the guinea pigs during the perinatal period.

B. Female Rats

Leavitt and Meismer (1968) reported that neonatal female rats given a single injection of 100 or 1000 μg of the triphenylethylene compound, clomiphene, or of the plant estrogen, coumestrol, showed persistent vaginal cornification and significantly smaller ovaries lacking corpora lutea accompanied by low levels of pituitary luteinizing hormone (LH). Gellert *et al.* (1971) reported that neonatal female rats of the Sprague-Dawley strain given a single injection of 100 μg clomiphene citrate at 3 days of age showed accelerated vaginal opening, constant vaginal cornification, ovaries with degenerating follicles and absence of corpora lutea, enlarged cleft clitoris, and uterine metaplasia at about 4 months of age. Female rats injected on day 1 of life with nafoxidine (1–100 μg per rat) or clomiphene citrate (10–500 μg per rat) showed early vaginal opening and a high incidence of estrous smears. These rats showed uterine hyperplasia, squamous metaplasia and tumors, hypertrophic and hyperplastic oviducts, polycystic atrophic ovaries lacking corpora lutea, and hilus cell tumors of the ovary at 100 days of age (Clark and McCormack, 1977, 1980). McCormack and Clark (1979) demonstrated similar abnormalities in 15-week-old rats given 2 mg/kg body weight of clomiphene citrate on days 0, 5, and 12 of pregnancy. Female rats treated neonatally with 5 μg of tamoxifen on days 1, 3, and 5 showed abnormalities in reproductive development, early vaginal opening and absent estrous cycles, atrophic uteri, and oviducal squamous metaplasia with abscess formation at 4 months of age (Chamness *et al.*, 1979).

Neonatal and prepubertal injections of 10 μg/day tamoxifen inhibited the formation of uterine glands in female rats (Branham *et al.*, 1985). Branham *et al.* (1988a,b) also examined the prepubertal growth of the uterine luminal epithelium, stroma, glands, and circular and longitudinal muscle after exposure of neonatal rats to 10 μg/day 17 β-estradiol, DES, ethynylestradiol, tamoxifen, or clomiphene citrate for 5 days from the day of birth. On postnatal day 26, the cross-sectional areas and total cell numbers of the luminal epithelium, endometrial stroma, and circular muscle were reduced after estrogen exposure compared with untreated controls, whereas the longitudinal muscle cross-sectional area was not affected. The synthetic estrogens, DES and ethynylestradiol, were more potent than estradiol with respect to reduction of uterine growth. Neonatal antiestrogen exposure caused major reductions only in the uterine glands and luminal epithelium. Little change in cell density occurred in any cell population exposed to antiestrogen. These results indicate that the decreased uterine growth resulting from estrogen exposure during early postnatal development is a consequence of combined atrophy and hypoplasia of all cell types except longitudinal muscle, whereas antiestrogen-induced morphological alterations were limited specifically to hypoplasia of the epithelial cell population.

Campbell and Satterfield (1988) examined effects of neonatal exposure to antiestrogens, nafoxidine and CI-628, on the uterus development in prepubertal rats. They injected 50 μg nafoxidine or CI-628 on day 3 of postnatal life or in combination with estradiol benzoate 24 hours later; 21–23 days later estrogen-stimulated glucose oxidation and cytoplasmic estrogen binding sites of the uteri were examined. Both antiestrogens administered alone produced defects that resembled the changes caused by neonatal estrogenization. The agonist property of each antiestrogen was differentially expressed. CI-628 directly reduced uterine production of estrogen receptor protein, whereas nafoxidine affected the development of the uterine phosphogluconate oxidative pathway indirectly through impaired ovarian function. However, antiestrogens blocked the neonatal estrogen-induced reduction of actomyosin in the adult uterine muscle. Therefore, although both antiestrogens are clearly agonists in the neonatal rat, each appears to exhibit cell-specific agonist and antagonist properties.

Döhler *et al.* (1986) also demonstrated that perinatal administration of tamoxifen (from 16 days of pregnancy to delivery, 200 μg/day to mothers, and 10 μg/day to neonates for 10 days) induced permanent anovulatory sterility. Ohta *et al.* (1989) showed that the administration of 100 or 200 μg/day tamoxifen during neonatal life induced sterility characterized by acyclicity and anovulation in female rats as previously reported in rats and mice exposed neonatally to tamoxifen (Chamness *et al.*, 1979, Forsberg, 1985; Taguchi and Nishizuka, 1985; Iguchi *et al.*, 1986a, 1988a). Ovaries of these rats exhibiting persistent vaginal mucification invariably contained various stages of follicle maturation without corpora lutea. This is in contrast to the findings of Chamness *et al.* (1979) and Iguchi *et al.* (1986a) who demonstrated that tamoxifen-treated rats and mice showing continued vaginal diestrus had atrophic ovaries with small follicles and degenerated oocytes; the vagina was lined by atrophic epithelium without mucus. The vagina of tamoxifen-injected rats ovariectomized at the ages of 10 and 60 days failed to respond to estrogen priming, showing no estrous smears (Ohta *et al.*, 1989). Therefore, continued vaginal diestrus in neonatally tamoxifen-treated rats may be accounted for by changed sensitivity of the ovary to gonadotropin and/or of the vagina to sex hormones. By contrast, a daily dose of 100 or 200 μg MER-25 had little effect on the female reproductive system when given neonatally. The MER-25–treated rats had regular estrous cycles; ovaries contained follicles and corpora lutea as reported in female mice given MER-25 neonatally (Forsberg, 1985).

The endometrium of neonatally tamoxifen-treated rats had a reduced capacity for proliferation and transformation into deciduoma in response to intraluminal oil instillation compared with that in the controls. Estrogen priming prior to the standard schedule of treatment failed to increase the uterine deciduoma response in tamoxifen-treated rats (Ohta *et al.*, 1989). These findings suggest that tamoxifen administered during neonatal life is primarily responsible for the sustained, low uterine responsiveness to the deciduogenic stimulus after the standard schedule of treatment. Ohta (1982, 1985) demonstrated that endogenous estrogen is not needed for the capacity of the uterus to form deciduoma within 10 days after birth. In addition, because MER-25 given neonatally had little effect on decidualization in adults (Ohta *et al.*, 1989), it is possible that in neonatally tamoxifen-treated rats, the lowered uterine responsiveness to the deciduogenic stimulus is a result not of its action as an estrogen antagonist but rather as an estrogen agonist on the developing uterus.

One of the benzothiophenes, LY117018, with a high affinity for the adult rat uterine estrogen receptor, has little uterotropic activity, and inhibits many estradiol-induced responses in the adult or immature uterus (Wakeling *et al.*, 1984; Black and Goode, 1980, 1981; Black *et al.*, 1981). LY117018 at 1.25 or 50 μg/fetus failed to decrease the 15–70% incidences of oviduct malformations and cleft phallus at 21 days of age induced by DES (2.5 μg/fetus) or estradiol (50 μg/fetus) when injected into day 19 rat fetuses. However, LY117018 alone (1–50 μg/fetus) was more potent than estradiol in eliciting these same urogenital malformations. LY117018 also failed to compete *in vitro* with plasma protein-bound ^{3}H-estradiol, and therefore, like DES, is more available than estradiol for uptake into fetal tissues (Henry and Miller, 1986). In the fetus, unlike in the adult, LY117018 was an estrogen agonist. These findings indicate that the fetus has a different sensitivity than the adult to estrogenic compounds and nominal antiestrogens and that the early estrogen receptor may have characteristics different from the adult receptor.

C. Female Mice

Treatment of neonatal C57BL/Tw mice with 100 μg MER-25/day resulted in prolonged estrous cycles and reduced number of corpora lutea at 2 and 8 months of age; however, no differences were found in the responsiveness of vaginal and uterine epithelia in these mice to exogenous estrogen (Mori *et al.*, 1977). Iguchi *et al.* (1979) showed that MER-25 (100 μg/day) had an inhibitory effect on the occurrence of permanent vaginal changes when given repeatedly before each of five daily neonatal 17β-estradiol (20 μg) injections. Neonatal treatment of female BALB/c mice with 5 μg clomiphene for 5 days resulted in heterotopic columnar epithelium (HCE) (vaginal adenosis) in the cervicovaginal region, but the incidence decreased with age from 30% (4 weeks) to 10% (24 weeks) (Gorwill *et al.*, 1982). Forsberg (1985) also demonstrated the occurrence of HCE in the same region of 8-week-old and 6-month-old NMRI female mice given five daily injections of 5 μg of tamoxifen, clomiphene, or nafoxidine from the day of birth. The regions with HCE in neonatally antiestrogen-treated mice were more extensive than those in mice treated neonatally with DES and 17β-estradiol. Adult females treated neonatally with antiestrogens had ovaries with or without corpora lutea, although all ovaries from adult females treated with DES or 17β-estradiol lacked corpora lutea. These findings suggest that in the neonatal period, the cervicovaginal epithelium is more sensitive to antiestrogens than the hypothalamohypophyseal system but less sensitive to DES and 17β-estradiol. Taguchi and Nishizuka (1985) showed that neonatal treatment of female NMRI mice with 20 μg tamoxifen for 3 days from the day of birth caused vaginal adenosis, hypospadias, cervical hypoplasia, uterine hypoplasia, and arrest of ovarian luteinization.

Three groups of female C57BL/Tw mice given daily injections of 2, 20, or 100 μg tamoxifen starting on the day of birth were examined at 35 and 150 days of age (Iguchi *et al.*, 1986a). Half of the mice killed at 150 days had been ovariectomized at 90 days. Uterine hypoplasia was found in neonatally tamoxifen-treated C57BL/Tw mice as was observed in similarly treated NMRI mice (Taguchi and Nishizuka, 1985). Uterine metaplasia and myometrial disorganization have been reported as major uterine abnormalities of mice treated perinatally with estrogen (Newbold and McLachlan, 1982; Ostrander *et al.*, 1985; Iguchi *et al.*, 1986b,e, 1987b). In contrast, no uterine metaplasia was found in neonatally tamoxifen-exposed mice, although the number of uterine glands significantly decreased in tamoxifen-treated mice. Branham *et al.* (1985) reported that neonatal and prepubertal tamoxifen injections inhibited the formation of uterine glands in female rats. Prenatal DES exposure (2 mg from day 15 to 18 of gestation) also caused a retardation of uterine gland genesis and reduced the number of glands in mouse uteri (Iguchi and Takasugi, 1987). Wordinger *et al.* (1991), however, indicated that prenatal DES exposure (10 μg/kg body weight on day 16 of gestation) resulted in premature formation of uterine glands during the first week of neonatal uterine development. Vaginal hypoplasia and hypospadias were common abnormalities in 150-day-old tamoxifen-treated mice. Vaginal adenosis was encountered in 35-day-old mice treated neonatally with 20 or 100 μg/day tamoxifen, but these lesions did not persist for more than 150 days. Because the vaginal epithelium of ovariectomized tamoxifen-treated mice was atrophic, neonatal tamoxifen treatment (2–100 μg/day) was not capable of inducing permanent vaginal epithelial proliferation, although these doses of estrogen induced permanent vaginal changes (Bern and Talamantes, 1981). Another study also demonstrated that neonatal treatment of female C57BL mice with MER-25 failed to induce permanent vaginal changes (Iguchi *et al.*, 1979). These results indicate that antiestrogens have little permanent effect on the vaginal epithelium of neonatal mice.

Female rats and mice treated neonatally with estrogen possess ovaries lacking corpora lutea, indicating noncyclic release of gonadotropins because of the impaired hypothalamohypophyseal function (Takasugi, 1976; Gorski *et al.*, 1977). In ovaries of tamoxifen-treated mice at 150 days, luteinization never occurred (Forsberg, 1985; Taguchi and Nishizuka, 1985). Ovaries of tamoxifen-treated 150-day-old mice contained small follicles, the oocytes of which frequently underwent degeneration. Frequency of oocyte death (14%, 77%, and 89%) rose with an increase in tamoxifen dose (2, 20, and 100 μg, respectively) given to the neonates. Follicular growth was markedly suppressed in ovaries of tamoxifen-treated mice, being similar to the ovary of hypophysectomized rats (Selye *et al.*, 1933). In addition, responsiveness of ovaries to prepubertally injected gonadotropin (hCG) was strikingly reduced in tamoxifen-treated mice. These findings imply that the hypothalamohypophyseal-ovarian system was impaired by neonatal tamoxifen treatment. We found that specific binding of follicle-stimulating hormone (FSH) to ovaries (both per milligram of tissue and per organ) of 40-day-old C57BL/Tw mice treated neonatally with 5 μg DES and 100 μg tamoxifen for 5 days from the day of birth to be significantly lower than that in the controls (T. Iguchi *et al.*, unpublished observations).

Vaginal adenosis and cervical ectropion are commonly encountered in women exposed *in utero* to DES during the first trimester of pregnancy for the prevention of threatened abortion (Herbst *et al.*, 1971; Herbst and Bern, 1981). Frequent occurrence of adenosis-like lesions in the cervicovaginal region has been reported in perinatally DES-exposed mice (Forsberg, 1969; Plapinger and Bern, 1979; Forsberg and Kalland, 1981; Iguchi *et al.*, 1986b, 1988b). Extensive adenosis-like lesions have also been reported in 2- to 6-month-old NMRI mice treated neonatally with antiestrogens including tamoxifen (Forsberg, 1985; Taguchi and Nishizuka, 1985). Only small lesions were induced in the vaginal fornix of 35-day-old C57BL mice treated neonatally with 20 or 100 μg/day tamoxifen (Iguchi *et al.*, 1986a); however, extensive lesions were seen in the upper vaginal region of 35-day-old ICR mice treated neonatally with 2–200 μg/day tamoxifen (22–100%), reflecting a probable strain difference (Iguchi *et al.*, 1989b). Adenosis-like lesions were found in the cervical and upper vaginal regions derived from the Müllerian duct, especially in the common cervical canal and the vaginal fornix, but not in the lower vagina derived from the urogenital sinus. This result is in accordance with that seen in prenatally DES-exposed mice (Iguchi *et al.*, 1986b,c). Neonatal treatment with 200 μg/day clomiphene or nafoxidine also induced lesions in the cervical and upper vaginal regions.

Adenosis-like lesions developed in prenatally DES-exposed mice in the presence of ovarian estrogen; ovariectomy diminished incidence of the lesions, and prepubertal daily injections of 10^{-4} to 1 μg estradiol restored the incidence to the level before ovariectomy (Iguchi *et al.*, 1986c). Ovariectomy performed at 10 days reduced the incidence of the lesions only in the vaginal fornix of tamoxifen- or clomiphene-injected (200 μg/day) mice, whereas the size of the lesions was decreased by ovariectomy in all vaginal regions of antiestrogen-treated mice. Prepubertal daily injections of 0.1 μg estradiol increased the lesion incidence only in the common cervical canal and the vaginal fornix of tamoxifen (20 μg/day)-exposed mice, although the size of the lesions was increased by the treatment.

In the uterus of prenatally DES-exposed mice, the stroma and the circular musculature showed disorganization but not involution (Iguchi and Takasugi, 1987). Uteri were heavier and uterine epithelial cells were taller in tamoxifen-injected (200 μg/day) mice than in the controls from 5 to 20 days of age. The uteri of tamoxifen-treated mice at 35 days had a significantly taller columnar epithelium and a smaller number of the glands than in the controls (Iguchi *et al.*, 1989b). These findings suggest that neonatal tamoxifen acts as an estrogen agonist on the uterovaginal epithelium and uterine glands.

Uteri of mice treated neonatally with tamoxifen, clomiphene, or nafoxidine showed a reduced number of glands, a disorganization of the circular musculature, and lowered responsiveness to estradiol given prepubertally (Iguchi *et al.*, 1989b). Nascent uterine glands in intact mice appeared at 5–7 days as pits of invaginated epithelium (Plapinger, 1982; Iguchi and Takasugi, 1987). In tamoxifen-treated mice, uterine epithelium began to invaginate at 5 days and formed the nascent glands at 10 days, suggesting a delay in the commencement of gland formation (Iguchi *et al.*, 1989b). Similar results were reported in both neonatally tamoxifen-treated rats and prenatally DES-exposed mice (Branham *et al.*, 1985; Iguchi and Takasugi, 1987).

In mice treated neonatally with tamoxifen, the organization of the stroma and the circular musculature was disrupted at ages of 5 and 10–15 days in ICR and C57BL strains, respectively (Iguchi *et al.*, 1986a, 1989b). Neonatal tamoxifen exposure gave rise to changes in the density and distribution of Types I and III collagen, fibronectin, and laminin (Iguchi *et al.*, 1989b). Irisawa and Iguchi (1990) found that Type I collagen and fibronectin were absent from the uterine stroma of the edematous region, whereas laminin disappeared from the tunica muscularis in neonatally tamoxifen-injected, 15-day-old C57BL mice. Thus, the suppression of gland formation and the lowered responsiveness to estrogen may result from changes in extracellular matrix proteins, and the increase in uterine weight in tamoxifen-treated mice during the prepubertal period is due to edema of the stroma. Tamoxifen induced a high incidence of PFs, but clomiphene and nafoxidine did not (Irisawa and Iguchi, 1990).

Triphenylethylene-derived antiestrogens tamoxifen, clomiphene, and nafoxidine increased weight, protein, and DNA in the uterus of ovariectomized adult mice. However, the benzothiophene-derived antiestrogen, keoxifene, had no stimulating effect on uterine weight gain or on protein and DNA synthesis in the ovariectomized adult mice. Neonatal injections of keoxifene (100 μg/day) induced infertility but not abnormalities in uterus and vagina in 60-day-old C57BL mice (Chou *et al.*, 1992).

D. Critical Period for Tamoxifen Action on Females

To examine the critical period for induction by tamoxifen of female genital organ abnormalities, female C57BL/Tw mice were given five daily injections of 100 μg tamoxifen starting at different early postnatal ages (Irisawa and Iguchi, 1990). Tamoxifen injections starting within 5 days of birth caused a high incidence of PFs in the ovary and aplasia of the tunica muscularis in the uterus at 60 days. The tamoxifen treatment starting within 7 days also induced atrophy of the uterine luminal epithelium. In mice given tamoxifen starting within 3 days, the vagina had a thinner epithelium than that in the controls. These findings suggest that the postnatal limit of the critical period for tamoxifen induction of female genital abnormalities is within 3–7 days after birth.

E. Female Humans

The potential estrogenicity and teratogenicity of triphenylethylene antiestrogens were examined in 54 genital tracts isolated from 4- to 19-week-old human female fetuses and grown for 1–2 months in untreated athymic nude mice or host mice treated with the antiestrogens, clomiphene or tamoxifen, or with the synthetic estrogen, DES (Cunha *et al.,* 1987b). Proliferation and maturation of the squamous vaginal epithelium were observed in specimens treated with these agents only when the fetuses were maintained to a gestational age equivalent to 16 weeks or more. The number of their endometrial and cervical glands was 87% of the control specimens maintained to a gestational age equivalent to 13 weeks or more in untreated hosts. By contrast, antiestrogen-treated specimens showed a 44% reduction of the glands seen in the control specimens. The developing uterus of untreated controls was separated into stroma and myometrium, whereas in drug-treated specimens, condensation and separation of the mesenchyme were greatly impaired. The oviduct was also affected by clomiphene and tamoxifen, resulting in hyperplasia and disorganization of its epithelium. These findings emphasize the unrecognized estrogenicity and potential teratogenicity of triphenylethylene antiestrogens on the developing human genital tract, indicating the need for caution to prevent inadvertent exposure of the developing fetus to these compounds (Heel *et al.,* 1978). Using a similar experimental system, progesterone was not associated with obvious teratogenic effects on the developing human female genital tract (Cunha *et al.,* 1988).

F. Male Rats and Mice

In male rats and mice, neonatal treatment with estrogens gave rise to a marked atrophy of the testis, resulting in a long-lasting arrest of spermatogenesis. The genital accessory organs also underwent varying degrees of atrophy (Takewaki and Takasugi, 1953; Mori, 1967; Takasugi, 1970; Ohta and Takasugi, 1974; Arai *et al.*, 1977; Jones, 1980). Perinatal exposure of rats and mice to DES brought about similar changes in male genital organs leading to sterility (Dunn and Green, 1963; McLachlan *et al.*, 1975; Warner *et al.*, 1979; Arai *et al.*, 1983). Three groups of male C57BL/Tw mice given five daily subcutaneous injections of 2, 20, and 100 μg tamoxifen starting within 24 hours after birth were examined at 160 days of age (Iguchi and Hirokawa, 1986). In all groups of mice, weights of testes, gubernacula, seminal vesicles, coagulating glands, and epididymides were significantly lower in tamoxifen-injected (20 and 100 μg/day) mice than in the controls. In 9 of 10 tamoxifen-injected (2 μg/day) mice, testes and the accessory organs were not different histologically from those of the controls. The tenth mouse possessed testes with extended intertubular spaces largely occupied by proliferated fibroblasts and showed a low spermatogenic index. In 13 mice given 20 μg tamoxifen, the mean spermatogenic index (67%) was lower than in the controls (88%). One mouse had testes lacking spermatids and spermatozoa; two mice had testes showing proliferation of fibroblasts in the intertubular spaces. In half of 100 μg tamoxifen-injected mice, spermatogenic indices were lowered (0–35%), whereas in the other half the indices remained unaffected. In all mice treated neonatally with tamoxifen, colloidal secretion in the lumina of the seminal vesicles was reduced in amount, the extent correlating with the dose, suggesting reduced androgen secretion. Previous studies have revealed that the suppressive effect of neonatally injected estrogen on testicular function is nullified or attenuated by simultaneous administration of androgens or gonadotropins (Takasugi and Furukawa, 1972; Takasugi and Mitsuhashi, 1972; Ohta, 1977). These findings suggest that the decline of spermatogenesis in neonatally estrogenized mice may be due to estrogen-induced impairment of the hypothalamohypophyseal gonadotropin axis with or without direct effect of estrogen on the testis. In castrated rats tamoxifen had no retarding effect on androgen-stimulated growth of genital accessory organs (Harper and Walpole, 1967). On the basis of these findings, it may be inferred that tamoxifen exerts an estrogenic action on testis and the accessory organs directly and/or indirectly through an altered function of the hypothalamohypophyseal system. Taguchi (1987) also demonstrated genital abnormalities, including testicular hypoplasia, suppression of spermatogenesis, intraabdominal testes, epididymal cysts, and squamous metaplasia of seminal vesicles and coagulating glands, in 8-month-old NMRI/Tg male mice given three daily injections of 20 μg tamoxifen from the day of birth. Daily injections of 100 μg tamoxifen, clomiphene, or nafoxidine for 5 days from the day of birth resulted in a decrease in spermatogenic index (22%, 69%, and 74%, respectively) in 60-day-old C57BL mice (Irisawa *et al.*, 1990). Neonatal injections of keoxifene (100 μg/day) also reduced the spermatogenic index (75%) in 60-day-old C57BL mice (Chou *et al.*, 1992). Tamoxifen caused more serious damage than did clomiphene, nafoxidine, or keoxifene. We found that specific binding of LH to the testis of 40-day-old C57BL/Tw mice given daily injections of 5 μg DES and 100 μg tamoxifen for 5 days from the day of birth is significantly reduced in mice injected neonatally with tamoxifen (15% of the controls) or with DES (58% of the controls) (T. Iguchi *et al.*, unpublished observations).

Brown and Chakraborty (1988) studied the effects of clomiphene citrate (0.05–5.0 mg/kg/day) on the reproductive physiology of adult (60 days), peripubertal (35 days), and prepubertal (10 days) male rats, and showed that testis weight was reduced only in prepubertal rats. Seminal vesicle and prostate weights, serum LH and testosterone levels, pituitary gonadotropin levels, and testicular LH receptors were decreased in all rats given clomiphene at certain ages, indicating that response to clomiphene is dependent on the age of administration.

G. Critical Period for Tamoxifen Action on Males

Male C57BL/Tw mice given five daily injections of 100 μg tamoxifen beginning at various early postnatal ages (0, 3, 5, 7, and 10 days) were examined at 60 days of age (Irisawa *et al.*, 1990). In all mice given tamoxifen, spermatogenic index and diameter of seminiferous tubules and Leydig cell nuclei were significantly less than in the controls. In mice given tamoxifen starting at 0 and 3 days, 70% had testes with spermatogenic indices lower than 25%. Epithelial cell height in the epididymis and seminal vesicle of mice given tamoxifen starting at 0 and 3 days was significantly lower than those in the controls. The postnatal limit of the critical period for tamoxifen-induced male genital organ dysfunction is 3 days after birth.

H. Changes in the Os Penis

The os penis of adult rats and mice is divided into two regions (Fig. 1A): (1) the proximal segment is comprised of membrane bone with hyaline cartilage having Type II collagen at its proximal end and (2) the distal segment is comprised of fibrocartilage having Type I collagen (Glucksmann and Cherry, 1972; Glucksman *et al.*, 1976; Yoshida *et al.*, 1980; Vilmann and Vilmann, 1983; Murakami and Mizuno, 1984a,b, 1986; Murakami, 1986; Yamamoto, 1987, 1989; Iguchi *et al.*, 1990b; Yamamoto *et al.*, 1990). The proximal segment becomes ossified in the early postnatal

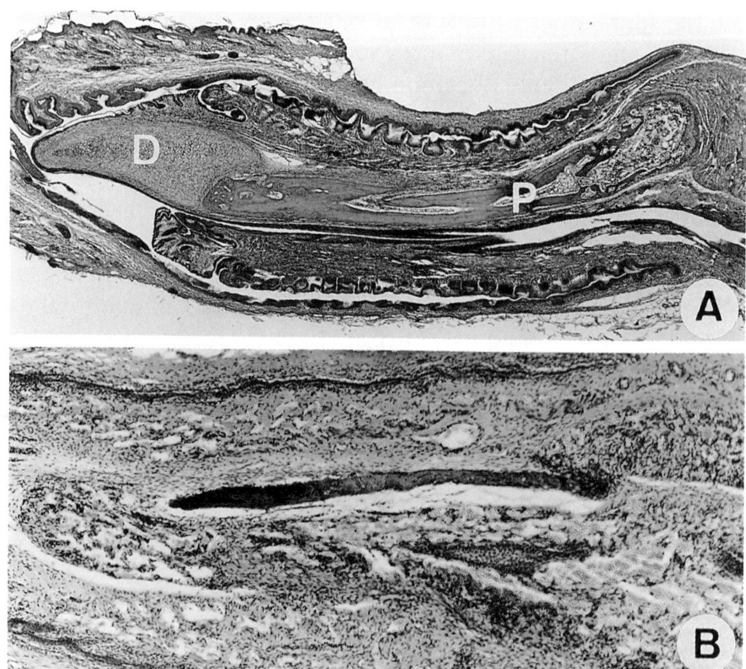

FIG. 1 Longitudinal section of penis of 60-day-old control (A) and neonatally tamoxifen (100 μg/day)-treated (B) male mice. There are well-developed distal (D) and proximal (P) segments in A. Atrophic P segment is seen, but D segment is not formed in B. A, ×20. B, ×64.

period, whereas the distal segment ossifies later and grows at a slower rate (Murakami, 1987a,b; Iguchi *et al.*, 1990b).

Hyaline cartilage, bone marrow, and trabeculae formed in the proximal segment of control mice at 5 days were lacking in mice treated neonatally with 100 μg/day tamoxifen for 5 days at the ages of 10 and 30 days. The size of the proximal segment in neonatally tamoxifen-treated mice at 10–60 days was significantly smaller than that in the age-matched controls. In mice treated neonatally with 100 μg/day clomiphene or nafoxidine, the size of the proximal segment was significantly smaller than in the controls at 60 days; however,the size was significantly larger than in tamoxifen-exposed mice at the same age, suggesting that tamoxifen is more active than other antiestrogens in inhibiting the growth of the proximal segment (Iguchi *et al.*, 1990b).

The distal segment of the os penis developed into fibrocartilage by 30 days in the controls, whereas such cartilage was undetectable at 30 days in neonatally tamoxifen-treated mice (Iguchi *et al.*, 1990b). Glucksmann

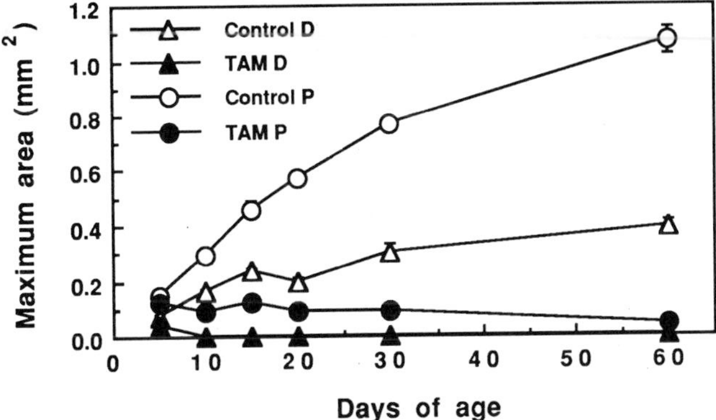

FIG. 2 Areas of the proximal and distal segments of the os penis in mice treated neonatally with tamoxifen (TAM). Values from tamoxifen-treated mice are significantly smaller ($p < .01$) than those from the controls except at 5 days of age. D, distal segment; P, proximal segment.

et al. (1976) reported that the growth and ossification of the distal segment are delayed in mice treated neonatally with the antiandrogen, cyproterone acetate, and that castration on the day of birth inhibited the growth and ossification more strongly than did treatment with antiandrogen. Howard (1959) showed aplasia of fibrocartilage in the distal segment of 40-day-old mice castrated at 4 days of age. These results suggest that the growth and formation of fibrocartilage in the distal segment are caused by androgen present neonatally.

In 60-day-old mice given tamoxifen injections starting within 5 days, hyaline cartilage was undetectable in the proximal segment (Fig. 1B). In mice given tamoxifen injections starting at 7 or 10 days, hyaline cartilage was formed at 60 days; however, the size of the proximal segment was significantly smaller than in the controls (Figs. 2 and 3). Hyaline cartilage was also found in the proximal segment of 60-day-old mice treated neonatally with nafoxidine or clomiphene (Iguchi *et al.*, 1990b). These results indicate that tamoxifen has a specific effect on suppression of hyaline cartilage in os penis and that there is a critical period for the suppression of cartilage formation by tamoxifen.

I. Changes in the Spines of the Glans Penis

Murakami (1987b) suggested that development of spines on the skin of mouse glans penis is dependent on androgens. Spines on the glans penis

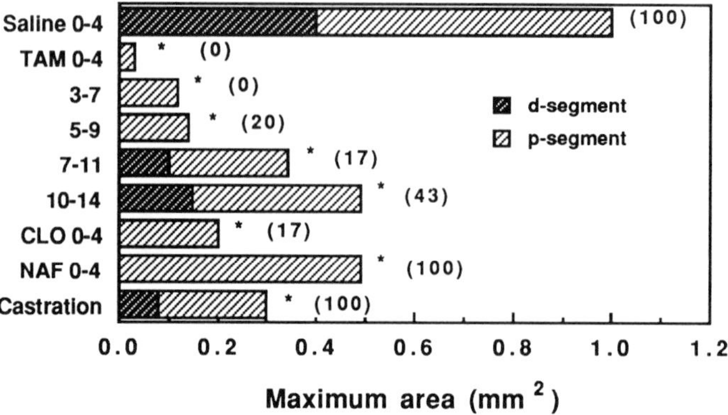

FIG. 3 Areas of proximal (P) and distal (D) segments in 60-day-old mice treated with antiestrogens. The numbers on the vertical axis indicate treatment periods for 5 days starting on 0 (day of birth), 3, 5, 7, and 10 days of age. Values in parentheses indicate percent incidence of mice having bone marrow in P segment of the os penis. TAM, tamoxifen; CLO, clomiphene; NAF, nafoxidine. $*p < .05$ versus saline.

skin began to form in both control and neonatally tamoxifen-treated mice between 5 and 10 days of age (Fig. 4), but the density was lower in neonatally tamoxifen-treated mice from 10 to 60 days than in the controls (Iguchi *et al.,* 1990b). The epidermis of the glans penis and the prepuce began to separate at 10 days in the controls. In neonatally tamoxifen-treated mice, however, even at 60 days they remained unseparated. Nafoxidine and clomiphene had no inhibitory effect on spine formation, whereas formation was significantly suppressed by tamoxifen injections starting within 5 days (Iguchi *et al.,* 1990b), suggesting that tamoxifen has a specific inhibitory effect on spine formation.

V. Nongenital Abnormalities Induced by Perinatal Exposure to Sex Hormones and Antihormones

A. Changes in the Sexually Dimorphic Brain Nuclei and Sexual Behavior

The volume of the sexually dimorphic nucleus in the preoptic area (SDN-POA) of the rat brain is severalfold larger in adult males than in adult females (Gorski *et al.,* 1978; Jacobson *et al.,* 1981; Döhler *et al.,* 1982b, 1984a). The sex difference in brain structure first appears at the later fetal stage depending on sex hormones during the critical period of sexual

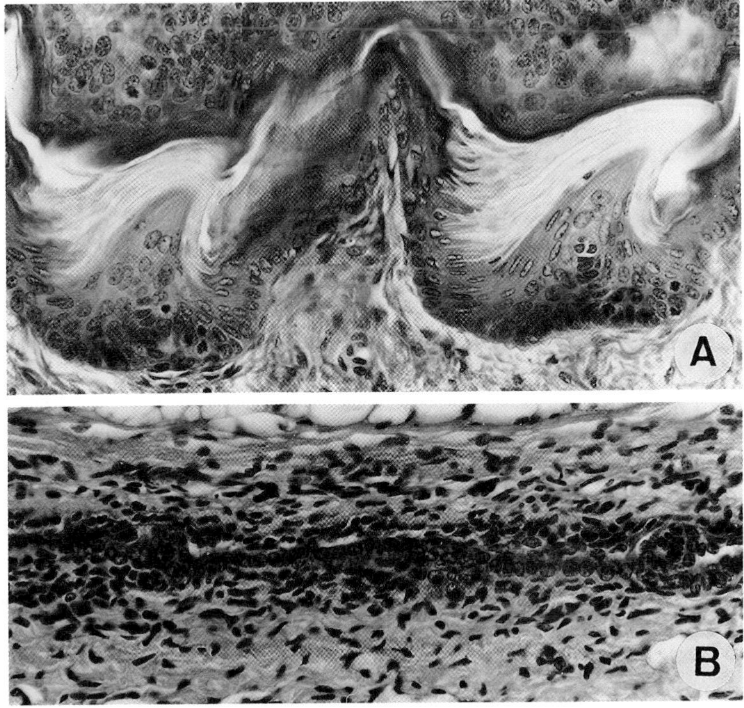

FIG. 4 Epidermis of the glans penis and the prepuce in 60-day-old control (A) and neonatally tamoxifen-treated (B) male mice. There are well-developed spines in A. ×310.

differentiation (Hsu *et al.*, 1980; Jacobson *et al.*, 1980; Jacobson and Gorski, 1981; Arai *et al.*, 1983). Perinatal treatment of female rats with an aromatizable androgen (Gorski *et al.*, 1978; Jacobson *et al.*, 1981; Döhler *et al.*, 1982b, 1984b) or with an estrogen (Döhler *et al.*, 1982a,1984a) increased SDN-POA volume. By contrast, neonatal castration (Gorski *et al.*, 1978; Jacobson *et al.*, 1981) or a single injection of tamoxifen (Döhler *et al.*, 1984b) permanently reduced the SDN-POA volume in male rats but not in female rats. Döhler *et al.* (1986) studied the effect of perinatal exposure to antiestrogen and antiandrogen on brain structure in rats. Pregnant rats were given daily injections of 200 μg tamoxifen or 10 mg cyproterone acetate from day 16 of pregnancy. After delivery the pups were given 10 daily injections of either 10 μg tamoxifen or 0.5 mg cyproterone acetate. Tamoxifen treatment did not alter serum levels of testosterone in male rats during the perinatal period but inhibited the development of SDN-POA. Perinatal treatment of male rats with cyproterone acetate injections resulted in the female phenotype but not in the feminization of

SDN-POA. Exposure of perinatal rats to tamoxifen resulted in permanent anovulatory sterility without influencing SDN-POA. These findings indicate that the development of SDN-POA is primarily under estrogenic control and that tamoxifen acts as an estrogen antagonist on this structure.

Hines *et al.* (1987) examined neural, behavioral, and ovarian development in female guinea pigs given daily injections of 3 μg DES dipropionate (DESDP) or 100 μg tamoxifen from days 29 to 65 of pregnancy. Prenatal exposure to DESDP resulted in masculinization (more mounting behavior and larger SDN-POA) and defeminization (delayed vaginal opening, impaired progesterone production, absence of corpora lutea, and impaired lordosis and mounting responses to estrogen and progesterone). Prenatal tamoxifen treatment caused less male-type behavior: diminished mounting, delayed vaginal opening, enhanced progesterone production, and impaired mounting in response to estrogen and progesterone. Lordosis behavior and the volume of SDN-POA were not affected. These findings suggest that estrogens play a substantial role in sexual differentiation in the guinea pig.

B. Effects on Enzyme Levels

Hepatic activity of dehydroepiandrosterone-16α-hydroxylase is higher in male rats than in females; however, the activity of the enzyme is reduced in the adult when animals are exposed to clomiphene or estradiol during the neonatal period (Tabei and Heinrichs, 1977). Therefore, the inherent estrogenicity of clomiphene is probably responsible for this phenomenon. Because the induction of this enzyme is mediated by androgen, neonatal clomiphene or estrogen antagonizes the action of androgen. Clomiphene treatment in male rats increases 3β-hydroxysteriod dehydrogenase activity in the liver, whereas estradiol treatment decreases it (Lax *et al.*, 1978).

Adult rats treated neonatally with 390 μg (1.45 μmol) DES have lower testis and uterine wet weights and lower circulating testosterone and estrogen concentrations. In addition, serum acetylcholinesterase and butyrylcholinesterase, hepatic histidase, and monoamine oxidase activity levels in rats are higher in adult female rats than in adult male rats. Both perinatal exposure to estrogens and neonatal castration alter the developmental patterns of some of these enzymes (Lamartiniere and Lucier, 1978; Dieringer *et al.*, 1979; Lamartiniere, 1979; Lamartiniere *et al.*, 1979, 1982; Illsley and Lamartiniere, 1980). Neonatal castration caused an increase in monoamine oxidase activity in adult male rats; however, administration of 1.45 μmol/day testosterone propionate or DES to these castrates at 2 days of age prevented the increase in the enzyme activity observed in the adult males after neonatal castration. Therefore, both sex-related differences in hepatic monoamine oxidase of adult rats and male pattern of the enzyme in response to androgen are imprinted during the neonatal period (Illsley and Lamartiniere, 1980). Neonatal treatment of rats with 1.45 μmol/day DES or 17β-estradiol from 2, 4, and 6 days after birth resulted in a decreased histidase activity in adult female rats, but no effect was seen when given to prepubertal females and adult males. In contrast, a similar neonatal dose of testosterone propionate had no effect on histidase. Neonatal estrogen treatment exerts a permanent and irreversible action on the hepatic activity (Lamartiniere, 1979). Neonatal exposure of rats to DES alters endogenous activity levels of enzymes involved in activation and detoxication of xenobiotics (Lamartiniere and Pardo, 1988; Lamartiniere, 1990). This alteration may be a consequence of altered imprinting mechanisms by DES, causing developmental modifications early in life.

Neonatal treatment of rats with 20 μg tamoxifen or 200 μg CI-628 on days 1,3, and 5 postpartum resulted in increased serum cholinesterase activity in adult females, whereas such treatment had no effect on hepatic histidase and monoamine oxidase in either sex (Lamartiniere *et al.*, 1986). Circulating estrogen levels in antiestrogen-treated female rats were not different from those in the controls. Neonatal tamoxifen treatment did not alter the uptake of estrogen in pituitary, preoptic anterior hypothalamic area, and median eminence–basal hypothalamus (Lamartiniere *et al.*, 1986).

C. Effects of Perinatal Treatment with 5α-Reductase Inhibitor

Testosterone, by virtue of its aromatization, is required for masculinization of the hypothalamohypophysiogonadal axis (Barraclough, 1967; McDonald and Doughty, 1972; Iguchi *et al.*, 1988c). Its 5α-reduced metabolite, 5α-DHT, in contrast, is involved in the development of the prostate, coagulating glands, and bulbourethral glands from the urogenital sinus (Cunha *et al.*, 1987a; George and Wilson, 1988).

Androgens at a critical period of morphogenesis cause a regression of the nipple anlage in male rats and mice (Kratochwil and Schwartz, 1976; Kratochwil, 1977, 1986), but this could be arrested by inhibiting the action of the androgen *in utero* by cyproterone acetate, an antiandrogen (Goldman and Neumann, 1969). Testosterone, DHT, or androstenediol inhibited normal nipple formation in rats (Greene *et al.*, 1941; Goldman *et al.*, 1976). Imperato-McGinley *et al.* (1986) showed that treatment of pregnant Sprague-Dawley rats from days 12 to 21 of gestation with a 5α-reductase inhibitor, 17β-N,N-diisopropylcarbamoyl-4-aza-5α-androstan-3-one (100 mg/kg/day), resulted in nipple development in male offspring and feminization of the external genitalia with urethral displacement to the base of the phallus. Imperato-McGinley *et al.* (1985) also reported that 5α-reductase inhibitor, 4-methyl-4-aza-5-pregnan-3-one-20[s] carboxylate (36 mg/kg/day) from days 12 to 21 of gestation caused complete feminization of external genitalia of male rats, with formation of a urogenital sinus and a pseudovagina, but had little inhibitory effect on prostatic bud formation in male rat offspring. George and Peterson (1988) demonstrated almost total inhibition of prostatic bud formation and development of external genitalia in male rats exposed prenatally to a 5α-reductase inhibitor L652,931 (50 mg/kg/day) from days 14 to 22 of gestation. Male rats given DHT (50 mg/kg/day) with the 5α-reductase inhibitor, in contrast, restored prostate development and anogenital distances of males. George (1989) revealed that the 5α-reductase *in utero* inhibited the normal growth rate of the gubernaculum. George *et al.* (1989) demonstrated that after neonatal treatment of male rats with a 5α-reductase inhibitor, N-(2-methyl-2-propyl)3-oxo-4-aza-5α-androst-1-ene-17β-carboxyamide (finasteride), 35 mg/kg/day for the first 4 weeks and 15 or 30 mg/kg/day for another 3 weeks, weights of prostate, penis, seminal vesicles, and epididymis were reduced to 30–50% of those of controls. However, DHT formation is not critical for postnatal development of preputial glands and spermatogenesis. Similar effects of finasteride on external genitalia, including hypospadias of male offspring, were reported when the agent was given orally to pregnant rats during days 6–20 of gestation (Anderson and Clark, 1990; Clark *et al.*, 1990).

Prenatal treatment of ICR/JCL mice with 5α-reductase inhibitor, 6-methylene-4-pregnene-3,20-dione (6-MP) (400 mg/kg/day) for 7 days from days 12 to 18 of gestation caused significantly shorter anogenital distance, feminization of the nipples, and hypospadias of the phallic urethra in male offspring. Development of prostate, coagulating gland, and bulbourethral gland was significantly suppressed; however, development of the testis and seminal vesicle was not affected. Simultaneous injections of DHT (20 mg/kg/day) with 6-MP prevented the regression of mammary gland, penis, prostate, coagulating gland, and bulbourethral gland in the male (Iguchi *et al.*, 1991b). These results further support the conclusion that DHT is necessary for the development of the urogenital sinus and penis and for the regression of the nipples in male mice. Reproductive abnormalities were not found in 90-day-old mice (both sexes) exposed to 6-MP *in utero* (Iguchi *et al.*, 1991b). The 6-MP-exposed male and female mice had normal reproductive capacity when mated with normal mice. These results show that 6-MP-induced growth retardation of reproductive organs is evident on day 19 of gestation but that such retardation is no longer apparent in the adult.

D. Sexual Dimorphism of the Mouse Pelvis

Sexual dimorphism of the pelvis has been described in pocket gophers, guinea pigs, and mice (Chapman, 1919; Todd, 1925; Gardner, 1936). A pair of innominate bones, the *ossa coxae*, are comprised of four separate units in mice: ilium, ischium, pubis, and acetabulum, which unite at the ventral midline as the pubic symphysis to form the pelvis. The innominate bone is connected dorsomedially with the sacrum by the iliosacral joint. Gardner (1936) reported that there is no difference in the shape of the innominate bone in young male and female mice; however, after sexual maturity the pubic bone in the females is thinner than in the males. Long-term administration of estrogenic hormones to male mice induces the female-type pelvis with thin pubic bones, indicating that sex hormones play a role in pelvic morphogenesis. Stein (1957) pointed out the difficulty in expressing this skeletal difference quantitatively. Festing (1972) emphasized, however, that morphometric analysis can be employed to demonstrate mandibular variations, and Bailey (1986) conducted a precise morphometric analysis of genetic differences in the shape of the mouse mandible. Computer analysis has revealed that genetic differences are also present in the shape of the innominate bones of different strains of mice (Lovell *et al.*, 1986).

Seven parameters of the innominate bone (Fig. 5) differentiated by a modified staining method for cartilage and calcified bone (Inoue, 1976; McLeod, 1980) were chosen for analysis (Iguchi *et al.*, 1989a): longitudinal length of the innominate bone (IL), distances from the upper edge of the pubic symphysis to its lower edge (SP), from the center of the acetabulum to the lower edge of the ischium (AI), from the lower ischium edge to the lower pubis edge (IP), and from the center of the acetabulum to the upper edge of the pubis (AP), and the widths of the ischium (IW) and of the pubis (PW). The lengths and widths measured by a Color Image Analyzer CIA-102 (Olympus, Tokyo) were calculated as ratios to IL and expressed as ratios (percentages) (Fig. 6).

IL increased with age, but no sexual difference in IL was found between intact male and female C57BL/Tw mice regardless of age. Sexual dimorphism of the innominate bone was found in C57BL mice as early as 1 day of age: in the females the ischium had a small spinous process of cartilage in the basal part until at least 30 days of age; in the males this process disappeared the day after birth (Iguchi *et al.*, 1989a). In male rats and mice higher levels of plasma testosterone were detected in the perinatal period than those detected in a later immature period, but this is not true of females (Pointis *et al.*, 1980; Slob *et al.*, 1980; Pang and Tang, 1984), suggesting that the early postnatal disappearance of the small spinous process in male mice results from the higher androgen levels during the

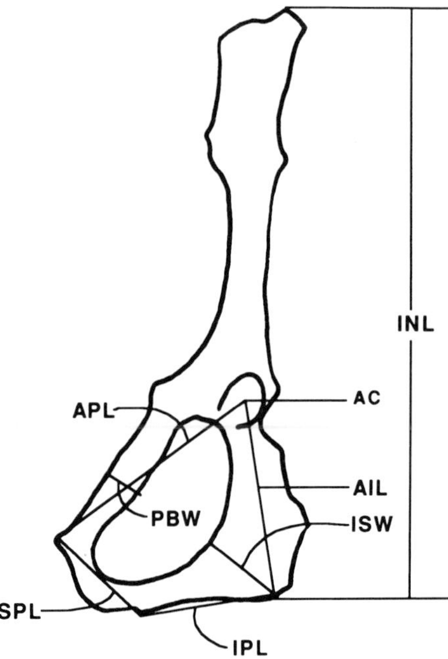

FIG. 5 Parameters of the left innominate bone. INL, longitudinal length of the innominate bone; SPL, distance from the upper edge of the pubis to the lower edge; AIL, distance from the center of acetabulum (AC) to the lower edge of ischium; IPL, distance from lower ischium edge to the lower pubis edge; APL, distance from AC to the upper pubis edge; ISW and PBW, widths of ischium and pubis, respectively.

perinatal period. Sexual dimorphism of the innominate bone was found in 90-day-old T-strain rats and Chinese hamsters (Uesugi *et al.*, 1992a).

Sexual differences in the pubis and the ischium appeared at 30 and 120 days, respectively (Iguchi *et al.*, 1989a). The pubis in female mice was longer and thinner than that in the males, and the ischium in male mice was shorter and thicker than that in the females of 14 strains of mice (Uesugi *et al.*, 1990, 1992b). No such sexual differences were found until 30 days in C57BL and ICR mice, indicating that some sexual differences in the pelvic bones appear during the prepubertal period. Serum androgen levels in male mice increase from 30 to 50 days of age (Sermanoff *et al.*, 1977), suggesting that in male mice the shape of pelvic bones is determined by prepubertally secreted androgens. However, Gardner (1936) showed that long-term administration of estrogen caused a reduction of the pubis width in male mice. Because the ratio of the ischium width to IL in 120-day-old female mice was significantly lower than in 30-day-old females, ovarian estrogen secreted postpubertally may participate in the formation

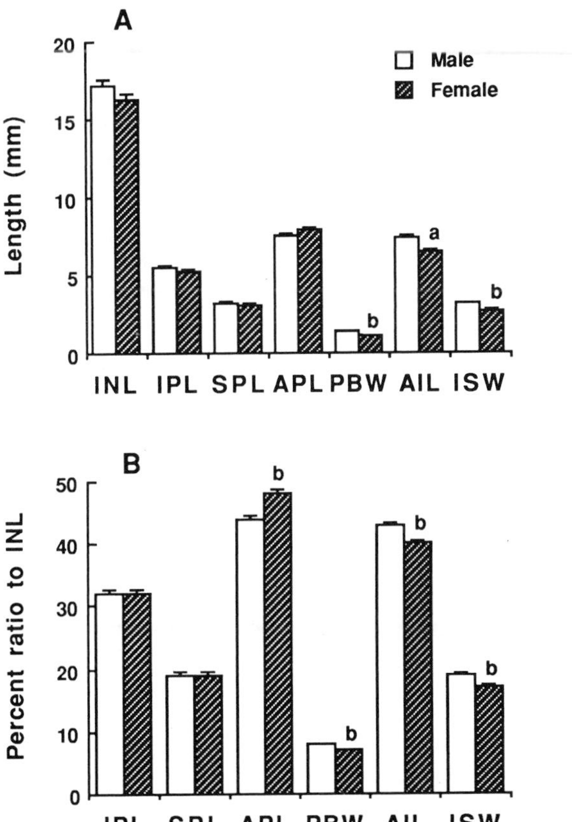

FIG. 6 Length and percentage ratio of parameters of the innominate bone to the longitudinal length (INL) of the bone in 120-day-old C57BL mice. The legend to Figure 5 provides the key to the abbreviations. $^{a}p < .002$, $^{b}p < .001$ versus male.

of the female pelvic bones (Iguchi *et al.*, 1989a). Estrogen receptors were immunohistochemically detected in mesenchymal cells surrounding pubis and ischium, periosteum, and osteocytes of pubis and ischium (Fig. 7), but not of the innominate bone, in both male and female newborn mice. Weak staining for the receptors was also observed in the chondrocytes of pubis and ischium (Uesugi *et al.*, 1992c).

E. Effects of Neonatal Treatment with Sex Hormones on Mouse Pelvis

Estrogen feminizes the bone structure of the pelvis and pubic symphysis in many mammals (De Fremery *et al.*, 1931; Gardner, 1936; Talmage,

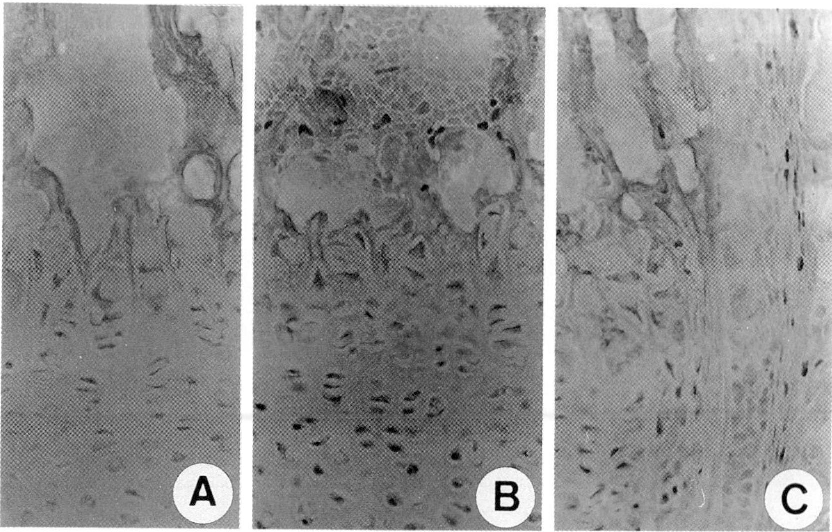

FIG. 7 Immunohistochemical localization of estrogen receptor in pubic bones of 15-day-old control female mice. (A) Pubic bone stained with control serum. (B and C) Pubic bones stained with monoclonal antibody against estrogen receptor. Estrogen receptor is present on nuclei of cells adjacent to trabeculae and cartilage cells (B) and periosteal cells (C). ×200.

1946, 1947). A breakdown of symphysial bone and cartilage and their replacement by connective tissue occurs, and in some species, the ventral pubic structures are completely resorbed coincidentally with the onset of puberty and estrogen secretion (Hisaw, 1925).

McLusty and Naftolin (1981) reported that in rats and mice neural sexual differentiation takes place during the critical period within 10 days after birth. Therefore, the innominate bones from male mice castrated on the day of birth and from female mice given daily injections of 20 μg testosterone and 5α-DHT for 5 days starting on the day of birth were examined at 30 days. The ratio of the pubis width to IL in neonatally androgen-treated females at 30 days was greater than in the age-matched untreated females, whereas this ratio was smaller in neonatally castrated 30-day-old males than in age-matched intact males (Iguchi *et al.*, 1989a). In adult testicular-feminized male (Tfm) mice lacking androgen receptors, ischium length and ischium width were significantly greater than that in the wild-type males. Pubis width in Tfm mice was intermediate between those of wild-type males and females. The ischium in females and castrated males was shorter and thinner than that in the males. The pubis in gonadectomized males and females was wider than in intact females and smaller than in intact males. The pubis in intact males and castrated males was shorter than in intact females. On the basis of these findings, the basic type of ischium is of the female type; postnatal endogenous androgen modifies the ischium to the male phenotype, and the pubis phenotype is intermediate between males and females; postnatal endogenous androgen induces the male type and postpubertal endogenous estrogen induces the female type (Fig. 8) (Uesugi *et al.*, 1990, 1992b). These results suggest that the shape of the innominate bone is transformed to the male type under the influence of early postnatal androgen.

Sex steroids are important for both the normal growth of the skeleton (Southwick and Crelin, 1969; Krabbe *et al.*, 1979; Turner *et al.*, 1987a) as

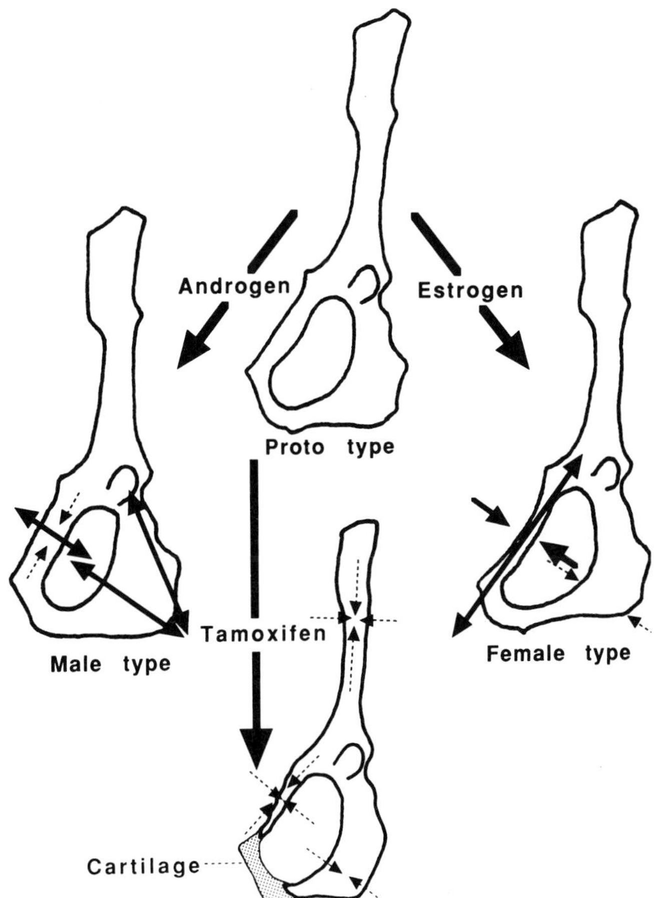

FIG. 8 Schematic drawings of mouse innominate bone modified by sex hormones and tamoxifen. Intermediate and small arrows show bone parts affected by physiological and supraphysiological amounts of hormones, respectively.

well as bone maintenance in adults (Riggs *et al.*, 1969; Aitken *et al.*, 1972; Johnston *et al.*, 1985; Wronski *et al.*, 1985; Slemenda *et al.*, 1987; Turner *et al.*, 1988). Intact gonadal function is essential for the sexual dimorphism of rat tibia (R. T. Turner *et al.*,1989) and mouse pelvis (Iguchi *et al.*, 1989a; Uesugi *et al.*, 1990). Reduction in serum estrogen during menopause in humans (Riggs *et al.*, 1969; Johnston *et al.*, 1985; Slemenda *et al.*, 1987) or by ovariectomy in both humans (Aitken *et al.*,1973) and rats (Wronski *et al.*, 1985, 1986, 1988; Turner *et al.*, 1987a, 1988) increases both bone formation and bone resorption. Estrogen treatment prevents these events in humans (Riggs *et al.*, 1969; Aitken *et al.*, 1972) and rats (Turner *et al.*, 1987a, 1988; Wronski *et al.*, 1988) by decreasing both bone formation and resorption, with the net effect of slowing the loss of bone mass. R. T. Turner *et al.* (1990) showed that DES treatment reduced periosteal bone formation and apposition rates in 8-week-old ovariectomized Sprague-Dawley rats. DES treatment had no effect on endosteal bone formation but suppressed endosteal bone resorption. Furthermore, DES treatment resulted in dramatic decreases in steady state mRNA levels for the bone matrix proteins: osteocalcin, prepro-α2(I) chain of Type I collagen, osteonectin, and osteopontin, as well as the osteoblast marker enzyme alkaline phosphatase. These results suggest that the inhibitory effects of estrogen on bone growth in rats are mediated, or at least accompanied, by the inhibition of the expression of bone matrix protein genes in periosteal cells.

Medroxyprogesterone acetate (MPA) is an effective, reversible, contraceptive agent (Rosenfield *et al.*, 1983). Carbone *et al.* (1990) studied embryofetal toxicity, teratogenicity, and effects on embryonic limb development of MPA administered at 5 and 50 mg/kg/day on gestational days 7–19 of C57BL/6J mice. MPA was embryotoxic at 50 mg/kg/day; however, it did not induce nongenital teratogenesis or limb defects.

F. Effects of Neonatal Treatment with Antihormones on Mouse Pelvis

The pubic bones in the mouse facilitate parturition by relaxation of the symphysis pubis under the influence of relaxin produced at parturition (Hall, 1947, 1957; Viell and Struck, 1987). Iguchi *et al.* (1986a) demonstrated that neonatal tamoxifen treatment caused a long-lasting inhibition of pubic bone calcification, suggesting that the elastic and cartilaginous nature of the symphysis region continued into adulthood (Figs. 9 and 10). Neonatally tamoxifen-exposed mice showed hernia of urinary bladder with or without descent of the cecum through the subpubic space. Although the mechanism of the bladder hernia is unknown, it may be related to the

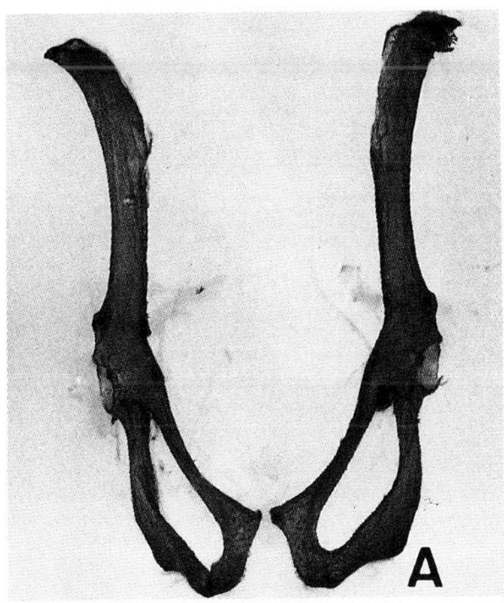

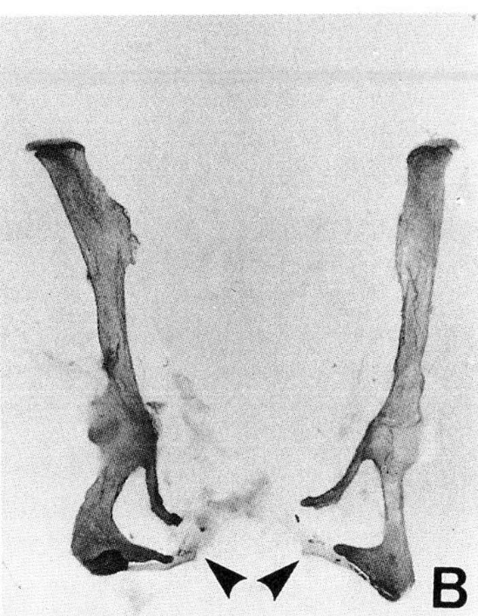

FIG. 9 Pelvis of 540-day-old control (A) and neonatally tamoxifen-treated (B) female mice. The junctional regions between a pair of pubic bones and between pubic and ischial bones are cartilaginous in B (arrowheads). ×2.5.

modified symphysis pubis. Mice treated with tamoxifen starting at 0–10 days had significantly longer pubic ligaments than did the corresponding controls (Figs. 11 and 12). However, mice treated neonatally with clomiphene and nafoxidine possessed normal pubic bones.

In 120-day-old female mice treated neonatally with 100 μg tamoxifen, the total area of the pelvis, and the individual areas of the ilium, ischium, and pubis were significantly smaller than in the controls. There was no significant difference in the length of ischium between tamoxifen-treated and control mice of both sexes. However, lengths of ilium and pubis, and widths of ilium, pubis, and ischium in tamoxifen-treated male and female mice were significantly smaller than in the respective controls. In contrast, neonatal treatment with 2 μg DES for 5 days from the day of birth did not affect the shape of pelvis of either sex. These results suggest that neonatally administered tamoxifen mainly retards the growth of ilium and pubis in mice (Fig. 8) (Uesugi *et al.*, 1990, 1992c).

Parameters of bone resorption were measured on the sections in the endosteal area of the pubic bone. The number of active osteoclasts was counted per unit area of bone section. In 15-day-old mice given neonatal injections of tamoxifen, the osteoclastic surface, the number of osteoclasts per unit area, and the number of nuclei per osteoclast were significantly smaller than those in the controls. Ratio of ossified area to pubic bone was significantly larger (Uesugi *et al.*, 1990). These findings indicate that neonatally injected tamoxifen suppressed osteoclastic activity. Inhibition

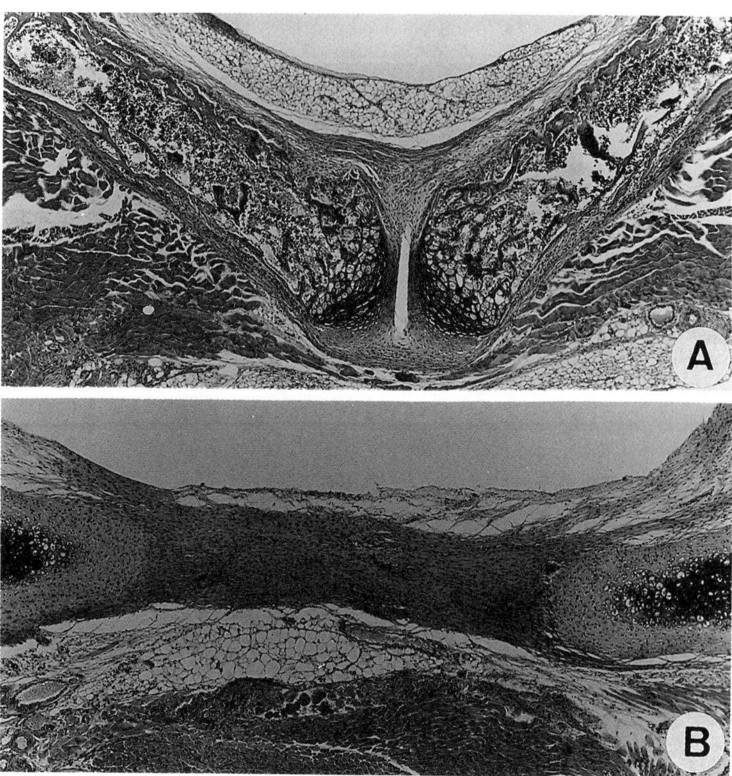

FIG. 10 Longitudinal sections of symphysis pubis in 30-day-old control (A) and neonatally tamoxifen-treated (B) female mice. Pubic ligament in B is expanded and pubic bones contain a large number of cartilage cells; marrow cavity is absent. ×46.

of ossification persisted in the junction of pubis and ischium of pelvis transplanted under the kidney capsule after treatment with tamoxifen *in vitro* (Uesugi *et al.*, 1992c). Further studies are needed to clarify alteration of gene expression of bone matrix proteins and alkaline phosphatase (R. T. Turner *et al.*, 1990) in the innominate bone of mice treated neonatally with tamoxifen.

Estrogen is required to maintain bone density and prevent osteoporosis (Takano-Yamamoto and Rodan, 1990). Tamoxifen is an antiestrogen, but there appears to be a target site specificity to its action (Jordan, 1990). Tamoxifen has an estrogen-like effect upon bone in ovariectomized adult rats (Turner *et al.*, 1987b, 1988); although it inhibits estrogen-stimulated increase in ovariectomized rat uterine wet weight, it has an additive estrogenic effect on bone density in rats (Jordan *et al.*, 1987). Fentiman *et al.*

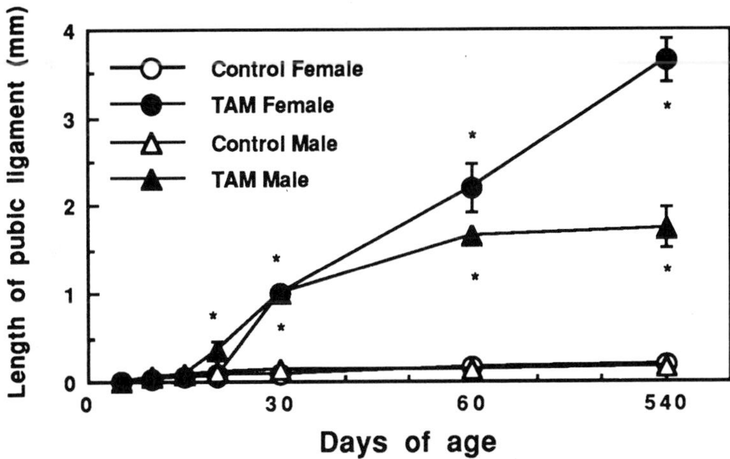

FIG. 11 Changes in length of pubic ligament in mice treated neonatally with tamoxifen. $*p < .005$ versus respective controls.

(1989) reported that tamoxifen administration (10 or 20 mg/day) for short periods (3–6 months) did not influence spinal or femoral bone density, osteocalcin, alkaline phosphatase, and electrolytes in humans.

VI. Nongenital Abnormalities Induced by Prenatal Exposure to Phytoestrogens

Zearalenone is an estrogenic mycotoxin produced by several species of Fusarium (Eugenio *et al.,* 1970; Kallela and Korpinen, 1973; Steele *et al.,* 1974). Enlargement of the uteri and mammary glands, vaginal prolapse, and testicular atrophy in rats and mice have been reported as its estrogenic effects (Mirocha *et al.,* 1968; 1974). Miller *et al.* (1973) have reported that prenatal injections of zearalenone to sows induced stillbirth and splayleg. Offspring of female rats treated with 1, 5, and 10 mg/kg/day zearalenone from days 6 through 15 of gestation showed skeletal defects, such as delayed or absent ossification in the rib, sternum, tarsus, and parietal. The incidence of skeletal defects increased with increasing doses: 12.8%, 26.1%, and 36.8% with 1, 5, and 10 mg/kg, respectively (Ruddick *et al.,* 1976). No obvious skeletal abnormalities, however, were noted in female mice treated neonatally with coumestrol and zearalenone (C. Burroughs and H. A. Bern, personal communication).

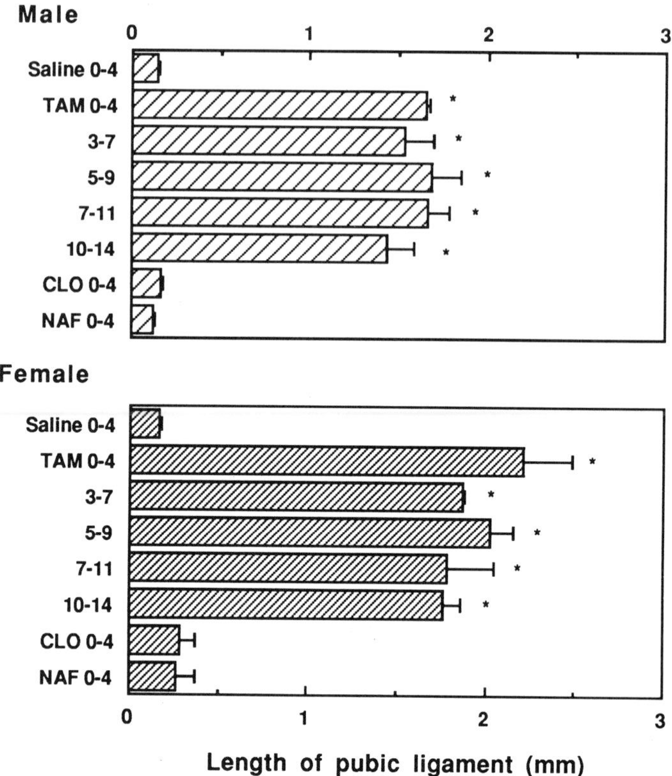

FIG. 12 Length of pubic ligament in 60-day-old mice treated neonatally with tamoxifen and other antiestrogens. $*p < .05$ versus saline.

VII. Possible Mechanisms of Action of Antihormones on Genital and Nongenital Organs

Interaction of tritium-labeled tamoxifen with rat and guinea pig uterine cytosol was described by Sutherland and Foo (1979), Faye *et al.* (1980, 1986), and Gulino and Pasqualini (1980). Ignar-Trowbridge *et al.* (1991) demonstrated that tritium-labeled tamoxifen exclusively binds to the protein in the nuclear fraction of mouse uterine epithelial and stromal cells. In the uterus tamoxifen and estradiol bind to a similar number of saturable binding sites,and estradiol could completely inhibit the binding of radiolabeled antiestrogens to these sites. Other studies (Sutherland and Murphy, 1980; Sutherland *et al.*, 1980) indentified antiestrogen binding sites in the cytosol of estrogen receptor-positive breast tumors and several estrogen target tissues, including immature rat uterus. The target site specificity, however, is controversial because estrogen receptor-negative tumors (Jordan *et al.,* 1981; Miller and Katzenellenbogen, 1983) and all human tissues tested have antiestrogen binding sites (Kon, 1983). In addition to binding directly to the estrogen receptor, antiestrogens interact with the binding sites with which estrogen does not compete (Sutherland *et al.,* 1980). Antiestrogen binding sites are present in a wide variety of tissues from several species of mammals: rat, mouse, guinea pig, and human (Gulino and Pasqualini, 1982; Kon, 1983; Sudo *et al.,* 1983). The antiestrogen binding sites in rats are highly concentrated in the liver, which shows about 10 times more binding than the uterus, ovary, and brain; even lower binding occurs in the lung and spleen (Sudo *et al.,* 1983). Antiestrogen binding sites in the uterus, vagina, and liver have been demonstrated *in vivo* in immature rats (Jordan and Bowser-Finn, 1982).

Miller and Katzenellenbogen (1983) have compared three breast cancer cell lines, MCF-7, T47D, and MDA-MB-231, containing similar levels of antiestrogen binding sites, although they have high, low, and intermediate levels of estrogen receptors, respectively. Tamoxifen inhibits the growth of the cells depending upon the presence of the estrogen receptors. This finding is similar to the results from human breast cancer patients (McGuire, 1979; DeSombre and Jensen, 1980). Sutherland *et al.* (1980) have shown that the presence of high affinity antiestrogen binding site distinct from estrogen receptor and high concentrations of antiestrogens that inhibit the growth of breast cancer cells cannot be reversed by estrogens. Antiestrogens inhibit the prolactin-responsive growth of rat Nb2 lymphoma cells, which are unresponsive to estrogen and devoid of estrogen receptors. Specific prolactin binding to these cells is inhibited by antiestrogens, and the potency of the various antiestrogens in this system parallels their ability to bind to the antiestrogen binding sites (Biswas and Vonderhaar, 1989). Biswas and Vonderhaar (1991) also showed that triphenylethylene antiestrogens, such as tamoxifen, 4-hydroxytamoxifen, and nafoxidine, selectively inhibited prolactin binding to microsomal membranes isolated from mammary glands of lactating mice and prevented prolactin-induced accumulation of caseins by cultured mouse mammary explants, acting through the antiestrogen binding sites present in the microsomal fraction. This represents a possible specific method for controlling tumor growth and casein production by a nonestrogen receptor-mediated mechanism.

Winneker and Clark (1983) also found antiestrogen binding sites in rat liver and uterus; in serum antiestrogens bind to a similar low density lipoprotein. The low density lipoprotein is involved in the control of cholesterol metabolism (Goldstein and Brown, 1979). Cholesterol synthesis is inhibited by antiestrogens (Clark and Markaverich, 1982). It is proposed, therefore, that the low density lipoprotein–antiestrogen binding sites may be involved in the control of cholesterol metabolism (Clark *et al.*, 1986).

Antiestrogens can bind to estrogen receptors (Borgna and Rochefort, 1980; Ignar-Trowbridge *et al.*, 1991), antiestrogen binding sites (Sutherland *et al.*, 1980; Sudo *et al.*, 1983; Watts *et al.*, 1984), and calmodulin (Lam, 1984). Tamoxifen inhibits calmodulin-mediated phosphodiesterase (Lam, 1984) by binding to calmodulin (Lopes *et al.*, 1990). The other recognized inhibitor of calmodulin is a major tranquilizer, trifluoperazine, which is an effective inhibitor of radiolabeled tamoxifen binding to rat liver "antiestrogen binding sites," resulting in the inhibition of colony formation by breast cancer cells (Wei *et al.*,1983). Calmodulin is believed to be involved in cell division (Chafouleas *et al.*, 1984; Willingham *et al.*, 1984), and its inhibitor, trifluoperazine, blocks the G1 phase of the cell cycle (Ito and Hidaka, 1983). Trifluoperazine has been shown to prevent the binding of iodinated EGF to neoplastic, but not to normal, cells in culture (Bodine and Tupper, 1984). This observation may explain the antiproliferative actions of tamoxifen on estrogen receptor-positive or -negative cells in culture (Green *et al.*, 1981).

Neonatal treatment of female C57BL/Tw mice with 20 μg trifluoperazine induced no abnormalities in reproductive tracts and bones except for reduction of spermatogenesis (Chou *et al.*, 1992). Brandes *et al.* (1986) demonstrated that a novel histamine antagonist, *N-N*-diethyl-2-[(4-phenylmethyl)-phenoxy]-ethanamine HCl, selectively binds with high affinity to the antiestrogen-binding site. This suggests that the antiproliferative properties of tamoxifen may be mediated by a histamine receptor.

Tamoxifen inhibits protein kinase C by interfering with the activity of the catalytic subunit of the enzyme (O'Brian *et al.*, 1985, 1986, 1988; Su *et al.*, 1985; Nakadate *et al.*, 1988). Protein kinase C has important roles in controlling normal and abnormal cell proliferation (Kikkawa *et al.*, 1983; Neidel *et al.*, 1983; Nishizuka, 1986, 1988). Tamoxifen inhibits proliferation of some estrogen receptor-negative cell lines (O'Brian *et al.*, 1986; Chouvet *et al.*, 1988) and human malignant glioma cell lines *in vitro* (Pollack *et al.*, 1990), acting through a mechanism independent of estrogen receptor binding, possibly by protein kinase C inhibition.

Tamoxifen also inhibits cell proliferation and voltage-dependent K^+ channels of neuroblastoma cells (Rouzaire-Dubois and Dubois, 1990), suggesting that cell mitosis is, in some way, controlled by functioning K^+ channels and that the antitumor actions of tamoxifen could be due to its interaction with K^+ channels.

Insulin-like growth factor I (IGF-I) is a potent mitogen for several types of cells, and IGF-I receptors are also present on these cells (Sara and Hall,

1990). Pollak *et al.* (1990) demonstrated that tamoxifen suppressed the growth of breast cancer cells by reducing the serum IGF-I level in patients.

Based on the literature on tamoxifen action at the cell level, the observed effects on neonatal genital and nongenital tissues could be mediated by (1) antiestrogen binding sites (Gulino and Pasqualini, 1982; Sudo *et al.*, 1983), (2) estrogen receptors (Borgna and Rochefort, 1980; Sutherland and Murphy, 1980), (3) growth factors (Pollack *et al.*, 1990), (4) calmodulin (Lam, 1984), (5) protein kinase C (O'Brian *et al.*, 1986; Nakadate *et al.*, 1988), and /or (6) K^+ channels (Rouzaire-Dubois and Dubois, 1990).

During long-term adjuvant tamoxifen therapy, the levels of tamoxifen and its metabolites are high and tissues will be saturated with antiestrogens. Even if tamoxifen were stopped immediately after the patient discovered her pregnancy, the long plasma half-life at steady state would result in the drug being present for at least 6 weeks (Furr and Jordan, 1984; Jordan, 1982, 1990). A fetus could thus be exposed to the drug throughout the first trimester. Therefore, as suggested by Jordan and Murphy (1990), tamoxifen should not be taken during pregnancy.

VIII. Conclusions

Sex hormones and related compounds are potent molecules, which can induce a wide variety of abnormalities in male and female mice when given during the critical developmental period of genital and nongenital target organs. Perinatal treatment of female mice with sex hormones, especially natural and synthetic estrogens and phytoestrogens, causes abnormalities in reproductive organs: estrogen-dependent or estrogen-independent persistent proliferation and cornification of vaginal epithelium, adenosis, oviducal hypertrophy, uterine metaplasia, hypospadias, aplasia of uterine glands, disorganization of uterine myometrium, PFs in ovaries, and infertility resulting from alterations of the hypothalamohypophyseal-ovarian axis. In male mice perinatal treatment with estrogens and androgens induces persistent suppression of spermatogenesis, atrophy of seminal vesicle and prostate, and gubernaculum degeneration resulting in cryptorchidism. Prostatic and Leydig cell tumors develop in aged prenatally DES-exposed mice. Estrogen receptors in vagina, uterus, and prostate, and androgen receptors in prostate of neonatally DES-exposed mice are reduced compared with the controls. Levels of FSH and LH receptors in the ovary and testis, respectively, of neonatally DES-exposed mice are also reduced. In addition, EGF receptors are reduced in the vagina, but not in the uterus and oviduct, of neonatally DES-exposed mice. Perinatal sex hormone treatment results in alterations of protein profiles in vagina, uterus, seminal vesicle, and prostate.

Neonatal treatment of mice with antiestrogens (tamoxifen, clomiphene, and nafoxidine) induces infertility by modification of the hypothalamo-hypophyseal-ovarian axis, vaginal adenosis, uterine hypoplasia, ovarian dysgenesis, PFs, and reduction of decidual response in rat uterus to an artificial stimulus in female mice, and persistent suppression of spermatogenesis and atrophy of seminal vesicle, epididymis, and gubernaculum in male mice. These so-called antiestrogens increase weight, protein, and DNA in the uterus of ovariectomized adult mice. Neonatal injections of keoxifene induce infertility and slight suppression of spermatogenesis in mice; however, this antiestrogen has no stimulating effect on uterine weight or on protein and DNA synthesis in the ovariectomized adult.

In addition to these abnormalities in reproductive organs, perinatal treatment with sex hormones affects the development of bones such as those of the pelvis. The pelvis (paired innominate bones) is comprised of four separate units: ilium, ischium, pubis, and acetabulum. The pubes are joined at the ventral midline by the pubic symphysis. The shape of the ischium is transformed to the male type under the influence of early postnatal androgen. The shape of the pubis is transformed to the female type under the presence of peripubertal estrogen. Estrogen receptors are present exclusively in the periosteum of the pubis on the day of birth; however, they appear in bone cells of the entire pelvis after 10 days of age. Permanent chondrification of the pubic and ischial bones is found in all mice given tamoxifen starting within 10 days of age. The pubic ligament of mice treated with tamoxifen within 5 days markedly expands, accompanied by bladder hernia with or without cecum hernia. Inhibition of ossification persists in the junction of pubis and ischium of pelvis treated with tamoxifen *in vitro*. In contrast, neonatal treatment of mice with other "antiestrogens," estrogen and androgen induces neither permanent chondrification in the pelvis nor expansion of the pubic symphysis. Neonatally administered tamoxifen mainly affects the pubis and some junctional regions in mice by changing activities of osteoclasts and osteoblasts, acting directly to inhibit ossification.

In neonatally tamoxifen-treated mice, fibrocartilage in the distal segment of the os penis, hyaline cartilage characterized by the presence of Type II collagen, and bone marrow in the proximal segment disappear by 30 days. In clomiphene- and nafoxidine-treated mice the distal segment lacks fibrocartilage. Neonatal castration does not suppress the formation of bone marrow and fibrocartilage in the os penis, although its size is smaller than in controls. Formation of spines on the glans penis skin is suppressed by tamoxifen given within 5 days of birth. Thus, neonatally administered tamoxifen inhibits the postnatal differentiation of the mouse os penis, resulting in aplasia of the distal segment and of bone marrow in the proximal segment and in reduced formation of spines on the glans penis.

The study of animals treated perinatally with sex hormones and related compounds provides an opportunity to analyze various factors influencing developmental and carcinogenic processes. The ovary-independent cornified vaginal epithelium of perinatally estrogen-treated mice progresses to epidermoid carcinoma; undifferentiated columnar cervicovaginal epithelium transforms into adenosis and adenocarcinoma or into epidermoid carcinoma. In the uterus metaplasia, cystic hyperplasia, and tumors develop in perinatally estrogen-exposed mice. Also, perinatal estrogen induces teratological abnormalities, including developmental arrest of the oviduct, hypoplasia of the uterus, clitoridal hypospadia, and retention of Wolffian structures. Some abnormalities induced by perinatal exposure to estrogen in rodents, such as the high incidence of adenosis and the occurrence of tumors, are similar to those that occur in human.

As Takasugi and Bern (1988) mentioned in their review:

> The central point is that the developing organism is subject to a variety of stimuli, some of which are still undefined, the influence of which during a critical period in development may result not only in teratological alterations that are readily discernible but also in subtle changes expressed much later in life. One expression of these delayed effects could be an alteration in tumor risk in tissues affected directly or indirectly by the initial stimulus.

Thus, the prenatal and neonatal mouse models continue to provide information of possible occurrence of genital abnormalities in human offspring exposed to sex hormones and related compounds, including antihormones, enzyme inhibitors, and tranquilizers. In further studies, however, as indicated in this review, more attention should be paid to nongenital organs exposed to various agents during fetal and early postnatal development in mammals including humans.

Acknowledgments

The author thanks Dr. Noboru Takasugi, President of Yokohama City University and Emeritus Professor Howard A. Bern, Department of Integrative Biology, University of California at Berkeley and Guest Professor at Yokohama City University, for their continuous encouragement, valuable advice, and critical reading of this review. Some studies described herein were supported by Grants-in-Aid from the Ministry of Education, Science and Culture of Japan (01540618, 04640690), a grant from the Kihara Foundation for Scientific Research, and a grant in Support of the Promotion of Research at Yokohama City University.

References

Aihara, M., Kimura, T., and Kato, J. (1980). *Endocrinology (Baltimore)* **107,** 224–230

Aitken, J. M., Armstrong, E., and Armstrong, J. B. (1972). *J. Endocrinol.* **55,** 79–87.

Aitken, J. M., Hart, D. M., and Lindsay, R. (1973). *Br. Med. J.* **3,** 515–518.

Allen, N., Mirocha, C. J., Weaver, G., Aakus-Allen, S., and Bates, F. (1981). *Poult. Sci.* **60,** 124–131.

Anderson, C. A., and Clark, R. L. (1990). *Teratology* **42,** 483–496.

Arai, Y., Suzuki, Y., and Nishizuka, Y. (1977). *Virchows Arch. A* **376,** 21–28.

Arai, Y., Mori, T., Suzuki, Y., and Bern, H. A. (1983). *Int. Rev. Cytol.* **84,** 235–268.

Arnold, L. (1912). *Anat. Rec.* **6,** 413–422.

Baggs, R. B. Miller, R. K., and Odoroff, C. L. (1991). *Cancer Res.* **51,** 3311–3315.

Bailey, D. W. (1986). *J. Hered.* **77,** 17–25.

Barraclough, C. A. (1967). *In* "Neuroendocrinology" (L. Martini and W. F. Ganong, eds.), pp. 61-99, Academic Press, New York.

Baum, M., and other members of Norvadex Adjuvant Trial Organization (1985). *Lancet* **1,** 836–840.

Bennett, G. A., and Shotwell, O. L. (1979). *J. Am. Oil Chem. Soc.* **56,** 812–819.

Bern, H. A. (1979). *Proc. Symp. Endocr.-Induced Neoplasia, Epply Inst. Res. Cancer, Univ. Nebraska Med. Center, Omaha* pp. 31–37.

Bern, H. A. (1992). *In* "Hormonal Carcinogenesis" (J. Li, S. Nandi, and S. A. Li, eds.). Springer-Verlag, New York. In press.

Bern, H. A., and Talamantes, F., Jr. (1981). *In* "Developmental Effects of Diethylstilbestrol (DES) in Pregnancy" (A. L. Herbst and H. A. Bern, eds.), pp. 129–147. Thieme-Stratton, New York.

Bern, H. A., Jones, L. A., Mills, K. T., Kohrman, A., and Mori, T. (1976). *J. Toxicol. Environ. Health* **1,** 103–116.

Bern, H. A., Mills, K. T., and Mori, T. (1984a). *Proc. Soc. Exp. Biol. Med.* **177,** 303–307.

Bern, H. A., Mills, K. T., Ostrander, P. L., Schoenrock, B., Graveline, B., and Plapinger, L. (1984b). *Teratology* **30,** 267–274.

Bern, H. A., Mills, K. T., and Edery, M. (1985). *In* "Estrogens in the Environment II, Influences on Development" (J. A. McLachlan, ed.), pp. 319–326. Elsevier, New York.

Bern, H. A., Edery, M., Mills, K. T., Kohrman, A. F., Mori, T., and Larson, L. (1987). *Cancer Res.* **47,** 4165–4172.

Bickoff, E. M., Booth, A. N., Lyman, R. C., Livingston, A. L., and Thompson, C. R. (1957). *Science* **126,** 969–970.

Bickoff, E. M., Livingston, A. L., Hendrickson, A. P., and Booth, A. N. (1958). *J. Agric. Food Chem.* **6,** 536–539.

Bickoff, E. M., Livingston, A. L., Hendrickson, A. P., and Booth, A. N. (1962). *J. Agric. Food Chem.* **10,** 410–412.

Biswas, R., and Vonderhaar, B. K. (1989). *Cancer Res.* **49,** 6295–6299.

Biswas, R., and Vonderhaar, B. K. (1991). *Endocrinology (Baltimore)* **128,** 532–538.

Black, L. J., and Goode, R. L. (1980). *Life Sci.* **26,**1453–1458.

Black, L. J., and Goode, R. L. (1981). *Endocrinology (Baltimore)* **109,** 987–989.

Black, L. J., Jones, C. D., and Goode, R. L. (1981). *Mol. Cell. Endocrinol.* **22,** 95–103.

Bodine, P. V., and Tupper, J. H. (1984). *Biochem. J.* **218,** 629–632.

Borgna, J. L., and Rochefort, H. (1980). *Mol. Cell. Endocrinol.* **20,** 71–85.

Boyd, P. A., and Wittliff, J. L. (1978). *J. Toxicol. Environ. Health* **4,** 1–8.

Brandes, L. J., Bogdanovic, R. P., Cawker, M. D., and Bose, R. (1986). *Cancer Chemother. Pharmacol.* **18,** 21–23.

Branham, W. S., Sheehan, D. M., Zehr, D. R., Medlock, K. L., Nelson, C. J., and Ridlon, E. (1985). *Endocrinology (Baltimore)* **117,** 2238–2248.

Branham, W. S., Zehr, D. R., Chen, J. J., and Sheehan, D. M. (1988a). *Teratology* **38,** 29–36.

Branham, W. S., Zehr, D. R., Chen, J. J., and Sheehan, D. M. (1988b). *Teratology* **38,** 271–279.

Brown, J. L., and Chakraborty, P. K. (1988). *Acta Endocrinol. (Copenhagen)* **118,** 437–443.

Bullock, B. C., Newbold, R. R., and McLachlan, J. A. (1988). *Environ. Health Perspect.* **77,** 29–31.

Burroughs, C. D., Bern, H. A., and Stokstad, E. L. R. (1985). *J. Toxicol. Environ. Health* **15,** 51–61.

Burroughs, C. D., Mills, K. T., and Bern, H. A. (1990a). *J. Toxicol. Environ. Health* **30,** 105–122.

Burroughs, C. D., Mills, K. T., and Bern, H. A. (1990b). *Reprod. Toxicol.* **4,** 127–135.

Campbell, P. S., and Satterfield, P. M. (1988). *J. Reprod. Fertil.* **83,** 225–231.

Carbone, J. P., Figurska, K., Bucks, S., and Brent, R. L. (1990). *Teratology* **42,** 121–130.

Chafouleas, J. G., Lagace, L., Boulton, W. E., Boyd, A. E., and Means, A. R. (1984). *Cell* **36,** 73–81.

Chamness, G. C., Bannayan, G. A., Landry, L. A., Jr., Scheridan, P. J., and McGuire, W. L. (1979). *Biol. Reprod.* **21,** 1087–1090.

Chang, M. C. (1959). *Endocrinology (Baltimore)* **65,** 339–342.

Chapmann, R. N. (1919). *Am. J. Anat.* **25,** 185–219.

Chou, Y.-C., Iguchi, T., and Bern, H. A. (1992). *Reprod. Toxicol.* (in press).

Chouvet, C., Vicard, E., Frappart, L., Falette, N., Lefebvre, M. F., and Saez, S. (1988). *J. Steroid Biochem.* **31,** 655–663.

Clark, J. H., and Markaverich, B. M. (1982). *Pharmacol. Ther.* **15,** 467–519.

Clark, J. H., and McCormack, S. (1977). *Science* **197,** 164–165.

Clark, J. H., and McCormack, S. A. (1980). *J. Steroid Biochem.* **12,** 47–53.

Clark, J. H., Markaverich, B. M., Guthrie, S. C., and Winneker, R. C. (1986). "Estrogen/Antiestrogen Action and Breast Cancer Therapy," pp. 115–126. Univ. of Wisconsin Press, Madison.

Clark, R. L., Antonello, J. M., Grossman, S. J., Wise, L. D., Anderson, C., Bagdon, W. J., Prahalada, S., MacDonald, J. S., and Robertson, R. T. (1990). *Teratology* **42,** 91–100.

Cole, M. P., Jones, C. T. A., and Todd, I. D. H. (1971). *Br. J. Cancer* **25,** 270–275.

Corner, G. W. (1923). *Am. J. Anat.* **31,** 523–545.

Csaba, G., Inczefi-Gonda, A., and Dobozy, O. (1986). *Acta Phys. Hung.* **67,** 207– 212.

Cunha, G. R., Donjacour, A. A., Cooke, P. S., Mee, S., Bigsby, R. M., Higgins, S. J., and Sugimura, Y. (1987a). *Endocr. Rev.* **8,** 338–362.

Cunha, G. R., Taguchi, O., Namikawa, R., Nishizuka, Y., and Robboy, S. J. (1987b). *Hum. Pathol.* **18,** 1132–1143.

Cunha, G. R., Taguchi, O., Sugimura, Y., Lawrence, W. D., Mahmood, F., and Robboy, S. J. (1988). *Hum. Pathol.* **19,** 777–783.

Davies, J., and Lefkowitz, J. (1987). *Acta Anat.* **130,** 351–358.

Davies, J., Russell, M., and Davenport, G. R. (1985). *Acta Anat.* **122,** 39–61.

Davis, D. E., and Hall, O. (1950). *Anat. Rec.* **107,** 187–192.

Dawson, A. B. (1951). *Anat. Rec.* **110,** 181–197.

Dederer, P. H. (1934). *Anat. Rec.* **60,** 391–403.

De Fremery, P., Kober, S., and Tausk, M. (1931). *Acta Brevia Neerl.* **1,** 146–148.

DeSombre, E. R., and Jensen, E. V. (1980). *Cancer (Philadelphia)* **46,** 2783–2790.

Dieckmann, W. E., David, M. E., Rynkiewicz, S. M., and Pottinger, R. E. (1953). *Am. J. Obstet. Gynecol.* **66,** 1062–1081.

Dieringer, C. S., Lamartiniere, C. A., and Lucier, G. W. (1979). *J. Steroid Biochem.* **13,** 1449–1453.

Dix, C. J., and Jordan, V. C. (1980). *Endocrinology (Baltimore)* **107,** 2011–2020.

Dodds, E. C., Goldberg, L., Lawson, W., and Robinson, R. (1938a). *Nature (London)* **141,** 247–248.

Dodds, E. C., Lawson, W., and Noble, R. L. (1938b). *Lancet* **i,** 1389–1391.

Döhler, K.-D., Coquelin, A., Davis, F., Hines, M., Shryne, J. E., and Gorski, R. A. (1982a). *Neurosci. Lett.* **33,** 295–298.

Döhler, K.-D., Hines, M., Coquelin, A., Davis, F., Shryne, J. E., and Gorski, R. A. (1982b). *Neuroendocr. Lett.* **4,** 361–365.

Döhler, K.-D., Coquelin, A., Davis, F., Hines, M., Shryne, J. E., and Gorski, R. A. (1984a). *Brain Res.* **302,** 291–295.

Döhler, K.-D., Srivastava, S. S., Shryne, J. E., Jarzab, B., Sipos, A., and Gorski, R. A. (1984b). *Neuroendocrinology* **38,** 297–301.

Döhler, K.-D., Coquelin, A., Davis, F., Hines, M., Shryne, J. E., Sickmöller, P. M., Jarzab, B., and Gorski, R. A. (1986). *Neuroendocrinology* **42,** 443–448.

Dunn, T. B., and Green, A. W. (1963). *J. Natl. Cancer Inst.* **31,** 425–455.

Edery, M., Mills, K. T., and Bern, H. A. (1989). *Biol. Neonate* **56,** 324–331.

Edery, M., Turner, T., Dauder, S., Young, G., and Bern, H. A. (1990). *Proc. Soc. Exp. Biol. Med.* **194,** 289–292.

Eiger, S., Mills, K. T., and Bern, H. A. (1990) *J. Steroid Biochem.* **135,** 617–621.

Engle, E. T. (1927). *Anat. Rec.* **35,** 341–343.

EORTC (European Organization for Research on Treatment of Cancer) (1972). *Eur. J. Cancer* **8,** 387–389.

Etienne, M., and Jemmali, M. (1982). *J. Anim. Sci.* **55,** 1–11.

Eugenio, C. P., Christensen, C. M., and Mirocha, C. J. (1970). *Phytopathology* **60,** 1055–1057.

Faye, J. C., Lasserre, B., and Bayard, F. (1980). *Biochem. Biophys. Res. Commun.* **93,** 1225–1231.

Faye, J. C., Fagin, A., and Bayard, F. (1986). *Mol. Cell. Endocrinol.* **47,** 119-124.

Fekete, E. (1946). *Cancer Res.* **6,** 263–269.

Fekete, E. (1950). *Anat. Rec.* **108,** 699–707.

Fentiman, I. S., Caleffi, M., Rodin, A., Murby, B., and Fogelman, I. (1989). *Br. J. Cancer* **60,** 262–264.

Festing, M. (1972). *Nature* (*London*) **238,** 351–352.

Fitzpatrick, D. W., Picken, C. A., Murphy, L. C., and Buhr, M. M. (1989). *Comp. Biochem. Physiol. C* **94,** 691–694.

Forsberg, J.-G. (1969). *Br. J. Exp. Pathol.* **50,** 187–195.

Forsberg, J.-G. (1979). *Natl. Cancer Inst. Monogr.* **51,** 41-56.

Forsberg, J.-G. (1985). *Biol. Reprod.* **32,** 427–441.

Forsberg, J.-G., and Kalland, T. (1981). *Cancer Res.* **41,** 721–734.

Forsberg, J.-G., Tenenbaum, A., Rydberg, C., and Sernvi, S. (1985). *In* "Estrogens in the Environment II, Influences on Development" (J. A. McLachlan, ed.), pp. 327–346. Elsevier, New York.

Fredericks, G. R., Kincaid, R. L., Bondioli, K. R., and Wright, R. W. (1981). *Proc. Soc. Exp. Biol. Med.* **167,** 237–241.

Furr, B. J. A., and Jordan, V. C. (1984). *Pharmacol. Ther.***25,** 127–205.

Gardner, W. U. (1936). *Am. J. Anat.* **59,** 459–483.

Gellert, R. J., Bakke, J. L., and Lawrence, N. L. (1971). *Fertil. Steril.* **22,** 244–250.

George, F. W. (1989). *Endocrinology* (*Baltimore*) **124,** 727–732.

George, F. W., and Peterson, K. G. (1988). *Endocrinology* (*Baltimore*) **122,** 1159–1164.

George, F. W., and Wilson, J. D. (1988). *In*"The Physiology of Reproduction" (E. Knobil and J. D. Neil, eds.), pp. 3–26. Raven, New York.

George F. W., Johnson, L., and Wilson J. D. (1989). *Endocrinology* (*Baltimore*) **125,** 2434–2438.

Gilloteaux, J., and Steegles, A. W. (1985). *Scanning Electron Microsc.* **1,** 303–309.

Gilloteaux, J., and Steegles, A. W. (1986), *Am. J. Anat.* **175,** 429–447.

Gilloteaux, J., Paul, R. J., and Steegles, A. W. (1982). *Virchows Arch. (Pathol.Anat.).* **398,** 163–183.

Glucksmann, A., and Cherry, C. P. (1972). *J. Anat.* **112,** 223–231.

Glucksmann, A., Ooka-Souda, S., Miura-Yasugi, E., and Mizuno, T. (1976). *J. Anat.* **121,** 363–370.

Goldman, A. S., and Neumann, F. (1969). *Proc. Soc. Exp. Biol. Med.* **132,** 237–241.

Goldman, A. S., Shapiro, B. H., and Neumann, F. (1976). *Endocrinology (Baltimore)* **99,** 1490–1495.

Goldstein, J. L., and Brown, M. S. (1979). *Annu. Rev. Genet.* **13,** 259–291.

Gorski, R. A., Harlan, R. E., and Christensen, L. W. (1977). *J. Toxicol. Environ. Health* **3,** 97–121.

Gorski, R. A., Gordon, J. H., Shryne, J. E., and Southam, A. M. (1978). *Brain Res.* **148,** 333–346.

Gorwill, R. H., Steele, H. D., and Sarda, I. R. (1982). *Am. J. Obstet. Gynecol.* **144,** 529–532.

Graham, C. E., and Bradley, C. F. (1971). *J. Reprod. Fertil.* **27,** 181–185.

Green, M. D., Whybourne, A. M., Taylor, I. W., and Sutherland, R. L. (1981). *In* "Nonsteroidal Antioestrogens" (R. L. Sutherland and V. C. Jordan, eds.), pp. 397–412. Academic Press, New York.

Greenblatt, R. B., Barfield, W. E., Jungek, E. C., and Roy, A. W. (1961). *J. Am. Med. Assoc.* **178,** 101–104.

Greenblatt, R. B., Roy, S., and Mahesh, V. B. (1962). *Am. J. Obstet. Gynecol.* **84,** 900–912.

Greene, A. R., Burrill, M. W., and Ivy, A. C. (1941). *J. Exp. Zool.* **87,** 211–232.

Greenman, D. L., Mehta, R. G., and Wittliff, J. L. (1979). *J. Toxicol. Environ. Health* **5,** 593–598.

Gulino, A., and Pasqualini, J. R. (1980). *Cancer Res.* **40,** 3821–3826.

Gulino, A., and Pasqualini, J. R. (1982). *Cancer Res.* **42,** 1913–1921.

Hall, K. (1947). *J. Endocrinol.* **5,** 174–185.

Hall, K. (1957). *J. Endocrinol.* **15,** 108–117.

Halling, A., and Forsberg, J.-G. (1989). *J. Steroid Biochem.* **32,** 439–443.

Halling, A., and Forsberg, J.-G. (1990a). *J. Reprod. Fertil.* **88,** 399–404.

Halling, A., and Forsberg, J.-G. (1990b). *Biol. Reprod.* **43,** 472–477.

Halling, A., and Forsberg, J.-G. (1991). *Biol. Reprod.* **45,** 157–162.

Haney, A. F., Newbold, R. R., and McLachlan, J. A. (1984). *Biol. Reprod.* **30,** 471–478.

Haney, A. F., Newbold, R. R., Fetter, B. F., and McLachlan, J. A. (1986). *Am. J. Pathol.* **124,** 405–411.

Harper, M. J. K., and Walpole, A. L. (1966). *Nature (London)* **212,** 87.

Harper, M. J. K., and Walpole, A. L. (1967). *J. Reprod. Fertil.* **13,** 101–119.

Hartman, C. G. (1926). *Am. J. Anat.* **37,** 1–51.

Heel, R. C., Brogden, R. N., Speight, T. M., and Avery, G. S. (1978). *Drugs* **16,** 1–24.

Henry, E. C., and Miller, R. K. (1986). *Teratology* **34,** 59–63.

Herbst, A. L., and Bern, H. A., eds. (1981). "Developmental Effects of Diethylstilbestrol (DES) in Pregnancy," p. 203 Thieme-Stratton, New York.

Herbst, A. L., and Scully, R. E. (1970). *Cancer* **25,** 745–757.

Herbst, A. L., Griffiths, C. T., and Kistner, R. W. (1964). *Cancer Chemther. Rep.* **43,** 39–41.

Herbst, A. L., Ulfelder, H., and Poskanzer, D. C. (1971). *N. Engl. J. Med.* **284,** 878–881.

Herbst, A. L., Poskanzer, D. C., Robboy, S. J., Friedlander, L., and Scully, R. E. (1975). *N. Engl. J. Med.* **292,** 334–339.

Hines, M., Alsum, P., Roy, M., Gorski, R. A., and Goy, R. W. (1987). *Horm. Behav.* **21,** 402–417.

Hisaw, F. L. (1925). *J. Exp. Zool.* **42,** 411–441.

Holtkamp, D. E., Greslin, S. C., Root, C. A., and Lerner, L. J. (1960). *Proc. Soc. Excp. Biol. Med.* **105,** 197–201.

Horwitz, C., Rozen, P., and Gilat, T. (1983). *Nutr. Cancer* **5,** 51–54.
Howard, E. (1959). *Endocrinology (Baltimore)* **65,** 785–801.
Hsu, H. K., Chen, F. N., and Peng, M. T. (1980). *Neuroendocrinology* **31,** 327–330.
Huppert, L. C. (1979). *Fertil. Steril.* **31,** 1–8.
Ignar-Trowbridge, D. M., Nelson, K. G., Ross, K. A., Washburn, T. F., Korach, K. S., and McLachlan, J. A. (1991). *J. Steroid Biochem. Mol. Biol.* **39,** 131–132.
Iguchi, T. (1984). *Proc. Jpn. Acad., Ser. B* **60,** 414–417.
Iguchi, T. (1985). *Proc. Jpn. Acad., Ser. B* **61,** 288–291.
Iguchi, T., and Hirokawa, M. (1986). *Proc. Jpn. Acad., Ser. B* **62,** 157–160.
Iguchi, T., and Ohta, Y. (1979). *Acta Anat.* **108,** 469–480.
Iguchi, T., and Takasugi, N. (1976). *Endocrinol. Jpn.* **23,** 327–332.
Iguchi, T., and Takasugi, N. (1979). *Anat. Embroyol.* **155,** 127–134.
Iguchi, T., and Takasugi, N. (1981). *Endocrinol. Jpn.* **28,** 207–213.
Iguchi, T., and Takasugi, N. (1983). *IRCS Med. Sci.* **11,** 696–697.
Iguchi, T., and Takasugi, N. (1986). *Anat. Embryol.* **175,** 53–55.
Iguchi, T., and Takasugi, N. (1987). *Biol. Neonate* **52,** 97–103.
Iguchi, T., Ohta, Y., and Takasugi, N. (1976). *Dev. Growth Differ.* **18,** 69–78.
Iguchi, T., Tachibana, H., and Takasugi, N. (1979). *IRCS Med. Sci.* **7,** 575.
Iguchi, T., Uchima, F.-D. A., Ostrander, P. L., and Bern, H. A. (1983). *Proc. Natl. Acad. Sci. U.S.A.* **80,** 3743–3747.
Iguchi, T., Iwase, Y., Kato, H., and Takasugi, N. (1985a). *Exp. Clin. Endocrinol.* **85,** 129–137.
Iguchi, T., Ostrander, P. L., Mills, K. T., and Bern, H. A. (1985b). *Cancer Res.* **45,** 5688–5693.
Iguchi, T., Uchima, F.-D. A., Ostrander, P. L., Hamamoto, S. T., and Berry, H. A. (1985c). *Proc. Jpn. Acad. Ser. B* **61,** 292–295.
Iguchi, T., Hirokawa, M., and Takasugi, N. (1986a). *Toxicology* **42,** 1–11.
Iguchi, T., Takase, M., and Takasugi, N. (1986b). *Proc. Soc. Exp. Biol. Med.,* **181,** 59–65.
Iguchi, T., Takase, M., and Takasugi, N. (1986c). *IRCS Med. Sci.* **14,** 187–188.
Iguchi, T., Takasugi, N., Bern, H. A., and Mills, K. T. (1986d). Teratology **34,** 29–35.
Iguchi, T., Takei, T., Takase, M., and Takasugi, N. (1986e). *Acta Anat.* **127,** 110–114.
Iguchi, T., Ohta, Y., Fukazawa, Y., and Takasugi, N. (1987a). *Med. Sci. Res.* **15,** 1407–1408.
Iguchi, T., Ostrander, P. L., Mills, K. T., and Bern, H. A. (1987b). *Med. Sci. Res.* **15,** 489–490.
Iguchi, T., Uchima, F.-D. A., and Bern, H. A. (1987c). *In Vitro Cell. Dev. Biol.* **23,** 535–540.
Iguchi, T., Irisawa, S., Uchima, F.-D. A., and Takasugi, N. (1988a). *Reprod. Toxicol.* **2,** 127–134.
Iguchi, T., Ostrander, P. L., Mills, K. T., and Bern, H. A. (1988b). *Cancer Lett.* **43,** 207–214.
Iguchi, T., Todoroki, R., Takasugi, N., and Petrow, V. (1988c). *Biol. Reprod.* **39,** 689–697.
Iguchi, T., Irisawa, S., Fukazawa, Y., Uesugi, Y., and Takasugi, N. (1989a). *Anat. Rec.* **224,** 490–494.
Iguchi, T., Todoroki, R., Yamaguchi, S., and Takasugi, N. (1989b). *Acta Anat.* **136,** 146–154.
Iguchi, T., Fukazawa, Y., Uesugi, Y., and Takasugi, N. (1990a). *Biol. Reprod.* **43,** 478–484.
Iguchi, T., Irisawa, S., Uesugi, Y., Kusunoki, S., and Takasugi, N. (1990b). *Acta Anat.* **139,** 201–208.
Iguchi, T., Kamiya, K., Uesugi, Y., Sayama, K., and Takasugi, N. (1991a). *In vivo* **5,** 359–364.
Iguchi, T., Uesugi, Y., Takasugi, N., and Petrow, V. (1991b). *J. Endocrinol.* **128,** 395–401.
Iguchi, T., Edery, M., Tsai, P.-S., Ozawa, S., Sato, T., and Bern, H. A. (submitted).
Illsley, N. P., and Lamartiniere, C. A. (1980). *Endocrinology (Baltimore)* **107,** 551–556.
Imagawa, W., Bandyopadhyay, G. K., and Nandi, S. (1990). *Endocr. Rev.* **11,** 494–523.

Imperato-McGinley, J., Binienda, Z., Arthur. A., Mininberg, D. T., Vaughan, E. D., Jr., and Quimby, F. W. (1985). *Endocrinology (Baltimore)* **116,** 807–812.
Imperato-McGinley, J., Binienda, Z., Gedney, J., and Vaughan, E. D., Jr. (1986). *Endocrinology (Baltimore)* **118,** 132–137.
Inoue, M. (1976). *Congr. Anom.* **16,** 171–173.
Irisawa, S., and Iguchi, T. (1990). *In Vivo* **4,** 175–180.
Irisawa, S., Iguchi, T., and Takasugi, N. (1990). *Zool. Sci.* **7,** 541–545.
Ito, H., and Hidaka, H. (1983). *Cancer Lett.* **19,** 215–220.
Jacobson, C. D., and Gorski, R. A. (1981). *J. Comp. Neurol.* **196,** 519–529.
Jacobson, C. D., Shryne, J. E., Shapiro, F., and Gorski, R. A. (1980). *J. Neurol.* **193,** 541–548.
Jacobson, C. D., Csernus, V. J., Shryne, J. E., and Gorski, R. A. (1981). *J. Neurosci.* **1,** 1142–1147.
Johnson, J. D. (1987). *In* "Monographs on Pathology of Laboratory Animals, Genital System" (T. C. Jones, U. Mohr, and R. D. Hunt, eds.), pp.84–109. Springer-Verlag, New York.
Johnson, L. D., Driscoll, S. G., Hertig A. T., Cole, P. T., and Nickerson, R. J. (1979). *Obstet. Gynecol.* **53,** 671–679.
Johnston, C. C., Hui, S. L., Witt, R. M., Appledorn, B., Baker, R. S., and Longcope, C. (1985). *J. Clin. Endocrinol. Metab.* **61,** 905–911.
Jones, L. A. (1980). *Proc. Soc. Exp. Biol. Med.* **165,** 17–25.
Jones, L. A., Verjan, R. P., Mills, K. T., and Bern, H. A. (1984). *Cancer Lett.* **23,** 123–128.
Jordan, V. C. (1982). *Breast Cancer Res. Treat.* **2,** 123–138.
Jordan, V. C. (1990). *Breast Cancer Res. Treat.* **15,** 125–136.
Jordan, V. C., and Bowser-Finn, R. A. (1982). *Endocrinology (Baltimore)* **110,** 1281–1291.
Jordan, V. C., and Gosden, B. (1982). *Mol. Cell. Endocrinol.* **7,** 291–306.
Jordan, V. C., and Koerner, S. (1976). *J. Endocrinol.* **68,** 305–310.
Jordan, V. C., and Murphy, C. S. (1990). *Endocr. Rev.* **11,** 578–610.
Jordan, V. C., Dix, C. J., Naylor, K. E., Prestwich, G., and Rowsby, L. (1978). *J. Toxicol. Environ.* **4,** 364–390.
Jordan, V. C., Fisher, A. M., and Rose, D. P. (1981). *Eur. J. Cancer* **17,** 121–122.
Jordan, V. C., Phelps, E. L., and Lindgren, J. U. (1987). *Breast Cancer Res. Treat.* **10,** 31–35.
Kallela, K., and Korpinen, E. L. (1973). *Nord. Vet.-Med.* **25,** 446–450.
Kang, Y. H., Anderson, W. A., and DeSombre, E. R. (1975). *J. Cell Biol.* **64,** 682–691.
Katzenellenbogen, B. S., Katzenellenbogen, J. A., and Mordecai, D. (1979). *Endocrinology (Baltimore)* **105,** 33–40.
Kent, H. A., Jr. (1959). *Anat. Rec.* **134,** 455–462.
Kent, H. A., Jr. (1960). *Anat. Rec.* **137,** 521–524.
Kiang, D. T., Kennedy, B. J., Pathre, S. V., and Mirocha, C. J. (1978). *Cancer Res.* **38,** 3611–3615.
Kikkawa, U., Takai, Y., Tanaka, Y., Miyake, R., and Nishizuka, Y. (1983). *J. Biol. Chem.* **258,** 11442–11445.
Kimura, T., and Nandi, S. (1967). *J. Natl. Cancer Inst.* **39,** 75–93.
Kimura, T., Nandi, S., and DeOme, K. B. (1967). *J. Exp. Zool.* **165,** 211–222.
Kincle, F. A. (1990). "Hormones and Toxicity in the Neonate," p. 334. Springer-Verlag, Berlin.
Kistner, R. W., and Smith, O. W. (1959). *Surg. Forum* **10,** 725–729.
Kistner, R. W., and Smith, O. W. (1961). *Fertil. Steril.* **12,** 121–141.
Kon, O. L. (1983). *J. Biol. Chem.* **258,** 3173–3177.
Krabbe, S., Christiansen, C., Rodlbro, P., and Transbol, I. (1979). *Arch. Dis. Child.* **54,** 950–953.

Kratochwil, K. (1977). *Dev. Biol.* **61,** 358–365.
Kratochwil, K. (1986). *In* "Hormones and Cell Regulation" (J. E. Nuez, J. E. Dumont, and R. J. B. King, eds.), pp. 9–14. John Libby, London.
Kratochwil, K., and Schwartz, P. (1976). *Proc. Soc. Exp. Biol. Med.* **73,** 4041–4044.
Kumagai, S., and Shimizu, T. (1982). *Arch. Toxicol.* **50,** 279–286.
Lam, P. H.-Y. (1984). *Biochem. Biophys. Res. Commun.* **118,** 27–32.
Lamartiniere, C. A. (1979). *Endocrinology* (*Baltimore*) **105,** 1031–1035.
Lamartiniere, C. A. (1990). *J. Biochem. Toxicol.* **5,** 41–46.
Lamartiniere, C. A., and Lucier, G. W. (1978). *J. Steroid Biochem.* **9,** 595–598.
Lamartiniere, C. A., and Pardo, C. (1988). *J. Biochem. Toxicol.* **3,** 87–103.
Lamartiniere, C. A., Dieringer, C. S., Kita, E., and Lucier, G. W. (1979). *Biochem. J.* **180,** 313–318.
Lamartiniere, C. A., Luther, M. A., Lucier, G. W., and Illsley, N. P. (1982). *Biochem. Pharmacol.* **31,** 647–651.
Lamartiniere, C. A., Nicholas, J. M., Hermann, J., and Whithworth, N. S. (1986). *J. Steroid Biochem.* **24,** 557–562.
Lax, E. R., Kreuzfelder, E., Ghraf, R., and Schriefers, H. (1978). *Acta Endocrinol.*(*Copenhagen*), *Suppl.* No. 215, 41–42.
Leavitt, W. W. (1965). *Endocrinology* (*Baltimore*) **77,** 247–254.
Leavitt, W. W., and Meismer, D. M. (1968). *Nature (London)* **218,** 181–182.
Lecerf, F., Nguyen, B.-L., and Pasqualini, J. R. (1988). *Acta Endocrinol.* (*Copenhagen*) **119,** 85–90.
Lerner, L. J., Holthaus, J. F., and Thompson, C. R. (1958). *Endocrinology* (*Baltimore*) **63,** 295–318.
Levay-Young, B. K., and Bern, H. A. (1989). *Proc. Soc. Exp. Biol. Med.* **192,** 187–191.
Lopes, M. C. F., Vale, M. G. P., and Carvalho, A. P. (1990). *Cancer Res.* **50,** 2753–2758.
Lovell, D. P., Johnson, F. M., and Willis, D. B. (1986). *Am. J. Anat.* **176,** 287–303.
Martin, P. M., Horwitz, K. B., Ryan, D. S., and McGuire, W. L. (1978). *Endocrinology* (*Baltimore*) **103,** 1860–1867.
McCormack, S., and Clark, J. H. (1979). *Science* **204,** 629–631.
McDonald, P. G., and Doughty, C. (1972). *J. Reprod. Fertil.* **30,** 55–62.
McGuire, W. L. (1979). *Adv. Intern. Med.* **24,** 127–140.
McLachlan, J. A., (1979). *Natl. Cancer Inst. Monogr.* **51,** 67–72.
McLachlan, J. A., Newbold, R. R., and Bullock, B. (1975). *Science* **190,** 991–992.
McLeod, M. J. (1980). *Teratology,* **22,** 229–301.
McLusty, N. J., and Naftolin, F. (1981). *Science* **211,** 1294–1303.
McNutt, S. H., Purivin, P., and Murray, C. (1928). *J. Am. Vet. Med. Assoc.* **73,** 484–492.
Menczer, J., Dulitzky, M., Ben-Baruch, G., and Modan, M. (1986). *Br. J. Obstet. Gynaecol.* **93,** 503–507.
Miller, J. K., Hacking, A., Harrison, J., and Gross, V. J. (1973). *Vet. Rec.* **93,** 555–559.
Miller, M. A., and Katzenellenbogen, B. S. (1983). *Cancer Res.* **43,** 3094–3100.
Mirocha, C. J., Christensen, C. M., and Nelson, C. H. (1968). *Cancer Res.* **28,** 2319–2322.
Mirocha, C. J., Schauerhamer, B., and Pathre, S. V. (1974). *J. Assoc. Off. Anal. Chem.* **57,** 1104–1110.
Mori, T. (1967). *J. Fac. Sci. Univ. Tokyo, Sect. 4* **11,** 244–254.
Mori, T., and Iguchi, T. (1988). *In* "Toxicity of Hormones in Perinatal Life" (T. Mori and H. Nagasawa, eds.), pp. 63–79. CRC Press, Boca Raton, Florida.
Mori, T., and Nagasawa, H. (eds.) (1988). "Toxicity of Hormones in Perinatal Life" p. 184. CRC Press, Boca Raton, Florida.
Mori, T., Komuro, M., and Nagasawa, H. (1977). *IRCS Med. Sci.* **5,** 384.
Mori, T., Nagasawa, H., and Bern, H. A. (1980). *J. Environ. Pathol. Toxicol.* **3,** 191–205.

Mori, T., Iguchi, T., and Takasugi, N. (1983). *J. Exp. Zool.* **225,** 99–105.

Murakami, R. (1986). *J. Anat.* **149,** 11–20.

Murakami, R. (1987a). *J. Anat.* **151,** 209–219.

Murakami, R. (1987b). *J. Anat.* **153,** 223–231.

Murakami, R., and Mizuno, T. (1984a). *Dev. Growth Differ.* **26,** 419–426.

Murakami, R., and Mizuno, T. (1984b). *C. R. Seances Soc. Biol. Ses Fil.* **178,** 576–579.

Murakami, R., and Mizuno, T. (1986). *J. Embryol. Exp. Morphol.* **92,** 133–143.

Nakadate, T., Jeng, A. Y., and Blumberg, P. M. (1988). *Biochem. Pharmacol.* **37,** 1541–1545.

Neidel, J. E., Kuhn, L. J., and Vandenbark, G. R. (1983). *Proc. Natl. Acad. Sci. U.S.A.* **80,** 36–40.

Nelson, K. G., Takahashi, T., Bossert, N. L., Walmer, D. K., and McLachlan, J. A. (1991). *Proc. Natl. Acad. Sci. U.S.A.* **88,** 21–25.

Newbold, R. R., and McLachlan, J. A. (1982). *Cancer Res.* **42,** 2003–2011.

Newbold, R. R., and McLachlan, J. A. (1985). *In* "Estrogens in the Environment II, Influences on Development" (J. A. McLachlan, ed.), pp. 288–318. Elsevier, New York.

Newbold, R. R., and McLachlan, J. A. (1988). *In* "Toxicity of Hormones in Perinatal Life"(T. Mori and H. Nagasawa, eds.), pp. 89–109. CRC Press, Boca Raton, Florida.

Newbold, R. R., Bullock, B. C., and McLachlan, J. A. (1983a). *Biol. Reprod.* **28,** 735–744.

Newbold, R. R., Tyrey, S., Haney, A. F., and McLachlan, J. A. (1983b). *Teratology* **27,** 417–426.

Newbold, R. R., Carter, D. B., Harris, S. E., and McLachlan, J. A. (1984). *Biol. Reprod.* **30,** 459–470.

Newbold, R. R., Bullock, B. C., and McLachlan, J. A. (1985a). *Cancer Res.* **45,** 5145–5150.

Newbold, R. R., Bullock, B. C., and McLachlan, J. A. (1985b). *Teratog. Carcinog. Mutagen.* **5,** 473–480.

Newbold, R. R., Bullock, B. C., and McLachlan, J. A. (1987a). *J. Urol.* **137,** 1446–1450.

Newbold, R. R., Bullock, B. C., and McLachlan, J. A. (1987b). *Teratog, Carcinog. Mutagen.* **7,** 377–389.

Newbold, R. R., Pentecost, B. T., Yamashita, S., Lum, K., Miller, J. V., Nelson, P., Blair, J., Kong, H., Teng, C., and McLachlan, J. A. (1989). *Endocrinology (Baltimore)* **124,** 2568–2576.

Newbold, R. R., Bullock, B. C., and McLachlan, J. A. (1990). *Cancer Res.* **50,** 7677–7681.

Nguyen, B. L., Giambiagi, N., Mayrand, C., Lecerf, F., and Pasqualini, J. R. (1986). *Endocrinology (Baltimore)* **119,** 978–988.

Nishizuka, Y. (1986). *Science* **233,** 305–312.

Nishizuka, Y. (1988). *Nature (London)* **334,** 661–665.

Normand, T., Jean-Faucher, C., and Jean, C. (1990). *J. Steroid Biochem.* **36,** 415–423.

O'Brian, C. A., Liskamp, R. M., Solomon, D. H., and Weinstein, I. B. (1985). *Cancer Res.* **45,** 2462–2465.

O'Brian, C. A., Liskamp, R. M., Solomon, D. H., and Weinstein, I. B. (1986). *J. Natl. Cancer Inst.* **76,** 1243–1246.

O'Brian, C. A., Ward, N. E., and Anderson, B. W. (1988). *J. Natl. Cancer Inst.* **80,** 1628–1633.

O'Donoghue, C. H. (1912). *Anat. Anz.* **41,** 353–368.

Ohta, Y. (1977). *Endocrinol. Jpn.* **24,** 287–294.

Ohta, Y. (1982). *Biol. Reprod.* **27,** 303–311.

Ohta, Y. (1985). *Zool. Sci.* **2,** 89–93.

Ohta, Y., and Iguchi, T. (1976). *Endocrinol. Jpn.* **23,** 333–340.

Ohta, Y., and Takasugi, N. (1974). *Endocrinol. Jpn,* **21,** 183–190.

Ohta, Y., Iguchi, T., and Takasugi, N. (1989). *Reprod. Toxicol.* **3,** 207–212.

Ostrander, P. L., Mills, K. T., and Bern, H. A. (1985). *J. Natl. Cancer Inst.* **74,** 121–135.

Ozawa, S., Iguchi, T., Sawada, K., Ohta, Y., Takasugi, N., and Bern, H. A. (1991a). *Cancer Lett.* **58,** 167–175.

Ozawa, S., Iguchi, T., Takemura, K., and Bern, H. A. (1991b). *Proc. Soc. Exp. Biol. Med.* **198,** 760–763.

Pang, S. F., and Tang, F. (1984). *J. Endocrinol.* **100,** 7–11.

Pasqualini, J. R., Sumida, C., and Gelly, C. (1976a). *Acta Endocrinol.* (*Copenhagen*) **83,** 811–828.

Pasqualini, J. R., Sumida, C., Gelly, C., and Nguyen, B.-L. (1976b). *J. Steroid Biochem.***7,** 1031–1038.

Pasqualini, J. R., and Lecerf, F. (1986). *J. Endocrinol.* **110,** 197–202.

Pasqualini, J. R., and Nguyen, B.-L. (1980). *Endocrinology* (*Baltimore*) **106,** 1160–1165.

Pasqualini, J. R., Nguyen, B.-L., Mayrand, C., and Lecerf, F. (1986a). *Acta Endocrinol.* (*Copenhagen*) **111,** 378–386.

Pasqualini, J. R., Nguyen, B.-L., Sumida, C., Giambiagi, N., and Mayrand, C. (1986b). *J. Steroid Biochem,* **25,** 853–857.

Pasqualini, J. R., Giambiagi, N., Sumida, C., Nguyen, B.-L., Gelly, C., Malluche, H. H., Faugere, M. C., Rush, M., and Friedler, R. (1986c). *Endocrinology* (*Baltimore*) **119,** 264–275.

Pentecost, B. T., and Teng, C. T. (1987). *J. Biol. Chem.* **262,** 10134–10139.

Pentecost, B. T., Newbold, R. R., Teng, C. T., and McLachlan, J. A. (1988). *Mol. Endocrinol.* **2,** 1243–1248.

Plapinger, L. (1982). *Biol. Reprod.* **26,** 961–972.

Plapinger, L., and Bern, H. A. (1979). *J. Natl. Cancer Inst.* **63,** 507–518.

Pollack, I. F., Randall, M. S., Kristofik, M. P., Kelly, R. H., Selker, R. G., and Vortosick, F. T., Jr. (1990). *Cancer Res.* **50,** 7134–7138.

Pollak, M., Costantino, J., Polychronakos, C., Blauer, S.-A., Guyda, H., Redmond, C., Fisher B., and Margolese, R. (1990). *J. Natl. Cancer Inst.* **82,** 1693–1697.

Pointis, G., Latreille, M.-T., and Cedard, L. (1980). *J. Endocrinol.* **86,** 483–488.

Powell-Jones, W., Raeford, S., and Lucier, G. W. (1981). *Mol. Pharmacol.* **20,** 34–42.

Riggs, B. L., Jowsey, J., Kelly, P. J., Jones, D. J., and Maher, F. T. (1969). *J. Clin. Invest.* **48,** 1065–1072.

Robson, J. M., and Schonberg, A. (1937). *Nature* (*London*) **140,** 196.

Rosenfield, A., Maine, D., Rochat, R., Shelton, J., and Hatcher, R. A. (1983). *J. Am. Med. Assoc.* **249,** 2922–2928.

Rothschild, T. C., Boylan, E. S., Calhoon, R. E., and Vonderhaar, B. K. (1987). *Cancer Res.* **47,** 4508–4516.

Rothschild, T. C., Calhoon, R. E., and Boylan, E. S. (1988). *Exp. Mol. Pathol.* **48,** 59–76.

Rouzaire-Dubois, B., and Dubois, J.-M. (1990). *Cell. Signall.* **2,** 387–393.

Ruddick, J. A., Scott, P. M., and Harwig, J. (1976). *Bull. Environ. Contam. Toxicol.* **15,** 678–681.

Sara, V. R., and Hall, K. (1990). *Physiol. Rev.* **70,** 591–614.

Segal, J. S., and Nelson, W. O. (1958). *Proc. Soc. Exp. Biol. Med.* **98,** 431–436.

Selye, H., Collip, J. B., and Thomson, D. L. (1933). *Endocrinology* (*Baltimore*) **17,** 494–500.

Sermanoff, M. K., Goldman, B. D., and Ginsburg, B. E. (1977). *Endocrinology* (*Baltimore*) **100,** 122–127.

Sheehan, D. M., Branham, W. S., Medlock, K. L., and Shanmugasunduram, E. R. B. (1984). *Teratology* **29,** 383–392.

Shyamala, G., Mori, T., and Bern, H. A. (1974). *J. Endocrinol.* **63,** 275–284.

Sigemoro, E. (1947). *J. Fac. Sci. Hokkaido Imp. Univ., Ser. 4* **9,** 233–242.

Sigemoro, E., and Makino, S. (1947). *Seibutsu, Suppl.* **1,** 63–67.

Slemenda, C., Hui, S. L., Witt, R. M., Longcope, C., and Johnston, C. C. (1987). *J. Clin. Invest.* **80,** 1261–1269.

Slob, A. K., Ooms, M. P., and Vreeburg, J. T. M. (1980). *J. Endocrinol.* **87,** 81–87.
Smith, O. W. (1948). *Am. J. Obstet. Gynecol.* **56,** 821–834.
Smith. O. W., and Kistner, R. W. (1963). *J. Am. Med. Assoc.* **184,** 122–130.
Southwick, F. S., and Crelin, E. S. (1969). *J. Biol. Med.* **41,** 446–447.
Steele, J. A., Lieberman, J. R., and Mirocha, C. J. (1974). *Can. J. Microbiol.* **20,** 531–534.
Stein, K. F. (1957). *J. Genet.* **55,** 313–324.
Stumpf, W. E., Narbeitz, R., and Sar, M. (1980). *J. Steroid Biochem.* **12,** 55–64.
Su, H. D., Mazzei, G. J., Vogler, W. R., and Kuo, J. F. (1985). *Biochem. Pharmacol.* **34,** 3649–3653.
Sudo, K., Monsma, F. J., and Katzenellenbogen, B. S. (1983). *Endocrinology (Baltimore)* **112,** 425–434.
Sumida, C. and Pasqualini, J. R. (1979). *Endocrinology (Baltimore)* **105,** 406–413.
Sutherland, R. L., and Foo, M. S. (1979). *Biochem, Biophys. Res. Commun.* **91,** 183–191.
Sutherland, R. L., and Murphy, L. C. (1980). *Eur. J. Cancer* **16,** 1141–1148.
Sutherland, R. L., Mester, J., and Baulien, E. E. (1977). *Nature (London)* **267,** 434–435.
Sutherland, R. L., Murphy, L. C., Foo, M. S., Green, M. D., Whybourne, A. M., and Krozowski, Z. S. (1980). *Nature (London)* **288,** 273–275.
Suzuki, Y., and Arai, Y. (1986). *Proc. Jpn. Acad. Ser. B* **62,** 412–415.
Tabei, T., and Heinrichs, W. L. (1977). *Horm. Res.* **7,** 227–231.
Tachibana, H., Iguchi, T., and Takasugi, N. (1984a). *Zool. Sci.* **1,** 777–785.
Tachibana, H., Iguchi, T., and Takasugi, N. (1984b). *Endocrinol. Jpn.* **31,** 645–650.
Taguchi, O. (1987). *Biol. Reprod.* **37,** 113–116.
Taguchi, O., and Nishizuka, Y. (1985). *Am. J. Obstet. Gynecol.* **151,** 675–678.
Takamatsu, Y., Iguchi, T., and Takasugi, N. (1990). *Zool. Sci.* **7,** 1137.
Takamatsu, Y., Iguchi, T., and Takasugi, N. (1992a). *In Vivo* **6,** 1–8.
Takamatsu, Y., Iguchi, T., and Takasugi, N. (1992b). *In Vivo* (in press).
Takamatsu, Y., Iguchi, T., and Takasugi, N. (1992c). *In Vivo* (in press).
Takano-Yamamoto, T., and Rodan, G. A. (1990). *Proc. Natl. Acad. Sci. U.S.A.* **87,** 2172–2176.
Takasugi, N. (1963). *Endocrinology (Baltimore)* **72,** 607–619.
Takasugi, N. (1970). *Endocrinol. Jpn.* **17,** 277–281.
Takasugi, N. (1976). *Int. Rev. Cytol.* **44,** 193–224.
Takasugi, N. (1979). *Natl. Cancer Inst. Monogr.* **51,** 57–66.
Takasugi, N., and Bern, H. A. (1964). *J. Natl. Cancer Inst.* **33,** 855–865.
Takasugi, N., and Bern, H. A. (1988). *In* "Toxicity of Hormones in Perinatal Life" (T. Mori and H. Nagasawa, eds.), pp. 1–7. CRC Press, Boca Raton, Florida.
Takasugi, N., and Furukawa, M. (1972). *Endocrinol. Jpn.* **19,** 417–422.
Takasugi, N., and Mitsuhashi, Y. (1972). *Endocrinol. Jpn.* **19,** 423–428.
Takasugi, N., Bern, H. A., and DeOme, K. B. (1962). *Science* **138,** 438–439.
Takasugi, N., Kimura, T., and Mori, T. (1970). *In* "Postnatal Development of Phenotype" (S. Kazda and L. H. Denenberg, eds.), pp. 229–251. Academia, Prague.
Takasugi, N., Tanaka, M., and Kato, C. (1983). *Endocrinol. Jpn.* **30,** 35–42.
Takasugi, N., Iguchi, T., Kurihara, J., Tei, A., and Takase, M. (1985). *Exp. Clin. Endocrinol.* **86,** 273–283.
Takewaki, K. (1937). *Zool. Mag.* **49,** 231–233.
Takewaki, K., and Takasugi, N. (1953). *Annot. Zool. Jpn.* **26,** 99–105.
Talmage, R. V. (1946). *Anat. Rec.* **96,** 528.
Talmage, R. V. (1947). *Anat. Rec.* **99,** 91–113.
Tanaka, M., and Takasugi, N. (1982). *Proc. Jpn. Acad., Ser. B* **58,** 311–314.
Tanaka, M., Iguchi, T., and Takasugi, N. (1984). *IRCS Med. Sci.* **12,** 814–815.
Telfer, E., and Gosden, R. G. (1987). *J. Reprod. Fertil.* **81,** 137–147.

Tenenbaum, A., and Forsberg, J.-G. (1985). *J. Reprod. Fertil.* **73,** 465–477.
Teng, C. T., Walker, M. P., Bhattacharyya, S. N., Klapper, D. G., DiAugustine, R. P., and McLachlan, J. A. (1986). *Biochem. J.* **240,** 413–422.
Teng, C. T., Pentecost, B. T., Chen, Y. H., Newbold, R. R., Eddy, E. M., and McLachlan, J. A. (1989). *Endocrinology (Baltimore)* **124,** 992–999.
Terenius, L. (1971). *Acta Endocrinol. (Copenhagen)* Suppl. No. 66, 431–447.
Todd, T. W. (1925). *Am. J. Anat.* **31,** 345–357.
Tomooka, Y., and Bern, H. A. (1982). *J. Natl. Cancer Inst.* **69,** 1347–1352.
Tomooka, Y., Bern, H. A. and Nandi, S. (1983). *Cancer Lett.* **20,** 255–261.
Tsai, P.-S., Uchima, F.-D. A., Hamamoto, S. T., and Bern, H. A. (1991). *In Vitro Cell. Dev. Biol.* **27A,** 442–446.
Turner, R. T., Vandersteenhoven, J. J., and Bell, N. H. (1987a). *J. Bone Miner. Res.* **2,** 115–122.
Turner, R. T., Wakley, G. K., Hannon, K. S., and Bell, N. H. (1987b). *J. Bone Miner. Res.* **2,** 449–456.
Turner, R. T., Wakeley, G. K., Hannon, K. S., and Bell, N. H. (1988). *Endocrinology (Baltimore)* **122,** 1146–1150.
Turner, R. T., Hannon, K. S., Demers, L. M., Buchanan, J., and Bell, N. H. (1989). *J. Bone Miner. Res.* **4,** 557–563.
Turner, R. T., Colvard, D. S., and Spelsberg, T. C. (1990). *Endocrinology (Baltimore)* **127,** 1346–1351.
Turner, T., and Bern, H. A. (1990). *Cancer Lett.* **52,** 209–218.
Turner, T., Edery, M., Mills, K. T., and Bern, H. A. (1989). *J. Steroid Biochem,* **32,** 559–564.
Turner, T., Bern, H. A., Young, P., and Cunha, G. R. (1990). *In Vitro Cell. Dev. Biol.* **26,** 722–730.
Uchima, F.-D. A., Edery, M., Iguchi, T., Larson, L., and Bern, H. A. (1987). *Cancer Lett.* **35,** 227–235.
Uchima, F.-D. A., Vallerga, A. K., Firestone, G. L., and Bern, H. A. (1990). *Proc. Soc. Exp. Biol. Med.* **195,** 218–223.
Uchima, F.-D. A., Edery, M., Iguchi, T., and Bern, H. A. (1991a). *J. Endocrinol.* **128,** 115–120.
Uchima, F.-D. A., Iguchi, T., Pattamakom, S., Mills, K. T., and Bern, H. A. (1991b). *Zool. Sci.* **8,** 713–719.
Uesugi, Y., Iguchi, T., and Takasugi, N. (1990). *Zool. Sci.* **7,** 1137.
Uesugi, Y., Ohta, Y., Asashima, M., and Iguchi, T. (1992a). *Anat. Rec.* (in press).
Uesugi, Y., Taguchi, O., Noumura, T., and Iguchi, T. (1992b). *Anat. Rec.* (in press).
Uesugi, Y., Sato, T., and Iguchi, T. (1992c). *Anat. Rec.* (in press).
Verdeal, K., and Ryan, D. S. (1979). *J. Food. Prot.* **42,** 577–583.
Viell, B., and Struck, H. (1987). *Horm. Metab. Res.* **19,** 415–418.
Vilmann, A., and Vilmann, H. (1983). *Acta Anat.* **117,** 136–144.
Wakeling, A. E., Valcaccia, B., Newboult, E., and Green, R. (1984). *J. Steroid Biochem.* **20,** 111–120.
Walker, B. E. (1984). *In* "Issues and Reviews in Teratology" (H. Katler, ed.), Vol. 2, pp. 157–187. Plenum, New York.
Warner, M. R., Warner, R. L., and Clinton, D. W. (1979). *Biol. Reprod.* **20,** 310–322.
Watts, C. K. W., Murphy, L. C., and Sutherland, R. L. (1984). *J. Biol. Chem.* **259,** 4223–4229.
Wei, J. W., Hickie, R. A., and Klaessen, D. J. (1983). *Cancer Chemother. Pharmacol.* **11,** 86–90.
Williams, B. A., Mills, K. T., Burroughs, C. D., and Bern, H. A. (1989). *Cancer Lett.* **46,** 225–230.
Willingham, M. C., Wehland, J., Klee, C. B., Richert, N. O., Rutherford, A. V., and Pastan, I. H. (1984). *J. Histochem, Cytochem.* **31,** 445–461.

Winneker, R. C., and Clark, J. H. (1983). *Endocrinology* (*Baltimore*) **112,** 1910–1915.

Wordinger, R. J., and Derrenbacker, J. (1989). *Acta Anat.* **134,** 312–318.

Wordinger, R. J., and Highman, B. (1984). *Virchows Arch. B* **45,** 241–253.

Wordinger, R. J., and Morrill, A. (1985). *Virchows Arch. B* **50,** 71–79.

Wordinger, R. J., Brown, D., Atkins, E., and Jackson, F. L. (1989). *J. Reprod. Fertil.* **85,** 383–388.

Wordinger, R. J., Nile, J., and Stevens, G. (1991). *J. Reprod. Fertil.* **92,** 209–216.

Wronski, T. J., Lowry, P. L., Walsh, C. C., and Ignaszewski, L. A. (1985). *Calcif, Tissue Int.* **37,** 324–328.

Wronski, T. J., Walsh, C. C., and Ignaszewski, L. A. (1986). *Bone* **7,** 119–123.

Wronski, T. J., Clintron M., Dohrty, A. L., and Dann, I. M. (1988). *Endocrinology* (*Baltimore*) **123,** 681–686.

Yamamoto, M. (1987). *Teikyo Med. J.* **10,** 303–313.

Yamamoto, M. (1989). *Arch. Histol. Cytol.* **52,** 529–541.

Yamamoto, M., Umekita, S., and Nishimura, M. (1990). *Teikyo Med. J.* **13,** 375–383.

Yang, J., and Nandi, S. (1983). *Int. Rev. Cytol.* **81,** 249–286.

Yoshida, H., Kadota, A., and Fukunishi, R. (1980). *Exp. Anim.* **29,** 39–43.

Biogenesis of the Vacuole in *Saccharomyces cerevisiae*

Christopher K. Raymond, Christopher J. Roberts, Karen E. Moore, Isabelle Howald, and Tom H. Stevens
Institute of Molecular Biology, University of Oregon, Eugene, Oregon 97403

I. Introduction

The complex structure of eukaryotic cells includes intracellular organelles of discrete composition and function. The biogenesis and maintenance of any one of these organelles requires unique arrays of molecular components and processes that are beginning to be characterized in some detail. In this review, various aspects of the biogenesis, maintenance, and function of the vacuole in the budding yeast *Saccharomyces cerevisiae* are described. This organism has proven to be tractable to genetic, biochemical, and cell-biological analysis, and many of the fundamental processes in these cells appear to be general to all eukaryotic cells. The vacuole in yeast shares many features in common with vacuoles in other fungi and simple eukaryotic cells, lysosomes in animal cells, and vacuoles/tonoplasts in plant cells. All possess acidic interiors, contain many different hydrolytic enzymes, and are derived in large part from a major branch of the secretory pathway. The biogenesis of the yeast vacuole is thus a reasonable model for organelle assembly in a diverse spectrum of more complex organisms.

The vacuole in *S. cerevisiae* occupies approximately 10–20% of the overall cell volume and is a prominent feature in most micrographs of yeast cells (Pringle *et al.*, 1989; Raymond *et al.*, 1990; Weisman *et al.*, 1987). The organelle undergoes dynamic changes in response to growth conditions, but it often appears as a complex, multilobed structure in many laboratory yeast strains that are actively growing (Gomes de Mesquita *et al.*, 1991; Pringle *et al.*, 1989; Raymond *et al.*, 1990; Weimken *et al.*, 1970). During cell growth, portions of the vacuole from the mother cell are transported into developing buds, indicating that specific cellular mechanisms exist that ensure partitioning of the organelle between dividing

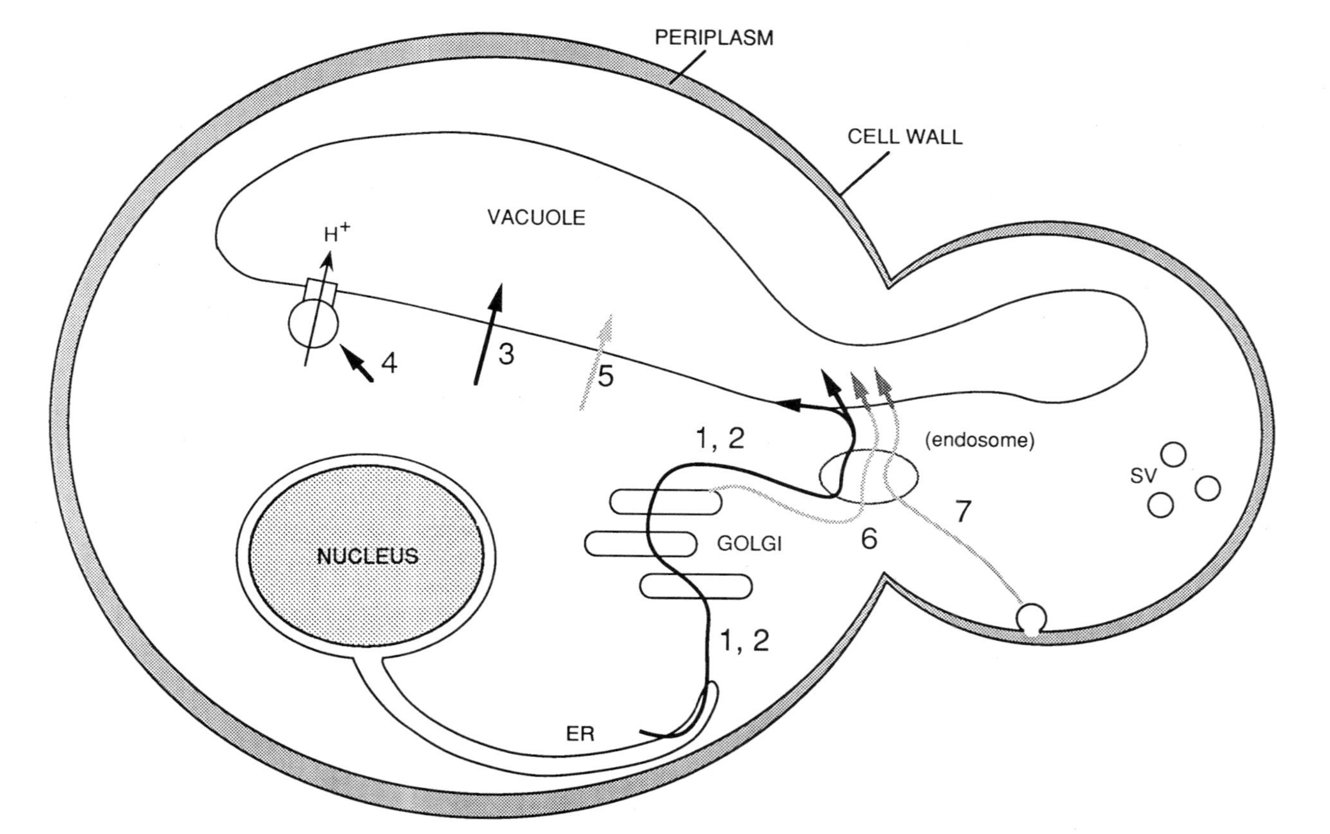
PERIPLASM
CELL WALL
VACUOLE
H+
4
3
5
1, 2
(endosome)
SV
NUCLEUS
GOLGI
6
7
1, 2
ER

cells (Gomes de Mesquita *et al.*, 1991; Raymond *et al.*, 1990; Shaw and Wickner, 1991; Weisman *et al.*, 1987, 1990; Weisman and Wickner, 1988). The vacuole contains a variety of glycoproteins, some of which have well-characterized hydrolytic activities toward specific biological substrates (for reviews see Achstetter and Wolf, 1985; Jones, 1984; Klionsky *et al.*, 1990). The pH optimum for most of these enzymes is below neutral, consistent with the fact that the vacuole is maintained at a pH of about 6.0 (Preston *et al.*, 1989). The acidic pH of the vacuole is generated and maintained by a multisubunit vacuolar H^+-ATPase, and the electrochemical gradient generated by proton pumping drives the import of several metabolites across the vacuolar membrane (Anraku *et al.*, 1989; Kane *et al.*, 1989a,b; Klionsky *et al.*, 1990; Ohsumi and Anraku, 1981, 1983; Uchida *et al.*, 1985). The vacuole is known to participate in several important physiological roles in response to changing growth conditions of yeast cells. It serves as a major storage reserve for basic amino acids, inorganic phosphate, and calcium ions, and these reserves are mobilized in response to nutrient limitation (for review see Klionsky *et al.*, 1990). Yeast mutants with impaired vacuolar function are sensitive to extremes of pH and osmotic strength, suggesting the vacuole is required for pH and osmotic homeostasis (Anraku *et al.*, 1989; Banta *et al.*, 1988; Klionsky *et al.*, 1990; Nelson and Nelson, 1990; Yamashiro *et al.*,1990). Protein turn-

---

FIG. 1 Protein traffic to the yeast vacuole. (1,2) Soluble vacuolar glycoproteins (Section II,B,C) and integral membrane proteins of the vacuolar membrane (Section II,E) are synthesized on polysomes bound to the surface of the ER, and are co-translationally translocated into the lumen or inserted into the membrane of this organelle, respectively. Proteins bound for the vacuole or the cell surface travel in common compartments through the Golgi apparatus, where the two classes of proteins are partitioned away from one another in a late-Golgi compartment (Section IV,A). Secreted proteins are packaged into secretory vesicles (SV) and exit through the bud. Vacuolar proteins may pass through an endosomal compartment en route to the vacuole. (3) α-Mannosidase is a peripheral membrane protein that resides on the luminal face of the vacuolar membrane (Section II,F). It appears to be sequestered directly from the cytoplasm into the vacuole by an unknown mechanism. (4) Several subunits of the vacuolar H^+-ATPase (Section III,E) assemble onto the cytoplasmic face of the vacuole, joining with a proton pore complex that spans the vacuolar membrane to form an ATP-dependent proton pump. (5) The cytoplasmic enzyme FBPase (Section III,C) is degraded in the vacuole in response to glucose addition to the growth medium. Like α-Mannosidase, vacuolar delivery of FBPase appears to occur by direct translocation across the vacuolar membrane. (6) Three integral membrane proteases that define a late-Golgi compartment, Kex1p, Kex2p, and DPAP A, are degraded in the vacuole (Section III,B). (7) The mating pheromone α-factor and Ste3p, the receptor for the pheromone **a**-factor are endocytosed and broken down in the vacuole (Section III,D); α-factor passes through a temperature-sensitive, endosome-like intermediate compartment en route to the vacuolar compartment.

over in yeast is greatly accelerated in response to nitrogen limitation, and the liberation of amino acids from intracellular proteins requires vacuolar proteases (Zubenko and Jones, 1981). The fact that vacuolar protease-deficient mutants exhibit drastically reduced viability upon nitrogen starvation or that diploids lacking vacuolar proteases fail to sporulate underscores the importance of vacuolar proteolysis for the survival and/or differentiation of the organism (Jones, 1984; Kaneko *et al.,* 1982; Teichert *et al.,* 1989; Zubenko and Jones, 1981). Degradation of a cytosolic protein, fructose 1,6-bis-phosphatase, and internalized mating pheromone α-factor also require vacuolar proteases (Chiang and Schekman, 1991; Dulic and Reizman, 1989), and these provide specific examples of more generalized vacuole-dependent protein turnover.

Although vacuolar function is required to regulate the physiology and growth of yeast cells in response to extremes in environmental growth conditions, many of these vacuolar functions are dispensable in the laboratory. As a consequence, mutations in genes required for vacuolar biogenesis generally do not compromise the overall viability of the organism (Bankaitis *et al.,* 1986; Banta *et al.,* 1988; Jones, 1977; Nelson and Nelson, 1990; Rothman and Stevens, 1986; Yamashiro *et al.,* 1990). This feature of yeast cells has permitted several aspects of vacuolar biogenesis to be elucidated. Most proteins that normally reside in the lumen of the vacuole are translocated across the endoplasmic reticulum (ER) membrane and glycosylated in the ER lumen (Fig. 1) (Klionsky *et al.,* 1990). These newly synthesized vacuolar proteins traverse the secretory pathway from the ER to the Golgi apparatus together with secreted proteins, and vacuolar proteins are sorted away from proteins destined for secretion in an ill-defined, *trans*-Golgi compartment (Ammerer *et al.,* 1986; Graham and Emr, 1991; Klionsky and Emr, 1989; Klionsky *et al.,* 1990; Roberts *et al.,* 1989; Stevens *et al.,*1982; Woolford *et al.,* 1986). Many vacuolar hydrolases move through the common compartments of the secretory pathway as inactive precursor proteins. Upon delivery to the vacuolar compartment, these proteins undergo proteolytic cleavages and assume their mature, enzymatically active conformation. Proteolytic activation of vacuolar protein precursors requires the activity of proteinase A, the product of the *PEP4* gene (Ammerer *et al.,* 1986; Hemmings *et al.,* 1981; Jones *et al.,* 1982; Woolford *et al.,* 1986; Zubenko *et al.,* 1983). Consequently, *pep4* strains of yeast lack a large number of vacuolar hydrolases, and protein turnover in these mutants is greatly retarded. In addition to *PEP4,* many additional genes required for vacuolar biogenesis have been identified, but the molecular activities of the corresponding gene products are less clear (Section IV) (Bankaitis *et al.,* 1986; Banta *et al.,* 1988; Jones, 1977; Klionsky *et al.,* 1990; Robinson *et al.,* 1988; Rothman *et al.,* 1986, 1989a).

The intent of this review is to focus on two broad themes of vacuolar biogenesis in *S. cerevisiae*. In the following two sections, the biosynthesis of endogenous vacuolar proteins, the intracellular routes taken by proteins that are degraded in the vacuole, and the assembly of the vacuolar H^+-ATPase are considered. Several distinct biosynthetic or degradative routes, as they are thought to occur, are illustrated in Fig. 1. Trafficking of these various proteins to the vacuole occurs through several, apparently distinct pathways. As such, these pathways, which converge at or in the vacuole, define a central aspect of the overall phrase vacuolar biogenesis. Sections IV and V describe the isolation and characterization of several classes of mutants with defects in vacuolar protein sorting, vacuolar assembly, and vacuolar segregation. Characterization of these mutants, and the wild-type genes affected in these strains, is providing clues as to the molecular activities required for assembly of this complex organelle.

II. Biosynthesis of Vacuolar Proteins

A. Distinct Sorting Pathways for Different Classes of Vacuolar Proteins

Several vacuolar hydrolytic activities have been described, and certain vacuolar enzymes and their corresponding structural genes have been characterized in great detail (Achstetter and Wolf, 1985; Jones, 1984; Klionsky *et al.*, 1990; Rothman and Stevens, 1988). Analysis of the biosynthesis of these proteins, from their site of translation in the cytoplasm to their final destination in the vacuole, has provided evidence that vacuolar proteins exploit diverse pathways to become localized to the vacuolar compartment. In this section, the biosynthesis of specific resident vacuolar proteins illustrate examples of these unique biosynthetic routes. Experiments to date suggest that soluble vacuolar glycoproteins possess targetting information within their primary amino acid sequences that specify interaction with an active vacuolar sorting machinery; missorted soluble precursors are secreted at the cell surface through the late stages of the secretory pathway (Sections II,B,C,D) (Bankaitis *et al.*, 1986; Robinson *et al.*, 1988; Rothman and Stevens, 1986; Stevens *et al.*, 1986; Valls *et al.*, 1987, 1990). By contrast, the mechanism by which vacuolar membrane proteins are localized to the vacuole is less clear, but available evidence is consistent with a model in which these proteins lack specific targetting information and are localized to the vacuole by "default" (Section II,E) (Roberts, 1991; Roberts *et al.*, 1992). The biosynthesis of the vacuolar enzyme α-mannosidase has provided evidence for a third vacuolar localization pathway (Section II,F) (Yoshihisa and Anraku, 1990). This protein appears to be sequestered directly from the cytoplasm into the vacuolar lumen, and its biosynthesis suggests that there are mechanisms for import of proteins directly across the vacuolar membrane.

B. Biosynthesis of Carboxypeptidase Y

The soluble vacuolar glycoprotein carboxypeptidase Y (CPY) is perhaps the best characterized vacuolar protein in yeast, and it provides a more or less typical example of a soluble vacuolar protein that traverses the secretory pathway during its biosynthesis and delivery to the vacuole. The primary sequence of the gene encoding CPY, *PRC1,* revealed that the protein is synthesized with three principle segments (Fig. 2) (Valls *et al.,* 1987). The amino-terminal 20 amino acids are largely hydrophobic residues characteristic of a typical ER signal sequence (Randall and Hardy, 1989). The following 91 amino acids constitute the pro region of the CPY precursor, and mature CPY is encoded in the 421 carboxy-terminal residues of the *PRC1* product. The biosynthesis of CPY has been examined in great detail (Hasilik and Tanner, 1976, 1978; Stevens *et al.,* 1982). The ER signal sequence directs translocation of CPY across the ER membrane via a common, genetically defined translocation apparatus (Deshaies and Schekman, 1987, 1989; Rothblatt *et al.,* 1989). The signal sequence of CPY is removed upon entry into the ER, and the precursor CPY molecule undergoes dolichol-mediated, asparagine-linked glycosylation at four Asn-X-Thr acceptor sites (Blachly-Dyson and Stevens, 1987; Hasilik and Tanner, 1978; Johnson *et al.,* 1987). The 67 kDa ER form of proCPY is referred to as p1 CPY. Movement from the ER to the Golgi apparatus is accompanied by additional modification of the carbohydrate side chains on proCPY, and this 69 kDa Golgi-modified form of the precursor is referred to as p2 CPY. CPY is sorted away from soluble proteins destined for secretion in a late-Golgi compartment and delivered instead to the vacuole (Graham and Emr, 1991; Stevens *et al.,* 1982). Delivery to the vacuole is accompanied by proteolytic removal of the pro segment of CPY, thereby yielding the mature, enzymatically active, 61 kDa CPY molecule. The processing event that yields mature CPY in wild-type cells requires the activity of both proteinase A and proteinase B (Mechler *et al.,* 1987), the products of the *PEP4* and *PRB1* gene, respectively (Ammerer *et al.,* 1986; Moehle *et al.,* 1987a,b; Woolford *et al.,* 1986). The biosynthesis of CPY requires an average of about 6 mintues (Hasilik and Tanner, 1978; Rothman and Stevens, 1986; Stevens *et al.,* 1982).

The pro region of the CPY precursor possesses information that is necessary and sufficient for delivery of CPY to the vacuole (Johnson *et*

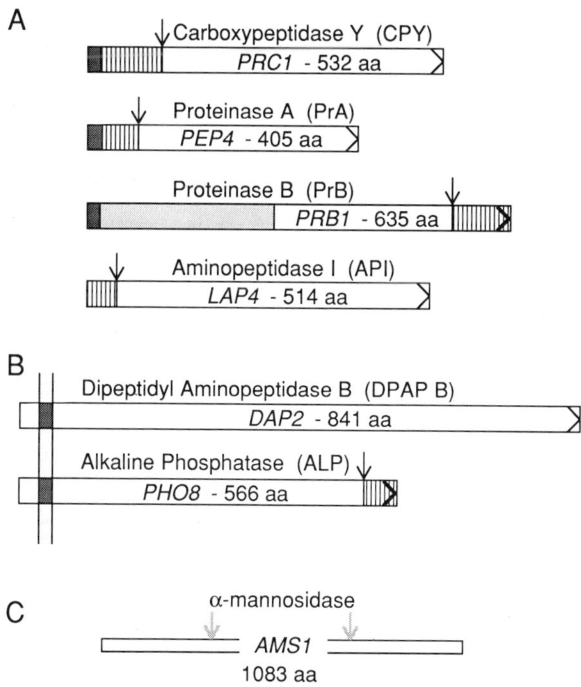

FIG. 2 Primary structures of vacuolar proteins (drawn to scale). (A) Soluble vacuolar glycoproteins (Sections II,B,C). Darkly shaded regions correspond to hydrophobic signal sequences; the additional shaded region in the *PRB1* sequence is removed prior to exit from the ER. Striped segments are propeptide portions that are removed in a *PEP4*-dependent reaction, as denoted by the vertical arrows. The precise position of the carboxy-terminal cleavage in the pro-PrB protein has not been determined. Glycosylation sites are not shown. (B) Vacuolar integral membrane proteins (Section II,E). The transmembrane segments are shown as shaded regions integrated into a schematic membrane, with cytoplasmic tails protruding to the left. The precise location of the *PEP4*-dependent cleavage in the PHO8 product is unknown. (C) Primary structure of α-mannosidase (Section II,F). Despite seven N-glycosylation acceptor sites in the sequence of α-mannosidase, this vacuolar peripheral membrane protein is not a glycoprotein. α-Mannosidase undergoes a slow, *PEP4*-dependent cleavage that liberates a 73 kDa and 31 kDa enzymatically active dimer from a 107 kDa enzymatically active monomer. However, the position of the cleavage site is unclear from the available data; the two possible cleavage sites are shown by grey arrows, only one of which is used.

al., 1987; Valls *et al.*, 1987, 1990). The pro region also maintains the precursor form of CPY in an inactive state as it migrates through the common compartments of the secretory pathway, and it may assist in the folding of the protein. Fusion of the first 50 amino acids of the *PRC1* coding sequence (and thus the first 30 residues of the pro region) to the coding region of the nonvacuolar enzyme invertase directs invertase to the vacuole (Johnson *et al.*, 1987). Furthermore, deletions and point mutations within the pro region compromise efficient vacuolar sorting of CPY and result in secretion of substantial percentages of the newly synthesized, mutant molecules (Valls *et al.*, 1987, 1990). These studies have demonstrated that four contiguous residues near the amino-terminus of the pro region are required for efficient sorting, and, as such, these four amino acids constitute a critical segment of the CPY sorting determinant (Valls *et al.*, 1990). These same studies indicate that CPY is sorted as a monomer. In cells that expressed both wild-type CPY and altered forms of CPY carrying sorting mutations in the pro region of the molecule, the wild-type protein was efficiently sorted to the vacuole, whereas the mutant proteins were largely secreted (Valls *et al.*, 1990). Unlike certain examples of lysosomal protein sorting in mammalian cells (Kornfeld and Melman, 1989), the recognition determinant on proCPY recognized by the vacuolar protein sorting apparatus is not generated by N-linked carbohydrate modification. Unglycosylated CPY is efficiently sorted to the vacuole in cells treated with tunicamycin, a drug that completely blocks the addition of N-linked carbohydrate to asparagine residues (Schwaiger *et al.*, 1982; Stevens *et al.*, 1982). Likewise, mutations that eliminate the four N-linked glycosylation acceptor sites have been engineered into the CPY molecule. This unglycosylated mutant form of CPY is both enzymatically active and, despite slowed kinetics of movement through the secretory pathway, efficiently sorted to the vacuole (Winther, 1989; Winther *et al.*, 1991).

Although carbohydrate side chains do not play a role in the sorting of CPY, the nature of carbohydrate modifications on vacuolar and secreted forms of both CPY and CPY-invertase fusions indicate that the process of sorting occurs distal to the last detectable Golgi-localized carbohydrate modification enzymes. CPY receives only limited carbohydrate modifications in the Golgi apparatus, and this led to the suggestion that it was sorted in an early- to middle-Golgi compartment (Hasilik and Tanner, 1978). However, vacuolar CPY-invertase hybrid proteins are hyperglycosylated in a manner indistinguishable from secreted forms of invertase (Johnson *et al.*, 1987). Likewise, secreted CPY receives the marginal modifications characteristic of vacuolar CPY despite evidence that the missorted molecules utilize the same late stages of the secretory pathway as normally secreted proteins (Stevens *et al.*, 1986; Rothman and Stevens, 1986). These data argue that differential carbohydrate modifications are a consequence of intrinsic properties of various proteins and not a reflection of passage through distinct compartments. More recent evidence suggests that CPY sorting occurs in or after transport through the most distal Golgi compartment yet characterized (Section IV,A) (Graham and Emr, 1991).

C. Biosynthetic Routes of Other Soluble Vacuolar Glycoproteins

The biosynthetic routes of several other soluble vacuolar hydrolases have been characterized (see Klionsky *et al.*, 1990, for a more extensive review), and in many respects the biosyntheses of these proteins have similar themes to the biosynthesis of CPY. The examples mentioned next are all glycoproteins that traverse the early and middle stages of the secretory pathway en route to the vacuole (Ammerer *et al.*, 1986; Chang and Smith, 1989; Klionsky *et al.*, 1988; Mechler *et al.*, 1988; Moehle *et al.*, 1989; Woolford *et al.*, 1986). These proteins are all synthesized as inactive precursors that, once committed to delivery to the vacuole, undergo *PEP4*-dependent proteolytic maturation to their respective mature, active species. Nonetheless, unusual features of the biosynthesis of proteinase A (PrA), proteinase B (PrB), and aminopeptidase I (API) underscore the diversity of posttranslational modifications soluble vacuolar glycoproteins can undergo during transit through the secretory pathway.

PrA, the 42 kDa product of the *PEP4* gene, is synthesized as a 405 amino acid precursor with a typical, cleaved ER signal sequence of 22 residues, an amino-terminal pro region of about 55 amino acids, and the mature protease domain consisting of the remaining 329 residues (Fig. 2) (Ammerer *et al.*, 1986; Klionsky *et al.*, 1988; Rothman *et al.*, 1986; Woolford *et al.*, 1986). The overall organization of the PEP4 product is thus similar to CPY; however, three features of its biosynthesis distinguish it from CPY. First, the vacuolar sorting information for the PrA precursor is encoded in the pro region of molecule (Klionsky *et al.*, 1988). Although this is similar to CPY, fusion studies with the amino-terminus of proPrA linked to the normally secreted enzyme invertase demonstrated that the entire pro region of PrA is required for efficient delivery of the fusion protein to the vacuole, and the implication from these studies is that an important vacuolar sorting determinant in proPrA resides in the carboxy-terminal region of the pro segment. Small deletions have been introduced into the pro region of the PrA precursor with the adverse effect of grossly destabilizing the protein (Klionsky *et al.*, 1988). Thus, it has not been possible to identify specific amino acids important for vacuolar sorting, but it is intriguing to note that there are no regions of significant sequence identity shared between the pro regions of CPY and PrA (Klionsky *et al.*, 1988; Valls *et al.*, 1987). These studies also indicate that the pro region of the PrA precursor may play additional roles in folding and/or stabilizing the protein during transport to the vacuole. The second distinction between CPY and PrA is that asparagine-linked carbohydrate side chains on PrA appear to be required for efficient sorting to the vacuole (Winther,

1989). The PEP4 gene product has two N-linked glycosylation acceptor sites, both of which are located in the mature region of the PrA molecule and both of which are utilized (Mechler *et al.,* 1982). Removal of these acceptor sites by introducing site-directed mutations in the *PEP4* coding region resulted in a reduced vacuolar pool of the mature protein, and about half of the newly synthesized, mutationally deglycosylated PrA was secreted in its precursor form (Winther, 1989). It has yet to be determined if the carbohydrate is itself a sorting determinant or if glycosyl modifications indirectly stabilize a sorting competent form of the precursor molecule. The third significant difference between CPY and PrA is that maturation of the 52 kDa Golgi form of proPrA to the mature, 42 kDa vacuolar enzyme appears to be autocatalytic. PrA activity is required for the activation of a large number of vacuolar hydrolases, and thus, *pep4* mutants lack many active vacuolar hydrolytic enzymes (Hemmings *et al.,* 1981; Jones *et al.,* 1982; Jones, 1984; Zubenko *et al.,* 1983). The mechanism of conversion of proPrA to PrA remains obscure. PrA is a member of the aspartyl protease family, as is pepsinogen, a protease precursor capable of autoactivation to pepsin under low pH conditions (James and Sielecki, 1986; Tang, 1979). This led to the proposal that conversion of proPrA to PrA was stimulated by the decrease in pH encountered when pro-PrA was was delivered to the vacuole (Ammerer *et al.,* 1986; Rothman and Stevens, 1988; Rothman *et al.,* 1989b; Woolford *et al.,* 1986). In support of this hypothesis, the pro-PrA species can undergo autocatalytic activation *in vitro* in a low pH environment containing sodium polyphosphate (Mechler *et al.,* 1987). However, the fact that rapid and efficient conversion is observed in a *vat2* yeast mutant, which is completely defective for acidification of the vacuole (Section III,E) (Yamashiro *et al.,* 1990), argues that a decrease in pH is not required for maturation of PrA. The switch that activates conversion of PrA precursor to its mature form *in vivo* remains unknown.

The biosynthesis of PrB initially appeared more or less typical of soluble vacuolar glycoproteins in *S. cerevisiae* (Mechler *et al.,* 1982). An inactive, 40 kDa precursor form of the protease passes through the secretory pathway and accumulates in the vacuoles of a *pep4* strain (Mechler *et al.,* 1982, 1988; Moehle *et al.,* 1989). Both the precursor and the mature, 30 kDa PrB molecules are glycosylated. Studies with tunicamycin, a drug that blocks dolichol-mediated, asparagine-linked glycosylation, indicate that the precursor peptide(s) of proPrB possesses one N-linked carbohydrate while the mature species carries tunicamycin-insensitive carbohydrate; the latter is presumed to result from O-linked modifications (Mechler *et al.,* 1982; Runge, 1988). This comparatively simple scheme was complicated when the sequence of the PrB structural gene, *PRB1,* revealed a 635 amino acid open reading frame with the capacity to encode a ~70 kDa protein species

(Fig. 2) (Moehle *et al.*, 1987a,b). This entire region is translated; a 70 kDa PrB-related protein species accumulates in yeast mutants defective for translocation of secretory pathway proteins across the ER membrane (Mechler *et al.*, 1988; Moehle *et al.*, 1989). The amino-terminal region encodes a large superpeptide, whereas the 40 kDa pro-PrB precursor and 30 kDa mature regions of PrB are encoded in the carboxy-terminal region of the *PRB1* gene (Hirsch *et al.*, 1991; Moehle *et al.*, 1987 a,b). The current model for PrB biosynthesis and maturation is that the large precursor is translocated across the ER membrane, the amino-terminal 281 amino acid superpeptide is cleaved from pro-PrB in the ER lumen, a carboxy-terminal segment not found in the mature enzyme receives a single N-linked carbohydrate side chain, and this carboxy-terminal extension is removed upon delivery to the vacuole in a two-step process requiring both PrA and PrB activity (Hirsch *et al.*, 1991; Klionsky *et al.*, 1990; Mechler *et al.*, 1988; Moehle *et al.*, 1987a,b, 1989). Cleavage of the large, amino-terminal segment in the ER lumen fails to occur in *prb1* mutants in which the active site serine is replaced by an alanine residue (Hirsch *et al.*, 1991). This suggests that the ER cleavage event is autocatalytic in nature. Both pro-PrB and the superpeptide are enriched in vacuoles isolated from *pep4* mutants, suggesting that the two proteins are targetted to the vacuole, possibly as a complex (Hirsch *et al.*, 1991). The superpeptide is rapidly degraded in wild-type cells. The location of vacuolar targetting information within the proPrB precursor (or the superpeptide or both) has not been established. Finally, genetic evidence suggests that vacuolar activation of PrB may involve formation of a stoichiometric molecular complex with PrA. Diploids that carry a single copy of the PrA structural gene, *PEP4*, have about half the amount of *PRB1* activity as diploids with both copies of *PEP4* (Jones, 1984; Rothman *et al.*, 1986; Rothman, 1988). Similarly, overexpression from the *PRB1* gene leads to only modest increases in PrB activity; however, when both the *PRB1* and *PEP4* are overexpressed, the levels of PrB activity increase significantly (Moehle and Jones, 1990).

Several additional soluble vacuolar hydrolases have been described (Achstetter and Wolf, 1985; Jones, 1984; Klionsky *et al.*, but they have not been analyzed in detail. API is a soluble glycoprotein that requires *PEP4*-dependent processing for activation (Fig. 2) (Trumbly and Bradley, 1983). The cloning and sequencing of the structural gene for API, *LAP4*, was recently reported (Chang and Smith, 1989). The surprising feature deduced from translation of the structural gene for this enzyme is that it lacks a conventional ER signal sequence. Rather, the amino-terminal 45 residues, which are cleaved from the mature enzyme, are predicted to form an amphipathic alpha helix. Given that API is a glycoprotein, this region of the protein must be sufficient to serve as an ER targeting sequence; this has not been tested directly, but the fact that a wide variety of diverse sequences serve as ER targetting sequences in yeast (Kaiser *et al.*, 1988) makes this a likely possibility.

D. Overproduced CPY and PrA Are Secreted

Overproduction of proCPY or proPrA by increasing the gene dosage of their respective structural genes results in secretion of substantial amounts of Golgi-modified precursor forms of these proteins (Rothman *et al.*, 1986; Stevens *et al.*, 1986). In both cases the expression level of these proteins was increased approximately eightfold, and about half of the overproduced, newly synthesized protein was secreted. Overexpression of pro-CPY did not result in the secretion of normally expressed proPrA, nor did overexpression of proPrA result in secretion of endogenous proCPY (Rothman *et al.*, 1986; Stevens *et al.*, 1986). These data are subject to at least two interpretations. Given that sorting of soluble vacuolar proteins is an active process, there is likely to be a receptor complex that recognizes and assists in the sorting of these proteins. Overproduction-indiced missorting could result from the saturation of such a receptor complex (Stevens *et al.*, 1986). By this model, the fact that overproduction of pro-CPY does not affect the sorting of proPrA, and vice versa, forces the conclusion that there are either different receptors for individual vacuolar proteins or that a single receptor complex has multiple, unique binding sites for individual precursors. An alternative explanation, namely, that overproduced precursors form some sort of homomultimers that fail to be recognized by the sorting machinery, would appear equally plausible. By this model, the overproduced vacuolar precursor forms multimers with itself, thereby sterically masking the normally exposed sorting determinant. Such masked multimers are secreted. Discrimination between these models awaits further experimentation.

E. Membrane Proteins Are Delivered to the Vacuole by Default

The biosynthesis of soluble vacuolar glycoproteins in yeast is similar to that of soluble lysosomal glycoproteins in animal cells (Kornfeld and Melman, 1989). In both cases newly synthesized precursors enter the ER and travel in common compartments with secreted proteins to a late-Golgi compartment. Furthermore, soluble hydrolases in both systems are actively sorted away from secreted soluble proteins in a trans-Golgi compartment, and in the absence of efficient sorting, hydrolase precursors are secreted from the cell (Graham and Emr, 1991; Johnson *et al.*, 1987;

Kornfeld and Melman, 1989; Rothman and Stevens, 1986; Stevens *et al.*, 1986; Valls *et al.*, 1987; von Figura and Hasilik, 1986). The situation with vacuolar and lysosomal membrane proteins is somewhat different. Human genetic diseases have been described that result in the aberrant secretion of soluble lysosomal proteins from several cell types including fibroblasts. By contrast, lysosomal membrane proteins are efficiently delivered to lysosomes in these mutant cells (Kornfeld and Melman, 1989). Similarly, yeast mutants unable to efficiently sort soluble proteins to the vacuole (Section IV) nonetheless target vacuolar membrane proteins to the vacuole with wild-type fidelity (Klionsky and Emr, 1989; Rothman and Stevens, 1988; Rothman, 1988; C. Raymond and T. Stevens, unpublished observations). Detailed analysis of the sorting determinants on vacuolar membrane proteins has revealed a radical new concept. In *S. cerevisiae* sorting of membrane proteins to the vacuole may occur by default just as the default pathway for soluble secretory pathway proteins appears to be secretion (Kelly, 1985; Roberts, 1991; Roberts *et al.*, 1992).

Two integral membrane glycoproteins of the yeast vacuole, dipeptidyl aminopeptidase B (DPAP B) and alkaline phosphatase (ALP), have been characterized in detail, and these proteins share many characteristics with regard to their structure and biosynthesis (Fig. 2) (Bordallo *et al.*, 1984; Clark *et al.*, 1982; Klionsky and Emr, 1989; Mitchell *et al.*, 1981; Roberts *et al.*, 1989; Suarez-Rendueles and Wolf, 1987). Antibodies directed against either protein decorate the vacuolar membrane when yeast cells are examined by indirect immunofluorescence microscopy (Raymond *et al.*, 1990; Roberts *et al.*, 1989; Roberts, 1991). Furthermore, DPAP B and ALP fractionate with vacuolar membranes and are resistant to extraction with alkaline buffers; the latter property is characteristic of integral membrane proteins (Fujiki *et al.*, 1982; Klionsky and Emr, 1989; Roberts *et al.*, 1989). Both proteins have been shown to traverse the secretory pathway and receive asparagine-linked carbohydrate during biosythesis and both proteins appear to be sorted to the vacuole after transport through the Golgi apparatus (Klionsky and Emr, 1989; Roberts *et al.*, 1989). ALP is synthesized as a zymogen that undergoes *PEP4*-dependent activation upon delivery to the vacuole (Jones *et al.*, 1982); DPAP B does not undergo proteolytic processing and is active in *pep4* yeast strains (Roberts *et al.*, 1989).

DPAP B is encoded by the *DAP2* gene and is a 120 kDa glycoprotein, approximately 97 kDa of which is protein (Roberts *et al.*, 1989; Suarez-Rendueles and Wolf, 1987). Sequencing of *DAP2* and translation of the DNA sequence revealed several features of the DPAP B protein that have been confirmed by subsequent experiments (Roberts *et al.*, 1989). The amino-terminal 29 amino acids of DPAP B are mainly hydrophilic residues that constitute a small cytoplasmic domain of the protein, the following

16 residues are hydrophobic amino acids that form a membrane spanning anchor, and the remaining 796 amino acids make up the enzymatic portion of the protein that protrudes into the vacuolar lumen (Fig. 2) (Roberts, 1991). These features of DPAP B topology relative to the vacuolar membrane are characteristic of Type II integral membrane proteins (Singer *et al.,* 1987). ALP is also a Type II integral membrane protein, and the apparent molecular weights of its precursor and mature forms are 76 and 72 kDa, respectively (Klionsky and Emr, 1989, 1990). The sequence of its structural gene, *PHO8,* demonstrated that the cytoplasmic tail and membrane anchor of ALP are similar in size but largely dissimilar in sequence to those of DPAP B (Fig. 2) (Kaneko *et al.*, 1987; Klionsky and Emr, 1989, 1990). The predicted topology of these proteins relative to the vacuolar membrane is supported by several observations. Both proteins receive N-linked glycosylation at several sites, and the Asn-X-Ser/Thr carbohydrate addition sites reside in the large regions of DPAP B and ALP predicted to be luminal (Klionsky and Emr, 1989; Roberts *et al.*, 1989). Furthermore, in-frame fusion of portions of *DAP2* or *PHO8* encoding the predicted cytoplasmic tail and transmembrane domains to the DNA encoding cytoplasmic invertase results in glycosylated fusion proteins that acquire the characteristics of integral membrane proteins (Klionsky and Emr, 1990; Roberts *et al.,* 1989; Roberts, 1991; C. J. Roberts and T. H. Stevens, unpublished observations).

To understand the cis-acting signals on DPAP B and ALP that direct these proteins to the vacuole, the experimental strategy has been to delete regions of their coding sequences and/or to fuse selected regions of *DAP2* or *PHO8* to genes encoding reporter proteins. The cellular distribution of these constructs is then followed by indirect immunofluorescence, subcellular fractionation, or vacuolar processing in the case of ALP (Klionsky and Emr, 1990; Roberts, 1991; Roberts *et al.,* 1992). It seems clear that the transmembrane domains are required as ER targetting sequences; deletion of this region from an ALP-invertase fusion protein construct results in the expression of a cytoplasmic protein (Klionsky and Emr, 1990). Substitution of a cleavable signal sequence onto the luminal domains of DPAP B or ALP results in glycosylated, enzymatically active proteins that are secreted from the cell, arguing that the luminal domains of these proteins lack vacuolar targetting information (Klionsky and Emr, 1990; Roberts, 1991; Roberts *et al.,* 1992). By contrast, proteins containing the cytoplasmic tail and transmembrane segments of DPAP B or ALP fused to invertase are delivered to the vacuole. Deletion of the cytoplasmic domain from DPAP B has no effect on its vacuolar targetting (Roberts, 1991; Roberts *et al.,* 1992). Finally, replacement of the transmembrane domain of DPAP B with the transmembrane domain from a protein that resides in the Golgi apparatus, DPAP A (Section III,B), also has no effect on vacuolar delivery (Roberts, 1991; Roberts *et al.*, 1992). Thus, other than the simple requirement for membrane anchorage, there is no evidence for a specific vacuolar targetting region in DPAP B. This suggests at least two models for vacuolar delivery of vacuolar membrane proteins. First, the default destination for unsorted membrane proteins lacking sorting information may simply be delivery to the vacuole. Second, there may be redundant vacuolar targetting information within both the cytoplasmic tail and transmembrane domains of both DPAP B and ALP.

Recent evidence favors the model in which membrane proteins within the secretory pathway that lack sorting information are delivered to the vacuole. The most compelling data come from work with two Golgi membrane proteins, DPAP A and Kex1p (Section III,B) (Cooper, 1990; Cooper and Bussey, 1989; Julius *et al.*, 1983). DPAP A is a 120 kDa Type II integral membrane protein similar in organization to DPAP B except that the cytoplasmic tail of DPAP A is 118 residues in length (Roberts *et al.*, 1989; Roberts, 1991). Kex1p is a 110 kDa Type I integral membrane protein; it has a cleavable signal sequence that directs entry of the majority of the protein into the lumen of the ER followed by a transmembrane domain and a short, carboxy-terminal cytoplasmic tail of about 100 residues (Cooper and Bussey, 1989; Dmochowska *et al.*, 1987). For both DPAP A and Kex1p, increased expression of the wild-type proteins in protease-deficient *pep4* yeast strains results in delivery of significant pools of these proteins to the vacuolar membrane (Cooper, 1990; Roberts, 1991; Roberts *et al.*, 1992). Likewise, mutations and deletions within the cytoplasmic tails of either protein lead to near-quantitative delivery of the mutant products to the vacuole (Cooper, 1990; Roberts, 1991; Roberts *et al.*, 1992). These data favor a model in which the cytoplasmic tails of these Golgi membrane proteins serve as retention domains; in the absence of retention, because of saturation of the retention mechanism(s) or mutational inactivation of the retention signal, these proteins enter the vacuole by default.

The notion that unsorted membrane proteins in the secretory pathway of *S. cerevisiae* are delivered to the vacuole by default is generally regarded with surprise. The default compartment for soluble proteins is firmly established to be secretion from the cell, and it was widely assumed that mislocalized membrane proteins would be delivered to the plasma membrane. Nonetheless, default delivery of membrane proteins to the vacuole may make biological sense. Whereas delivery to the cell surface of aberrant soluble proteins that are within the secretory pathway results in their expulsion from the cell, delivery of aberrant membrane proteins to the cell surface would only result in their residence on the plasma membrane. A more expedient mechanism for the elimination of mislocalized membrane proteins would potentially result from delivery to and subsequent degradation in the vacuole.

F. α-Mannose Defines a New Import Pathway to the Vacuole

The vacuolar enzyme α-mannosidase has frequently been used as a membrane marker for the vacuole in *S. cerevisiae* (Opheim, 1978). The enzyme is composed of 107, 73, and 31 kDa polypeptides, which co-purify with variable stoichiometry depending on the growth phase of the cells from which it is harvested (Fig. 2) (Yoshihisa *et al.,* 1988). Characterization of the biosynthesis of α-mannosidase demonstrated that all three polypeptides are the product of a single gene, *AMS1* (Kuranda and Robbins, 1987; Yoshihisa *et al.,* 1988; Yoshihisa and Anraku, 1989). Only the 107 kDa species, which itself is enzymatically active, is present in *pep4* yeast strain (Jones *et al.,* 1982; Yoshihisa and Anraku, 1990). The 73 and 31 kDa species are derived from the 107 kDa species by proteolytic cleavage in the vacuole (Yoshihisa and Anraku, 1990). Several unusual features of α-mannosidase have been elucidated. Purification of the enzyme revealed that it is not an integral membrane protein but, rather, that it is peripherally associated with the luminal face of the vacuolar membrane (Yoshihisa *et al.,* 1988). Furthermore, the enzyme is not glycosylated despite the presence of seven potential N-glycosylation sites in the *AMS1* sequence (Yoshihisa *et al.,* 1988; Yoshihisa and Anraku, 1989). Lastly, there were no apparent ER targetting sequences within the *AMS1* sequence (Yoshihisa and Anraku, 1989).

Characterization of the biosynthesis of α-mannosidase suggests that this enzyme is delivered to the vacuole by a novel mechanism; a tenable hypothesis at present is that the protein is sequestered directly from the cytoplasm into the vacuolar lumen (Fig. 1). Attachment of a cleavable ER signal sequence to the amino-terminus of α-mannosidase resulted in the synthesis of a glycoprotein (Yoshihisa and Anraku, 1990). Hence the N-glycosylation sites in α-mannosidase are competent to receive carbohydrate despite the fact that the native protein is not glycosylated. A chimeric protein in which invertase (normally a cytoplasmic protein and a secreted glycoprotein; Carlson and Botstein, 1982) was fused to the carboxy-terminus of α-mannosidase was delivered to the vacuolar lumen; yet the fusion protein was not glycosylated (Yoshihisa and Anraku, 1990). These data are consistent with a direct cytoplasm to vacuole import mechanism. The signal on the α-mannosidase protein that specifies uptake into the vacuole has not been precisely defined. A series of carboxy-terminal deletions of α-mannosidase attached to invertase were tested for vacuolar import, and only those fusions possessing more than 927 of the 1083 residues of the α-mannosidase were efficiently sequestered into the vacuole (Yoshihisa and Anraku, 1990). This may suggest that a targetting sequence is located in the extreme carboxy-terminal region of α-mannosidase, but it may also imply that proper folding of the α-mannosidase domain in the fusion protein is required for import. Finally, the import machinery for α-mannosidase appears to be saturable. Increased expression of the protein resulted in a significant decrease in the percentage of vacuole-associated α-mannosidase activity; the overexpressed protein was not secreted (Yoshihisa and Anraku, 1990).

III. Proteins Degraded in the Vacuole, Endocytosis, and Vacuolar Acidification

A. Protein Degradation and Vacuolar Acidification

Evidence has indicated that proteins that are targetted to a variety of intracellular destinations undergo turnover in the vacuole. Vacuolar degradation of these proteins augments the long held supposition that the vacuole is an organelle involved in the normal degradation of nonvacuolar proteins, and it suggests that transport pathways exist with the capacity to shuttle proteins into the vacuole from their normal location in other subcellular compartments. Specific protein turnover events have provided evidence for receptor-mediated endocytosis in yeast. In addition, there are data suggesting that an inducible pathway for uptake of a cytoplasmic protein into the vacuole exists. These examples support the idea that the vacuole is connected to several diverse intracellular compartments through specific transport pathways.

The vacuole is maintained at an acidic pH by a multisubunit vacuolar H^+-ATPase. The complex is comprised of both integral membrane components and peripheral subunits attached to the cytoplasmic face of the vacuolar membrane. Vacuolar H^+-ATPase function is not required for cell viability, but the poor growth of ATPase-deficient cells indicates that complex plays important physiological roles that have yet to be precisely defined. Additional research challenges involving the vacuolar H^+-ATPase include defining the composition of the active proton pump, understanding the functions of various subunits, and determining the site of biosynthesis and sequence of assembly of this complex from newly synthesized components.

B. Turnover of Golgi-Localized Processing Enzymes in the Vacuole

In *S. cerevisiae,* the Golgi apparatus is a morphologically indistinct organelle. In immunofluorescence experiments, antibodies directed against proteins that function in the Golgi label several discrete patches distributed throughout the cytoplasmic compartment (Cleves *et al.,* 1991; Cooper, 1990; Franzusoff *et al.,* 1991; Redding *et al.,* 1991; Roberts *et al.,* 1992). By other criteria the Golgi apparatus in yeast is well characterized. It was first defined by the yeast *sec* mutants as a compartment(s) in which elaboration of the outer chains on asparagine-linked carbohydrates and extensive processing of the mating pheromone α-factor and killer toxin occur (Bussey, 1988; Julius *et al.,* 1984b; Novick *et al.,* 1980; Runge, 1988; Schekman, 1985). Subcellular fractionation has been used to demonstrate that the enzymes characteristic of the yeast Golgi reside in enclosed compartments separable from other cell membranes (Bowser and Novick, 1991; Cunningham and Wickner, 1989). Biochemical and genetic evidence for vesicle-mediated transport between the yeast ER and Golgi apparatus further supports the contention that the Golgi complex is physically separate from other secretory pathway organelles (Baker *et al.,* 1988; Groesch *et al.,* 1990; Kaiser and Schekman, 1990; Rexach and Schekman, 1991; Ruohola *et al.,* 1988). The Golgi apparatus appears to be comprised of at least three functionally distinct compartments (Graham and Emr, 1991). Soluble vacuolar glycoproteins are sorted from secreted soluble proteins in (or after transport through) the most distal of these (Graham and Emr, 1991). The Golgi-localized proteolytic processing of the precursor to the yeast mating pheromone α-factor occurs in this same distal compartment and requires the sequential action of three proteases encoded by the *KEX2, KEX1,* and *STE13* genes, respectively (Fuller *et al.,* 1988; Graham and Emr, 1991). Kex2p is an endopeptidase that cleaves between pairs of basic amino acid residues (Fuller *et al.,* 1989; Julius *et al.,* 1984a), Kex1p is a carboxypeptidase (Cooper, 1990; Cooper and Bussey, 1989; Dmochowska *et al.,* 1987), and *STE13* encodes DPAP A, a dipeptidyl aminopeptidase (Julius *et al.,* 1983; Roberts *et al.,* 1989). The structural genes for these enzymes have been cloned and sequenced (Dmochowska *et al.,* 1987; Fuller *et al.,* 1989; Roberts, 1991; Roberts *et al.,* 1989), and all three proteases are integral membrane proteins (Cooper and Bussey, 1989; Julius *et al.,* 1983, 1984a; Roberts *et al.,* 1992).

There is mounting evidence that turnover of Kex2p, Kex1p, and DPAP A occurs in the vacuole and that the cytoplasmic domains of these proteins are required for retention in the Golgi apparatus. In the case of Kex2p, the half-life of the protein is extended in a *pep4* mutant relative to an isogenic, wild-type yeast strain (C. Wilcox and R. Fuller, personal communication). Overexpression of any one of these proteins in a *pep4* yeast strain resulted in its accumulation on the vacuolar membrane (Cooper, 1990; Roberts, 1991; K. Redding and R. Fuller, personal communication; Roberts *et al.*, 1992). Vacuolar labeling was less frequently observed in *PEP4* yeast overexpressing Kex2p, Kex1p, or DPAP A, and the turnover of Kex2p was accelerated in a Kex2p overproducing strain (Cooper, 1990; C. Wilcox *et al.*, personal communication; Roberts *et al.*, 1992). These data indicate that retention of these membrane proteins in the Golgi is saturable, that failure to be retained results in delivery to the vacuole, and that vacuolar degradation of these proteins is rapid. Deletions of the cytoplasmic domain from Kex1p or DPAP A resulted in exclusive vacuolar localization of the mutant proteins (Cooper, 1990; Roberts, 1991; Roberts *et al.*, 1992). Furthermore, replacement of the cytoplasmic domains on the vacuolar membrane proteins DPAB B or ALP (Section II,E) with the cytoplasmic domain from DPAP A resulted in apparent Golgi localization of the fusion proteins (Roberts, 1991; Roberts *et al.*, 1992). These data are consistent with a model in which the cytoplasmic tails of Kex2p, Kex1p, or DPAP A serve to retain these proteases in the appropriate Golgi compartment and that compromise of this retention leads to vacuolar delivery by default (Section II,E). The retention apparatus has not been extensively investigated, but evidence suggests that clathrin may play an important role in maintaining the spatial integrity of the Golgi compartments (Payne and Schekman, 1989).

C. Vacuolar Degradation of a Cytoplasmic Enzyme Occurring in the Vacuole

Fructose 1,6-bis-phosphatase (FBPase) is a key regulatory enzyme of gluconeogenesis. Upon transfer to glucose-rich media, FBPase synthesis is repressed and existing enzyme is both phosphorylated and degraded (Guerra *et al.*, 1988; Rittenhouse *et al.*, 1987; Schafer *et al.*, 1987). FBPase is normally a cytoplasmic enzyme; yet a recent report provided compelling evidence that catabolite-induced degradation of the enzyme occurs in the vacuole (Chiang and Schekman, 1991). The authors found that the rapid, glucose-induced turnover of FBPase observed in wild-type *S. cerevisiae* did not occur in an isogenic *pep4* yeast strain. Moreover, subcellular fractionation indicated that the enzyme shifted from the cytoplasm to a vacuole-enriched fraction upon transfer of *pep4* mutants from poor carbon sources to glucose-rich media; protease protection data demonstrated that the enzyme was sequestered within the lumen of membrane-enclosed organelles once redistribution had occurred. Finally, indirect immunofluorescence using anti-FBPase antibodies showed that the enzyme moved from the cytoplasm to the vacuole over a period of about 60 minutes in *pep4* cells shifted into glucose-rich broth. The cytoplasmic signal observed in wild-type, *PEP4* cells grown on poor carbon sources simply disappeared upon shift into glucose, presumably as a consequence of rapid vacuolar degradation of the FBPase (Chiang and Schekman, 1991).

Degradation of a cytoplasmic enzyme in the vacuole raises an obvious transport question: how is the enzyme transferred across a membrane barrier? This question was partially addressed using temperature-conditional yeast *sec* mutants (Deshaies and Schekman, 1989; Franzusoff and Schekman, 1989; Kaiser and Schekman, 1990; Novick *et al.*, 1981). Degradation of FBPase was blocked under restrictive conditions by *sec* mutations that prevent translocation into the ER*(sec62)*, by mutations that block multiple stages in transport throughout the secretory pathway (Graham and Emr, 1991), the first being movement from the ER to the Golgi compartment *(sec18)*, or by mutations that impair movement of proteins through the Golgi (*sec7;* Chiang and Schekman, 1991). The *sec1* mutation, which inhibits fusion of secretory vesicles with the plasma membrane (Novick *et al.*, 1981), displayed normal, glucose-stimulated turnover. Importantly, FBPase remained in the cytoplasm in *sec18* mutants shifted simultaneously to the restrictive temperature and glucose-rich media. The collective interpretation of these data is that FBPase itself does not traverse the secretory pathway. Rather, a component required for import of FBPase into the vacuole is induced upon glucose addition, and this component requires the secretory pathway to become localized to the compartment where FBPase is normally translocated (Chiang and Schekman, 1991).

The regulated degradation of FBPase shares features with the turnover of specific cytoplasmic proteins in serum-deprived animal cell cultures (Chiang *et al.*, 1989; Dice, 1990) and with uptake of α-mannosidase into vacuoles described in Section II,F (Yoshihisa and Anraku, 1990). In all three cases specific cytoplasmic proteins appear to bypass early stages of the secretory pathway and to be translocated in a posttranslational manner directly into the vacuole/lysosome. The import of specific proteins into vacuoles/lysosomes appears to be a fundamentally different process from the bulk turnover of cytoplasmic proteins that occurs in the vacuole upon nitrogen starvation of *S. cerevisiae* (Zubenko and Jones, 1981; Teichert *et al.*, 1989) or upon serum deprivation of cultured liver cells (Dice, 1990; Dunn, 1990). The aforementioned examples of import are specific for certain proteins, and the uptake of α-mannosidase and FBPase occurs in well-nourished yeast cells. By contrast, in liver cells deprived of amino acids, apparently random cytoplasmic contents are encircled by double membranes that are most likely derived from the rough ER (Dunn, 1990). These "autophagic vacuoles" then fuse with preexisting lysosomes and degradation of the vacuole contents ensues; this bulk turnover process is termed "autophagy" (Dice, 1990; Dunn, 1990; Mortimere, 1987). Whether the bulk turnover of protein in nitrogen starved yeast cells occurs by a similar autophagic mechanism has not been established. Regardless, the *PEP4*-dependent turnover of bulk protein and presumed recycling of constituent amino acids in yeast is requisite for a variety of processes including viability of yeast cells under starvation conditions (Teichert *et al.*, 1989) and sporulation of diploid cells (Jones, 1984). The mechanism by which bulk protein is delivered to the vacuole awaits further investigation.

D. Endocytosis in *Saccharomyces cerevisiae*

In animal cells the term "endocytic pathway" is used to describe the complex, vesicle-mediated traffic routes that interconnect the *trans*-Golgi network, endosomes, lysosomes, and the cell surface (for reviews see Goldstein *et al.*, 1985; Kornfeld and Mellman, 1989). A wide variety of macromolecules are sequestered from the extracellular milieu via recognition and binding to their cognate receptors on the plasma membrane. One form of endocytosis is the process whereby such receptor–ligand complexes are internalized in clathrin-coated pits that invaginate to form vesicles; internalized complexes proceed to a wide variety of fates in the endocytic pathway. Similarly, unicellular organisms, such as *Paramecium,* capture and internalize foodstuffs via endocytic and/or phagocytic mechanisms (Fok and Allen, 1990). Evidence for an endocytic pathway in *S. cerevisiae* has been somewhat more elusive. The thick cell wall of the yeast cell precludes the penetration of large molecules, and it can be argued *a priori* that small molecules could be internalized via specific transport proteins in a manner similar to that employed by bacteria. Nonetheless, recent data suggest that yeast cells possess an endocytic pathway.

Early investigation of endocytosis in yeast has proven to be inconclusive. Studies suggesting that fluorescein isothiocyanate-dextran (FITC-dextran) was internalized via fluid-phase endocytosis (Makarow and Nevalainen, 1987) were cast in doubt by the demonstration that low molecular weight impurities were responsible for vacuolar labeling (Preston *et al.*, 1987). Separate studies on the internalization of the sulfonate dye lucifer yellow were also interpreted as evidence for endocytosis (Chvatchko *et al.*, 1986; Pringle *et al.*, 1989; Riezman, 1985). However, there has not been a clear demonstration that uptake of lucifer yellow occurs by fluid-phase endocytosis.

Evidence that yeast cells possess an endocytic pathway has been generated by studies into the mechanism of conjugation between haploid yeast cells. Haploid yeast are one of two mating types, **a** or α, and these cells can mate with one another to form an **a**/α diploid cell (for review see Sprague *et al.*, 1983). Mating is initiated in response to a peptide pheromone stimulus that channels through specific receptors; **a** cells secrete **a**-factor that binds to the **a**-factor receptor on α cells, and α cells secrete α-factor that binds to α-factor receptor on **a** cells. The α-factor receptor, expressed exclusively in **a** cells, is a 431 amino acid integral membrane protein encoded by the *STE2* gene (Jenness *et al.*, 1983). Likewise, the **a**-factor receptor is a 470 residue protein encoded by the *STE3* gene and expressed only α cells (Hagen *et al.*, 1986; Nakayama *et al.*, 1985). The two receptors share little if any significant sequence similarity; yet they both possess seven membrane-spanning domains in the amino terminal regions followed by hydrophilic, cytoplasmic domains in their carboxy-termini. Both receptors are found on the plasma membrane (Jenness and Spatrick, 1986; J. Horecka and G. F. Sprague, Jr., personal communication). When radiolabeled α-factor, which is a 13 amino acid, water-soluble peptide, was added to **a** cells, it was internalized and subsequently degraded (Chvatchko *et al.*, 1986; Dulic and Riezman, 1989; Konopka *et al.*, 1988; Singer and Riezman, 1990). Degradation was dependent on the *PEP4* gene product. In *pep4* mutants, which lack most vacuolar protease activities, internalized α-factor co-fractionated with the precursor form of the vacuolar glycoprotein CPY (Section II,B,C) (Singer and Riezman, 1990). Similarly, upon addition of α-factor, the Ste2p α-factor binding sites disappeared from the cell surface, but the fate of Ste2p, which was presumably internalized, was not reported (Jenness and Spatrick, 1986); α-factor internalization was found to require energy but not new protein synthesis, and the loss of α-factor binding sites was greatly accelerated by the addition of α-factor (Jenness and Spatrick, 1986). These data support the model that α-factor bound to its receptor, the STE2 product, is transported to and degraded in the vacuole.

Two lines of evidence suggest that the **a**-factor receptor, Ste3p, is also degraded in the vacuole. Steady-state turnover of radiolabeled Ste3p (in the absence of **a**-factor) was investigated in isogenic *PEP4* and *pep4* yeast strains, and the lifetime of the protein was greatly extended in *pep4* mutants (N. Davis and G. F. Sprague, Jr., personal communication). In further experiments, the *STE3* coding region was placed behind a regulated promoter, and receptor synthesis was briefly induced then repressed. A stable pool of the **a**-factor receptor accumulated in a *pep4* strain but not in an isogenic *PEP4* strain. This pool of receptor was unresponsive to exogenously added pheromone (N. Davis and G. F. Sprague, Jr., personal communication). The accumulated Ste3p was found to be localized to the vacuole using indirect immunofluorescence techniques (J. Horecka and G. F. Sprague, Jr., personal communication).

Studies with mutant Ste2p and Ste3p receptors have shown that the carboxy-terminal, cytoplasmic domains of these proteins are required for efficient receptor internalization but not for pheromone-induced receptor signaling (Konopka *et al.*, 1988; Reneke *et al.*, 1988; N. Davis *et al.*, personal communication). Remarkably, clathrin function is not required for α-factor uptake in *S. cerevisiae* (Payne *et al.*, 1988). Yeast mutants lacking the heavy chain of clathrin internalized the pheromone at about half the rate of wild-type strains. Three possibilities consistent with these data are as follows: (1) clathrin is not absolutely required for endocytosis of α-factor, which occurs through a single pathway; (2) there are multiple pathways for α-factor endocytosis, some of which are clathrin-independent, or (3) clathrin plays no role in endocytosis and the observed decrease in the rate of α-factor uptake is an indirect effect. It is as yet unclear whether any of the yeast *SEC* genes, required for the integrity of the secretory pathway (Novick *et al.*, 1980, 1981), or *VPS* genes, required for vacuolar protein sorting (Robinson *et al.*, 1988; Rothman *et al.*, 1989a), are also required for internalization and vacuolar delivery of α-factor.

Taken together, the data on internalization and degradation of pheromone and pheromone receptor in yeast are consistent with a model in which pheromone–receptor complexes on the plasma membrane are endocytosed and delivered to the vacuole in membrane-enclosed compartments. Evidence for an intermediate, endosomal compartment was obtained through examination of pheromone internalization and degradation at reduced temperatures (Singer and Riezman, 1990). At 30°C, α-factor was internalized with a $\tau_{1/2}$ of 4 minutes and degraded with a $\tau_{1/2}$ of 10 minutes. At 15°C the rate of uptake decreased by two-fold, but the rate of degradation decreased by five-fold. Several lines of evidence argue that pheromone accumulated in an endosome-like compartment at this temperature. Degradation by vacuolar hydrolases did not appear to be rate-limiting at the low temperature, and the delivery of internalized pheromone from the intermediate compartment to the vacuole was reversibly blocked by metabolic inhibitors. The 15°C-accumulated α-factor had sedimentation behavior distinct from plasma membrane or vacuolar marker proteins, it floated in a density gradient with a density characteristic of a membrane-enclosed compartment, and it was largely protected from exogenously added protease (Singer and Riezman, 1990). It will be of interest to determine whether newly synthesized vacuolar proteins also pass through an endosome-like compartment.

E. Function and Assembly of the Vacuolar H^+-ATPase

The lumen of the vacuole in *S. cerevisiae* is maintained at a mildly acidic pH of about 6.0 (Preston *et al.*, 1989). The proton gradient is generated and maintained by a vacuolar proton-translocating ATPase, a multisubunit enzyme complex situated on the vacuolar membrane (Anraku *et al.*, 1989; Bowman and Bowman, 1986; Kane *et al.*, 1989a,b; Klionsky *et al.*, 1990; Nelson, 1989). Vacuolar H^+-ATPases are a ubiquitous feature of eukaryotic cells, and they are responsible for acidification of a variety of organelles including lysosomes, endosomes, regulated secretory vesicles, and the *trans*-Golgi network (for review see Mellman *et al.*, 1986). Vacuolar H^+-ATPase purified from a variety of sources share similarities in subunit composition, subunit sequences, inhibitor sensitivity, and enzymatic properties. By these same criteria, vacuolar-type H^+-ATPases are distinct from the mitochondrial F_1F_0-H^+-ATPase and the plasma membrane E_1E_2-H^+-ATPases (Bowman and Bowman, 1986; Kane *et al.*, 1989a,b; Klionsky *et al.*, 1990; Nelson, 1989). Vacuolar H^+-ATPases are thought to influence a wide variety of physiologically important events. For example, in yeast cells the vacuole serves as a major storage repository for amino acids, ions, and polyphosphate (for review see Klionsky *et al.*, 1990). Transport of ions and metabolites across the vacuolar membrane is thought to be driven through specific proton antiporter systems that utilize the proton gradient generated by the vacuolar H^+-ATPase; this is inferred from the fact that several import reactions into intact vacuolar vesicles are blocked by the addition of protonophores (Klionsky *et al.*, 1990; Ohsumi and Anraku, 1981, 1983; Sato *et al.*, 1984). These and other potential functions of the vacuolar H^+-ATPase are considered next.

The reported subunit composition of purified vacuolar H^+-ATPases varies depending on the sources and procedures used to isolate and characterize the enzyme (for reviews see Kane *et al.*, 1989b; Nelson, 1989). At least three basic subunits of approximately 70, 60, and 17 kDa are shared among all vacuolar H^+-ATPases isolated to date. The 70 kDa subunit is thought to possess the catalytic site for ATP hydrolysis, the 60 kDa polypeptide has been proposed to be a regulatory ATP-binding subunit, and the 17 kDa is a hydrophobic protein(s) that forms an integral portion of the transmembranous proton pore (Kane *et al.*, 1989a,b; Manolson *et al.*, 1985; Sun *et al.*, 1987). In *S. cerevisiae,* the initial characterization of the enzyme partially purified from vacuolar membranes suggested it had a simple, three-subunit composition (Uchida *et al.*, 1985). Subsequent analysis has demonstrated that the yeast vacuolar H^+-ATPase, like the vacuolar ATPases purified from bovine tissues, possesses several other polypeptides (Kane *et al.*, 1989a,b). In addition to the 70, 60, and 17 kDa species, proteins of 100, 42, 36, 32, and 27 kDa were found associated with the enzyme (Kane *et al.*,1989b). Several lines of evidence argue that these subunits are required for assembly and/or function of the complex. First, these eight polypeptides co-purify with vacuolar ATPase activity on glycerol gradients (Kane *et al.*, 1989b). Second, immunoprecipitation of native complex from detergent-solubilized vacuolar membranes with a mono-

TABLE I
Structural Genes Encoding Subunits of the *S. cerevisiae* Vacuolar H$^+$-ATPase Complex

Subunit designation (kDa)	Structural gene	Source
100	*a*	
70	*VMA1, TFP1, CLS1*	Anraku *et al.* (1989); Hirata *et al.* (1989); Kane *et al.* (1990); Ohya *et al.* (1991); Shih *et al.* (1988)
60	*VMA2, VAT2*	Anraku *et al.* (1989); Nelson *et al.* (1989); Yamashiro *et al.* (1990)
42	*VMA5*	M. Ho *et al.* (unpublished observations)
36	*VMA6*[a,b]	C. Bauerle and T. H. Stevens (unpublished observations)
32	*a*	
27	*VMA4*	Foury (1990)
17	*VMA11, CLS9*	Ohya *et al.* (1991); Umemoto *et al.* (1992)
16	*VMA3, CLS7*	Anraku *et al.* (1989); Ohya *et al.* (1991); Nelson and Nelson (1989, 1990); Umemoto *et al.* (1990)

[a] These polypeptides co-purify with the yeast vacuolar H$^+$-ATPase complex, but it has not been determined if they are required for activity and/or assembly of the complex.

[b] The structural gene for encoding the 36 kDa subunit has tentatively been identified, but the phenotypes of mutants carrying null mutations in the *VMA6* gene have yet to be fully characterized.

clonal antibody directed against the 70 kDa subunit co-precipitates all of the peptides that co-purify with vacuolar H$^+$-ATPase activity (Kane *et al.*, 1989b; the 17 kDa species was not apparent, but see below). Third, disruption of the structural genes encoding the 70, 60, 42, 27, 17, and 16 kDa subunits results in complete inactivation of vacuolar H$^+$-ATPase activity *in vivo* and *in vitro* (Anraku *et al.*, 1989; Foury, 1990; Hirata *et al.*, 1989; Kane *et al.*, 1990; Nelson and Nelson, 1989, 1990; Noumi *et al.*, 1991; Umemoto *et al.*, 1990, 1992; Yamashiro *et al.*, 1990; M. Ho *et al.*, unpublished observations). There appear to be two different 16–17 kDa proteolipid species encoded by separate structural genes, both of which are required for vacuolar H$^+$-ATPase activity (Umemoto *et al.*, 1992). The various subunits and the names of the corresponding structural genes are displayed in Table I. Fourth, the predicted protein sequences of the yeast 70, 60, 42, 36, 27, 17, and 16 kDa subunits share sequence similarities with various subunits from vacuolar-type H$^+$-ATPases isolated from *Neurospora crassa* and/or from bovine sources (Foury, 1990; Hirata *et al.*, 1989; Kane *et al.*, 1990; Nelson and Nelson, 1989; Nelson *et al.*, 1989; Umemoto *et al.*, 1990, 1992; Yamashiro *et al.*, 1990; M. Ho *et al.*, unpublished observations). Finally, polypeptides of 116 and 38 kDa from clathrin-coated vesicle proton ATPase (a vacuolar H^+-ATPase purified from bovine brain tissue) have been shown to be required for Mg^{2+}-activated ATP hydrolysis by the purified complex (Xie and Stone, 1988); assuming that the yeast proteins of 100 and 36 kDa are the equivalent subunits, these data suggest an essential role for these polypeptides in the proper function of the proton pump. In support of this, vacuolar H^+-ATPase-deficient mutants lacking the 100 kDa protein subunit have been obtained; however, it has yet to be established whether any of the corresponding mutations fall in the structural gene for this polypeptide (K. Hill *et al.*, unpublished observations). There may be other, unidentified subunits of the vacuolar H^+-ATPase, and subunit composition of the minimum complex required for proton pumping has not been established.

As implied, vacuolar H^+-ATPase activity is not required for the viability of *S. cerevisiae* (Nelson and Nelson, 1990; Yamashiro *et al.*, 1990). Yeast mutants carrying disruptions in any one of the structural genes shown in Table I have indistinguishable phenotypes. Vacuolar H^+-ATPase activity is completely abolished, the lumen of the vacuole assumes a neutral pH, the growth rate of the mutants is slowed substantially, they exhibit sensitivity to extremes in pH and Ca^{2+} concentrations, they are unable to grow on nonfermentable carbon sources, they grow poorly or not at all at suboptimal temperatures tolerated by wild-type yeast, they fail to accumulate *ade2* pigment in their vacuoles (a metabolic intermediate in adenine biosynthesis that accumulates in the vacuoles of *ade2* yeast mutants), and peripheral membrane subunits of the vacuolar H^+-ATPase complex fail to assemble onto the vacuolar membrane *in vivo* (Anraku *et al.*, 1989; Foury, 1990; Kane *et al.*, 1990, 1992; Nelson and Nelson, 1990; Noumi *et al.*, 1991; Ohya *et al.*, 1991; Umemoto *et al.*, 1992; Yamashiro *et al.*, 1990). These deficiencies underscore the importance of reduced vacuolar pH and vacuolar H^+-ATPase activity on the physiological regulation of yeast cell growth. One facet of this regulation appears to be Ca^{2+} homeostasis. Vacuolar ATPase-deficient mutants were identified among mutants screened for a calcium-sensitive (*cls*) phenotype (Table I) (Ohya *et al.*, 1986, 1991). The mutants have approximately six-fold elevated levels of cytosolic Ca^{2+}, and vacuoles isolated from these strains have no detectable ATP-dependent Ca^{2+} uptake (Ohya *et al.*, 1991). These observations highlight the importance of the vacuole as a storage organelle and the critical role of the vacuolar H^+-ATPase in the transport of various compounds across the vacuolar membrane.

In contrast, vacuolar H^+-ATPase activity and, by inference, reduced vacuolar pH play relatively modest roles in the sorting of vacuolar proteins, in the proteolytic maturation of vacuolar hydrolase precursors, in the degradation of proteins, and in the overall morphological organization of the vacuole. ATPase-deficient mutants, or wild-type yeast strains treated with bafilomycin A_1, a drug that specifically abolishes vacuolar H^+-ATPase activity (Bowman *et al.*, 1988), were found to sort approximately 80% of newly synthesized soluble vacuolar proteins CPY and PrA (Section II,B,C) to the vacuole (Banta *et al.*, 1988; Yamashiro *et al.*, 1990). The vacuolar membrane protein ALP (Section II,E) was also delivered to the vacuole quantitatively (Klionsky and Emr, 1989; Yamashiro *et al.*, 1990). *PEP4*-dependent proteolytic maturation of hydrolase precursors to enzymatically active mature species was unaffected by loss of vacuolar H^+-ATPase function (Banta *et al.*, 1988; Klionsky and Emr, 1989; Yamashiro *et al.*, 1990). Glucose-induced vacuolar degradation of the cytoplasmic enzyme FBPase proceeded normally in a *vma2* mutant (Section III,C) (Chiang and Schekman, 1991) and internalization and degradation of α-factor occurred with nearly wild-type kinetics in bafilomycin-treated cells (Singer and Riezman, 1990). Finally, ATPase mutants assemble vacuoles that are both normal in appearance and in cell growth-dependent segregation into developing buds (Section IV,C) (Yamashiro *et al.*, 1990). These observations have been greeted with surprise because a large body of evidence has suggested that acidification of the vacuolar network in animal systems is essential for lysosomal protein sorting, zymogen maturation, selective packaging of proteins into secretory granules, and endocytosis (for reviews see Forgac, 1989; Kornfeld and Mellman, 1989; Mellman *et al.*, 1986; Robbins, 1988; von Figura and Hasilik, 1986).

Studies with ATPase-deficient mutants have demonstrated that the vacuolar H^+-ATPase is required for several aspects of vacuolar function; consequently, this has provided further insights into the participation of the vacuole in cellular physiology. Beyond understanding the role of the vacuolar H^+-ATPase in cell growth, there is significant interest in the biogenesis of this multisubunit complex. At least four issues regarding the structure, function, and assembly of the vacuolar H^+-ATPase are under consideration. First, what is the subunit structure of the complex, and what is the topology and stoichiometry of these subunits in the functional proton pump? Second, where is the site of proton pump assembly and what signals are required for the assembly event to occur? Third, what role do the individual subunits of the complex play in promoting assembly? Finally, are there "chaperone" proteins, not themselves subunits of the ATPase, that nonetheless play specific roles in ATPase assembly?

The available data on the topology and subunit organization of the vacuolar H^+-ATPase are strikingly similar to the ball-and-stalk arrangement of the mitochondrial F_1/F_0-H^+-ATPase. Indeed the 70 and 60 kDa subunits of the vacuolar H^+-ATPases share limited sequence identity with the α-and β-subunits of the mitochondrial enzyme, respectively (Anraku *et al.*, 1989; Bowman *et al.*, 1988; Gogarten *et al.*, 1989; Nelson, 1989;

Shih *et al.*, 1988). This has led to the proposal, supported by microscopy data, that the vacuolar H^+-ATPases also have a ball-and-stalk structure with a cytoplasmic catalytic domain attached to a transmembranous proton channel (Bowman *et al.*, 1989; Kane *et al.*, 1989a; Taiz and Taiz, 1991). Consistent with this model, several of the identified subunits of the yeast vacuolar H^+-ATPase complex appear to be peripheral membrane proteins that assemble onto the cytoplasmic side of vacuolar membranes. The 70, 60, and 42 kDa subunits were stripped from vacuoles by KNO_3 or cold treatment in the presence of Mg^{2+}-ATP, a procedure that did not affect the integrity of the vacuolar membrane (Kane *et al.*, 1989b; Noumi *et al.*, 1991). Lesser amounts of the 36, 32, and 27 kDa species were released by the same treatment (Kane *et al.*, 1989a). Moreover, the 70 and 60 kDa subunits appeared as cytoplasmic proteins, as judged by indirect immunofluorescence and subcellular fractionation, in yeast mutants that fail to assemble a functional ATPase complex (Kane *et al.*, 1992; Rothman *et al.*, 1989b; Umemoto *et al.*, 1992; Yamashiro *et al.*, 1990). Furthermore, the structural genes for the 70, 60, 42, and 27 kDa subunits (Table I) do not encode ER targetting sequences at their amino-termini (Foury, 1990; Hirata *et al.*, 1989; Kane *et al.*, 1990; Nelson *et al.*, 1989; Shih *et al.*, 1988; Yamashiro *et al.*, 1990). All of these data are consistent with the interpretation that the catalytic domain of the vacuolar H^+-ATPase faces the cytoplasm. The subunit stoichiometry of this peripheral membrane component has not been investigated for the *S. cerevisiae* enzyme; however, studies of the *N. crassa* vacuolar H^+-ATPase and clathrin-coated vesicle H^+-ATPase from bovine agree on a ratio of 3:3:1 for the 70 and 60 kDa species relative to the other peripheral membrane subunits (Arai *et al.*, 1988; Bowman *et al.*, 1989).

The peripheral subunits of the vacuolar H^+-ATPase are anchored onto integral membrane proteins that form the proton channel. The membrane portion of the complex is comprised in part of 16–17 kDa, DCCD-binding, proteolipid species that are probably encoded by the *VMA3* and *VMA11* genes (Table 1) (Kane *et al.*, 1992; Umemoto *et al.*, 1992). Both genes encode extremely hydrophilic proteins of 160 and 164 amino acids, respectively, and both genes are required for functional vacuolar H^+-ATPase assembly (Nelson and Nelson, 1990;Noumi *et al.*, 1991; Ohya *et al.*, 1991; Umemoto *et al.*, 1990, 1992). The predicted products of *VMA3* and *VMA11* share 57% identity, and loss of either gene resulted in the disappearance of all proteolipid species from isolated vacuolar membranes (Umemoto *et al.*, 1992). The 100 kDa subunit also behaves as an integral membrane protein because it resists solubilization with sodium carbonate (Kane *et al.*, 1992). This is consistent with a recent report that the structural gene encoding the 116 kDa subunit from bovine clathrin-coated vesicle H^+-ATPase (isolated from a rat cDNA library) encodes a protein with at least six potential membrane spanning regions in its carboxy-terminal half (Perin *et al.*, 1991). Interestingly, the levels of 100 kDa subunit are drastically reduced in *vma3* mutants (Kane *et al.*, 1992). One hypothesis consistent with the data is that the 100 kDa protein and the proteolipid proteins form a stable complex; in the absence of one of the components of the complex, the remaining components are destabilized and degraded.

The *S. cerevisiae* 70 kDa subunit is produced through a novel biosynthetic pathway. In *N. crassa,* the 70 kDa subunit is encoded by the *vma1* gene, and the predicted molecular weight of the 608 amino acid translation product agrees cosely with the observed apparent molecular weight of the subunit protein (Bowman *et al.*, 1988). In *S. cerevisiae* the 70 kDa subunit is encoded by the *TFP1* gene (also called the *VMA1* and *CLS1* genes; Table I) (Anraku *et al.*, 1989; Hirata *et al.*, 1989; Kane *et al.*, 1990; Ohya *et al.*, 1991; Shih *et al.*, 1988). The *TFP1* open reading frame is 1071 amino acids and has a predicted molecular mass of 119 kDa (Hirata *et al.*, 1989; Kane *et al.*, 1990; Shih *et al.*, 1988). Alignment of the *TFP1* predicted translation product with the *Neurospora vma1* sequence shows that the amino-terminal and carboxy-terminal regions of the *TFP1* product share about 75% identity with the *Neurospora* protein, but there appears to be a 454 amino acid insertion in the central region of the *TFP1* protein (Anraku *et al.*, 1989; Hirata *et al.*, 1990; Kane *et al.*, 1990). All available evidence suggests that the *TFP1* primary translation product undergoes an unusual protein splicing event that simultaneously joins the amino- and carboxy-terminal fragments to form the functional 70 kDa vacuolar H^+-ATPase subunit and releases a 50 kDa spacer protein of unknown function (Kane *et al.*, 1990). Protein splicing of the *TFP1* product is observed when the gene is expressed in *Escherichia coli* and when *TFP1* messenger RNA is translated in an *in vitro* translation mixture. This suggests that splicing is an intrinsic property of the *TFP1* primary protein product and that if exogenous factors are required for splicing, they are ubiquitous. The biological relevance of this novel protein splicing event is difficult to ascertain, but it is not required for functional assembly of the vacuolar H^+-ATPase. A genetically engineered construct of the *TFP1* gene that lacks the coding region of the spacer protein and therefore encodes a colinear 70 kDa subunit can fully substitute for the full length *TFP1* gene in supplying wild-type levels of vacuolar H^+-ATPase activity (Kane *et al.*, 1990).

The mechanism of assembly of the vacuolar H^+-ATPase remains obscure. In higher eukaryotic cells the vacuolar network is acidified while the rest of the secretory pathway remains at neutral pH (for reviews see Kornfeld and Mellman, 1989; Mellman *et al.*, 1986; Orci *et al.*, 1987). Similarly, in *S. cerevisiae* the vacuole was labeled with probes for acidic compartments while the remaining secretory pathway organelles appeared to be neutral (Rothman *et al.*, 1989b; Weisman *et al.*, 1987). Furthermore, peripheral membrane components of the vacuolar H^+-ATPase are found predominantly on the vacuolar membrane in wild-type cells (Anraku *et al.*, 1989; Yamashiro et al., 1990; Kane *et al.*, 1992). These data suggest that the assembly and distribution of the vacuolar H^+-ATPase in yeast are constrained to the vacuolar network. Given the probable structure and topology of the complex, it would seem most likely that membrane components traverse the secretory pathway, whereas peripheral cytoplasmic proteins are synthesized in the cytoplasm and joined to the membrane portion on the surface of the appropriate organelle. The 116 kDa integral membrane subunit of the clathrin-coated vesicle H^+-ATPase is apparently glycosylated, consistent with the notion that membrane components traverse the early stages of the secretory pathway (Adachi *et al.*, 1990). The factors and/or conditions that promote assembly are unknown, but it is not a consequence of a *PEP4*-dependent processing event; *pep4* mutants possess completely normal levels of vacuolar H^+-ATPase activity, and the lumen of the vacuole in *pep4* strains is acidified normally (Rothman *et al.*, 1989b; Yamashiro *et al.*, 1990).

In *S. cerevisiae* mutants lacking structural genes encoding vacuolar H^+-ATPase subunits, other ATPase components are expressed, but stable association of the peripheral membrane subunits with the vacuolar membrane fails to occur. In mutants lacking the 60 kDa protein, the 70 kDa subunit was found at wild-type levels in total cell extracts, but the protein no longer fractionated with vacuolar membranes (Yamashiro *et al.*, 1990). Similarly, the 60 kDa subunit failed to associate with vacuoles in mutants lacking the 70 kDa protein, and the 60 kDa protein appeared evenly distributed throughout the cytoplasm by indirect immunofluorescence under these conditions (Kane *et al.*, 1992). Both the 60 and 70 kDa subunits failed to associate with vacuoles in mutants lacking the 17 kDa proteolipid (Kane *et al.*, 1992; Umemoto *et al.*, 1992). A monoclonal antibody that recognized the 100 kDa subunit was used to examine the distribution of the protein by indirect immunofluorescence (Kane *et al.*, 1992). In wild-type cells little signal was observed. By contrast, intense labeling of the vacuolar membrane was observed in mutants that lack either the 70 or the 60 kDa subunits (Kane *et al.*, 1992). Furthermore, the 100 kDa subunit was shown to be stable in mutants lacking the 70 or 60 kDa subunits. One interpretation of these experiments is that the epitope on the 100 kDa subunit that is recognized by the antibody is shielded by peripheral subunits in the wild-type complex; failure of these components to associate with the vacuolar membrane allows antibodies to gain access to this region of the 100 kDa protein. One long-range goal of future experiments is to explore the possibility that partially assembled ATPase complexes accumulate in mutants lacking certain subunits. For example, it will be of interest to determine whether the 70 and 60 kDa subunits form a stable, soluble complex in mutants lacking the proteolipid and, if so, to determine whether other vacuolar H^+-ATPase subunits are associated with this complex.

It seems reasonable to suspect that there may be specific factors, not themselves subunits of the functional vacuolar H^+-ATPase, that promote assembly of the complex *in vivo*. The *vma12* and *vma13* mutants are ATPase-deficient; yet the *VMA12* and *VMA13* genes did not appear to encode a subset of previously identified vacuolar H^+-ATPase subunits (Ohya *et al.*, 1991). Likewise, a screen for vacuolar acidification mutants yielded the *vph1* mutant in which the vacuolar pH that was neutral, yet protein sorting and zymogen activation proceeded normally (Preston *et al.*, 1989; Yamashiro *et al.*, 1990). The wild-type *VMA12, VMA13,* or *VPH1* genes may encode additional subunits of the vacuolar H^+-ATPase, but they may also regulate some critical feature vacuolar H^+-ATPase assembly. In addition, several of the vacuolar protein sorting mutants considered in Section IV fail to assemble normal levels of the vacuolar H^+-ATPase. Vacuolar membranes isolated from *vps3* and *vps6* mutants are grossly deficient in ATPase activity, and the 70 and 60 kDa subunits do not associate with the vacuolar membrane despite being present at wild-type levels (Rothman *et al.*, 1989b; C. K. Raymond *et al.*, unpublished observations). Mutations in as many as eight additional *VPS* genes render yeast strains with a similar phenotype. Unlike mutant strains specifically defective for vacuolar H^+-ATPase function, these *vps* mutants display additional defects in protein sorting, zymogen activation, and vacuolar morphology. Hence, the products of these genes are unlikely to encode subunits of the vacuolar H^+-ATPase complex, yet they exert a critical influence over vacuolar H^+-ATPase activity. Clearly, there is much that remains to be learned about the assembly and proper localization of the vacuolar H^+-ATPase complex.

IV. Isolation and Characterization of Mutants Defective for Vacuolar Biogenesis

A. Protein Sorting in the Eukaryotic Secretory Pathway

Most newly synthesized secretory pathway proteins enter the lumen of the ER. Localization to the appropriate final compartment then requires both bulk flow interorganellar transport reactions and selective segregation of specific proteins from traffic bound for different destinations. The molecular details of transport reactions have enjoyed increasing definition and clarity with the marriage of classical genetic approaches and *in vitro* transport systems that faithfully mimic the *in vivo* processes. In *S. cerevisiae* temperature-conditional *sec* mutants were identified that were incapable of protein secretion under restrictive conditions (Novick *et al.*, 1980). Phenotypic characterization of these mutants revealed groups of genes that encode products required for transport of proteins between various secretory pathway organelles (Novick *et al.*, 1981). Further analysis of mutants blocked in ER to Golgi transport demonstrated that there were a subset of genes required for the formation of transport vesicles and a second set that accumulated transport vesicles, suggesting the mutations blocked vesicle fusion with an acceptor compartment (Kaiser and Schekman, 1990). Meanwhile, biochemical analysis had shown that transport between organelles generally involves the formation of a coated vesicle from a donor compartment, transport to and recognition by an acceptor compartment, and uncoating followed by fusion of the vesicle with the acceptor membrane (Balch *et al.*, 1984a,b; Malhotra *et al.*, 1988; Orci *et al.*, 1989; Pfanner *et al.*, 1990; Wattenberg *et al.*, 1986, 1990). The two lines of investigation converged when it was found that two of the yeast *SEC* genes required for fusion of ER-derived transport vesicles with the Golgi compartment encode the yeast homologs of *N*-ethylmaleimide–sensitive fusion protein (NSF) and soluble NSF attachment protein (SNAP), proteins identified by biochemical analysis as required for the vesicle fusion reaction (Clary *et al.*, 1990; Wilson *et al.*, 1989). Similarly, the *SEC23* product is required for ER to Golgi transport *in vivo*, and a permeabilized cell system reconstituted from temperature-conditional *sec23* mutant cells displayed temperature-conditional ER to Golgi transport *in vitro* (Baker *et al.*, 1988). Finally, the observation that the *SEC4* gene encodes a GTP-binding protein required for fusion of secretory vesicles with the plasma membrane (Goud *et al.*, 1988; Salminen and Novick, 1987; Walworth *et al.*, 1989) has led to the discovery that GTP-binding proteins mediate several unique stages in protein transport as defined by mutants *in vivo* or by various reconstituted transport reactions *in vitro* (Baker *et al.*, 1990; Bourne, 1988; Chavrier *et al.*, 1990a,b; Gorvel *et al.*, 1991; Melancon *et al.*, 1987; Nakano and Muramatsu, 1989; Plutner *et al.*, 1990; Segev *et al.*, 1988, 1991). These studies underscore the important fact that genes identified by genetic means have proven to encode proteins that play pivotal roles in reconstituted transport reactions, and the strength of the combined analysis supports the validity of both genetic and biochemical approaches to protein transport.

Unfortunately, the molecular mechanisms that underlie protein sorting in the secretory pathway remain fairly obscure. There appear to be several different types of sorting reactions. For example, evidence from *S. cerevisiae* and other organisms has shown that membrane proteins are segregated away from one another in a manner independent of the sorting of soluble proteins, and the default destination for unsorted membrane proteins (vacuole/lysosome) may be entirely different from the default compartment for soluble proteins (cell surface; considered in detail in Section II). For either class of protein there are different kinds of sorting reactions. Proteins moving through the secretory pathway may be segregated away from one another and funneled to specific destinations; the sorting of soluble vacuolar glycoproteins away from secreted proteins is an obvious instance of this. Alternatively, there are specific classes of proteins that are retained in organelles that other proteins pass through. For example, there are several soluble resident ER proteins that are retained in the lumen of the organelle by virtue of a carboxy-terminal tetrapeptide sequence (normally KDEL in animal cells, HDEL in *S. cerevisiae;* for review see Pelham, 1989). Soluble proteins retained in the ER nonetheless acquire carbohydrate modifications characteristic of early-Golgi compartments (Dean and Pelham, 1990; Pelham, 1988). Current evidence favors the hypothesis that soluble ER proteins that exit the ER are retrieved from an early-Golgi compartment and targetted back to the ER in specialized vesicles or tubules (Lippincott-Schwartz *et al.,* 1990; Pelham, 1989). Still other proteins negotiate complex traffic patterns among several distinct compartments. The mammalian cation-independent mannose-6-phosphate receptor is known to cycle among the *trans*-Golgi network, the endosome, and the cell surface, but it is rarely if ever found in lysosomes (for review see Kornfeld and Mellman, 1989). The latter example illustrates a fundamental barrier in our understanding of protein sorting reactions. Numerous examples of protein sorting or selective retention have been documented, and in many cases, the *cis*-acting signals that direct these proteins to various locations have been deciphered. In far fewer cases, the receptors that recognize sorting and/or retention signals have been identified. However, the complicated cycling of the mannose-6-phosphate receptor illustrates the point that receptors themselves are targetted to various destinations. Here again, *cis*-acting signals in the cytoplasmic tail of the receptor are crucial for its correct localization (Kornfeld and Mellman, 1989), but the molecular components that interpret this information have yet to be identified. There have been suggestions that proteins found within clathrin-coated pits, the so-called adaptins, recognize and package specific proteins into clathrin-coated vesicles (Glickman *et al.,* 1989, Hurtley, 1991). Alternatively, the recent revelation that the vesicle coats on intra-Golgi transport vesicles are comprised in part of proteins with a high degree of similarity to the adaptins may argue that adaptins play a more generic role in vesicle formation (Donaldson *et al.,* 1990; Duden *et al.,* 1991; Serafini *et al.,* 1991; Waters *et al.,* 1991). The important point is that there are no clear molecular descriptions of any protein sorting reaction.

Several laboratories have focused on the biosynthesis and sorting of CPY to the vacuole in *S. cerevisiae* as a model system for protein sorting in the eukaryotic secretory pathway (Bankaitis *et al.*, 1986; Rothman and Stevens, 1986; Stevens *et al.*, 1982). An outline of the biosynthesis of this soluble vacuolar glycoprotein and the *cis*-acting signals necessary for its efficient delivery to the vacuole were considered in Section II,B. The system has the distinct advantage that mutants that lack most aspects of vacuolar function are viable, and this has allowed for isolation of a large collection of mutants defective for the efficient sorting of CPY, many of which are simultaneously defective in other aspects of vacuolar biogenesis. To understand how these mutants influence protein sorting, it is first necessary to obtain a better understanding of the compartments involved in the sorting process. A lucid description of CPY sorting to the yeast vacuole was recently provided by Graham and Emr (1991). These investigators examined the requirement for *SEC18* function in movement of proteins through various secretory pathway compartments, and they attempted to define the Golgi compartment in which the sorting of CPY occurs. *SEC18* encodes the yeast equivalent of NSF (Wilson *et al.*, 1989). NSF was identified as a protein required for fusion of transport vesicles derived from the *cis*-Golgi with medial Golgi compartments in a cell-free system (Malhotra *et al.*, 1988), and it has since been shown to promote vesicle fusion in *in vitro* ER to Golgi transport reactions (Beckers *et al.*, 1989), in medial to *trans*-Golgi movement (Rothman, 1987), and in fusion of endocytic vesicles (Diaz *et al.*, 1989). A temperature-conditional yeast *sec18* mutant was radiolabeled under permissive conditions such that all forms of the precursors to CPY and the secreted mating pheromone α-factor were present (Graham and Emr, 1991). The α-factor precursor undergoes several discrete and sequential modifications during transport through the secretory pathway, many of which are readily discernible as mobility shifts on SDS-PAGE gels (Fig.3) (Fuller *et al.*, 1988; Julius *et al.*, 1984b). Transfer of the labeled *sec18* strains to restrictive conditions froze all forms of α-factor precursors (Graham and Emr, 1991). The interpretation of these data is that the yeast Golgi can be differentiated into at least three functional compartments. Furthermore, the fact that all stages of α-factor maturation were blocked, including secretion of the mature pheromone from the cell, argues that the *SEC18* protein promotes membrane fusion at multiple points in the secretory pathway. Surprisingly, a substantial fraction of the Golgi-localized proCPY (Section II,B) was matured in the absence of *SEC18* function. These data are consistent with a model (Fig. 3) in which the final step in transport of proCPY to the vacuole does not require the *SEC18* product. Importantly, Graham and Emr (1991) used a chimeric CPY–α-factor–invertase fusion protein to show that sorting of CPY away from secreted proteins is likely to occur in or after passage

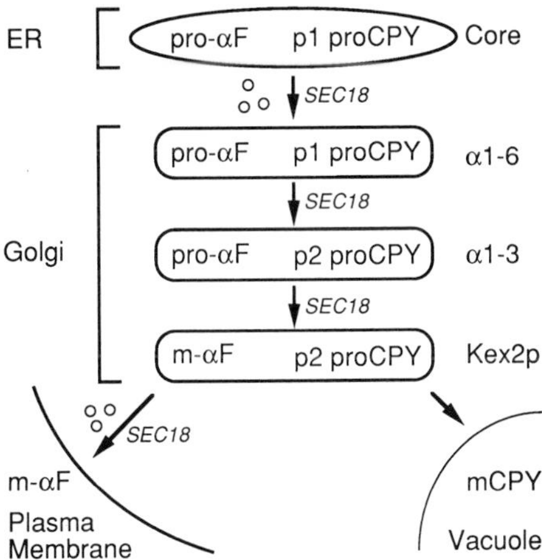

FIG. 3 Maturation of α-factor and CPY within the yeast secretory pathway (after Graham and Emr, 1991). The yeast mating pheromone α-factor undergoes several discrete modifications during its biosynthesis, all of which can be distinguished electrophoretically. The pro–α-factor precursor protein (pro-αF) receives core glycosylation in the ER. Movement from the ER to the *cis*-Golgi can be detected by the addition of α1-6-mannose linkages to the core oligosaccharide. In a subsequent compartment, mannose residues are added with α1-3 linkages. In the most distal Golgi compartment, pro-αF is cleaved by Kex2p protease, and additional processing yields mature α-factor (mαF; Section III,B) (Julius *et al.*, 1984b). Mature pheromone is secreted from the cell via the late secretory pathway. The *SEC18* product is required for progression through all stages of maturation of α-factor, as indicated (Graham and Emr, 1991). The biosynthesis of CPY is described in Section II,B. Core glycosylated CPY and α1-6-mannose–modified CPY co-migrate on SDS–PAGE gels (Franzusoff and Schekman, 1989), and both forms are referred to as p1 pro CPY. The experiments of Graham and Emr (1991) indicate that movement of α1-3-modified p2 proCPY from the Kex2p compartment (or a later compartment) to the vacuole (as defined by processing of proCPY to the mature species) is independent of the *SEC18* product.

through the Kex1p/Kex2p/DPAP A-containing compartment, the most distal Golgi compartment yet identified in yeast (Fig. 3; Section III,B). However, it was not possible to establish whether *SEC18*-independent transfer to the vacuole occurred from this late-Golgi compartment or from a more distal prevacuolar/endosome-like organelle.

B. Genes Required for Vacuolar Protein Sorting and Vacuolar Assembly

The observation that unsorted CPY was secreted from the cell via the later stages of the secretory pathway (Stevens *et al.*, 1986; Valls *et al.*, 1987)

TABLE II

Genetic Overlaps between Vacuolar Biogenesis Mutants[a]

vps mutant	*vpl* mutant	*vpt* mutant	*pep* mutant[b]	Other[c]
vps1[d]	*vpl1*	*vpt26*	—	*spo15*
vps2	*vpl2*	—	—	—
vps3	*vpl3*	*vpt17*	*pep6*	—
vps4	*vpl4*	*vpt10*	—	—
vps5	*vpl5*	*vpt5*	*pep10*	—
vps6	*vpl6*	*vpt13*	*pep12*	—
vps7[e]	*vpl7*	*vpt29*	*pep15*	—
vps8	*vpl8*	*vpt8*	—	—
vps9	*vpl31*	*vpt9*	—	—
vsp10	—	*vpt1*	—	—
vps11	*vpl9*	*vpt11*	*pep5*	*end1, vam1, spo-T7, cls13*
vps12	—	*vpt12*	—	—
vps13	—	*vpt2*	—	—
vps14	—	*vpt14*	—	—
vps15	*vpl19*	*vpt15*	—	—
vps16	—	*vpt16*	—	*vam9, cls17*
vps17	—	*vpt3*	*pep21*	—
vps18	—	*vpt18*	*pep3*	*vam8, cls18*
vps19	*vpl21*	*vpt19*	*pep7*	—
vps20	*vpl10*	*vpt20*	—	—
vps21	—	*vpt21*	—	—
vps22	*vpl14*	*vpt22*	—	—
vps23	*vpl15*	*vpt23*	—	—
vps24	*vpl26*	*vpt24*	—	—
vps25	*vpl12*	*vpt25*	—	—
vps26	—	*vpt4*	*pep8*	—
vps27	*vpl23*	*vpt27*	—	—
vps28	*vpl13*	*vpt28*	—	—
vps29	—	*vpt6*	—	—
vps30	—	*vpt30*	—	—
vps31	—	*vpt31*	—	—
vps32	—	*vpt32*	—	—
vps33	*vpl25*	*vpt33*	*pep14*	*cls14, slp1, vam5*
vps34	*vpl7*	*vpt29*	—	—
vps35	—	*vpt7*	—	—
vps36	*vpl11*	—	—	—
vps37	*vpl16*	—	—	—
vps38	*vpl17*	—	—	—
vps39	*vpl18*	—	—	—
vps40[e]	*vpl19*	*vpt15*	—	—
vps41	*vpl20*	—	—	*vam2*
vps42	*vpl22*	—	—	—
vps43	*vpl24*	—	—	—
vps44	*vpl27*	—	—	—
vps45	*vpl28*	—	—	—
vps46	*vpl30*	—	—	—

(*continued*)

TABLE II (*continued*)

[a] In most cases genetic overlap was determined by complementation testing (Robinson *et al.*, 1988; Rothman and Stevens, 1986). Genetic overlap in some complementation groups was verified by using standard genetic techniques to demonstrate that certain mutations are allelic; these were *vpl3* and *vpt17*, *vpl5* and *pep10*, *vpl9* and *end1*, *vpl9* and *vpt11*, *vpt3* and *pep21*, *vpl14* and *vpt22*, *and vpt4* and *pep8*. In addition, the following pairwise combinations of mutants failed to sporulate, indicating they are also probably allelic mutations: *vpl6* and *vpt13*, *vpl6* and *pep12*, *vpt13* and *pep12*, *vpl7* and *pep5*, *end1* and *pep5*, *end1* and *vpt11* and *pep5* and vpt11. The isolation of the *vpl20* through *vpl31* complementation groups and their genetic overlap with other vacuolar biogenesis mutants was only recently established (I. Howald-Stevenson and T. H. Stevens, unpublished observations).

[b] Six other *pep* mutants (Jones, 1977) display vacuolar biogenesis defects (such as secretion of CPY; Rothman *et al.*, 1989a), but genetic overlap with the *vps* mutants was not detected (Klionsky *et al.*, 1990); these were *pep1*, *pep2*, *pep9*, *pep11*, *pep13*, and *pep16*.

[c] Overlaps between the *vps* mutants and other *cls*, *vam*, *slp*, or *spo* mutants have not been tested. The identity between *VPS11* and *SPO-T7* was established on the basis that genes map to similar positions on chromosome XIII and the restriction maps of the cloned genes are basically identical (Tanaka and Tsuboi, 1985; Woolford *et al.*, 1990).

[d] The identify between *VPS1* and *SPO15* was established by linkage to CEN11, and the restriction maps and sequences of the two genes are one in the same (Raymond, 1990; Yeh *et al.*, 1986, 1991).

[e] In some cases intragenic complementation is observed and, thus, mutations in the same gene appear as different complementation groups. Such was the case for *vpl7* and *vpt15* or *vpl19* and *vpt29*, which have been reported previously as defining separate complementation groups (Klionsky *et al.*, 1990; Rothman *et al.*, 1989a); cloning of the *VPL7* and *VPL19* genes revealed they are identical to *VPT15*/*VPS15* and *VPT29*/*VPS34*, respectively (Herman and Emr, 1990; Herman *et al.*, 1991).

provided an avenue for isolation of vacuolar protein sorting mutants. In two cases selections for secretion of CPY were imposed (Bankaitis *et al.*, 1986; Rothman and Stevens, 1986; for review see Klionsky *et al.*, 1990). One selection relied on the fact that when CPY was secreted under aberrant conditions (always as the Golgi-modified, pro-CPY precursor form), a small amount of the enzyme was cleaved by extracellular proteases to an active species (Rothman and Stevens, 1986; Stevens *et al.*, 1986). This extracellular CPY activity, present in mutants that secreted CPY, liberated leucine from a dipeptide substrate present in the medium, permitting growth of leucine auxotrophs. Eight vacuolar protein localization (*vpl*) complementation groups were identified in the initial report (Rothman and Stevens, 1986). A second screen exploited the fact that extracellular invertase activity is required for utilization of sucrose as a carbon source under certain growth conditions. Yeast strains that lack the invertase structural gene yet express a CPY::invertase fusion protein are phenotypically Suc^- because invertase activity is sequestered in the vacuole. Selection of Suc^+ mutant colonies from such a strain allowed the isolation of eight vacuolar protein targetting *(vpt)* complementation groups (Bankaitis *et al.,* 1986). More exhaustive isolation and genetic analysis of these protein sorting mutants have shown that mutations in many of the same complementation groups were identified in both selection procedures and that at least 46 separate genes are required for the efficient delivery of CPY to the vacuole (Table II) (Robinson *et al.,* 1988; Rothman *et al.,* 1989a). The extensive genetic overlap of *vpl* and *vpt* mutants prompted a consolidation of nomenclature and renaming of the corresponding genes as vacuolar protein sorting *(VPS)* genes (Table II) (Robinson *et al.,* 1988; Rothman *et al.,* 1989a).

Several screens for mutants defective in other aspects of vacuolar function have also been reported, and many of the genes identified have proven to be genes found in the *vps* mutant selections. Peptidase, or *pep* mutants, were identified on the basis of deficient CPY activity (Jones, 1977). Mutations in the *PEP4* gene identified a key factor required for the proteolytic maturation of several vacuolar zymogen precursors (Section II,C) (Hemmings *et al.,* 1981; Jones *et al.,* 1982; Jones, 1984; Zubenko *et al.,* 1983). The remaining *PEP* genes were necessary for proper CPY biosynthesis, and they were later shown to secrete substantial amounts of newly synthesized CPY; many of the *pep* mutants overlap genetically with the *vps* mutants (TableII) (Rothman *et al.,* 1989a). Additional screens for mutations in genes required for endocytosis (*END*; Chvatchko *et al.,* 1986; Dulic and Riezman, 1989), wild-type vacuolar morphology (*VAM*; Wada *et al.,* 1988), resistance to high levels of calcium (*CLS*; Ohya *et al.,* 1986), or lysine (*SLP;* Kitamoto *et al.,* 1988) have also identified genes present in the *VPS* collection (Table II). The fact that diverse screens have identified common genes underscores the pleiotropic nature of many of the *vps* mutations.

All of the mutants shown in Table II share certain similar characteristics. They secrete between 20% and 95% of newly synthesized, Golgi-modified, pro-CPY. In several mutants the secreted pro-CPY was shown to traverse the later stages of the secretory pathway with the kinetics of secretion being similar to the normally secreted protein invertase (Rothman and Stevens, 1986; Robinson *et al.,* 1988). None of the mutants release substantial amounts of cytoplasmic proteins. These facts imply that the *vps* mutants externalize CPY because of defective sorting within the secretory pathway and not because of cell lysis or nonspecific release of vacular contents into the media. All of the mutants tested also secrete precursor forms of the soluble vacuolar glycoproteins PrA and PrB (Section II,C), suggesting that the *vps* mutations influence a common sorting apparatus (Robinson *et al.,* 1988; Rothman and Stevens, 1986). The percentage of newly synthesized PrA or PrB that is secreted is generally less than the amount of mislocalized CPY (Robinson *et al.,* 1988; Rothman and Stevens, 1986). The reason for this is unclear. Perhaps the sorting of CPY is more sensitive to perturbations in the sorting apparatus than is the sorting of PrA or PrB. Alternatively, vacuolar precursors may exhibit substantial variations in stability in the extracellular milieu. Extracellular pro-PrA has recently been shown to be unstable (Winther, 1989). Rapid degradation of external protein would have the trivial effect of diminishing apparent mislocalization phenotypes.

None of the *vps* mutants display detectable levels of vacuolar membrane proteins on the cell surface. External DPAP B activity (Section II,E) was assayed in wild-type and *vps* mutant cells, and the mutants displayed little if any external activity in excess of background levels (Rothman and Stevens, 1988; Rothman, 1988). The cellular distribution of ALP (Section II,E) was examined in all of the *vps* mutants using indirect immunofluorescence (Section IV,C Fig. 4). None of the mutants exhibited detectable cell surface labeling. These observations are consistent with observations that the sorting of soluble proteins is fundamentally distinct from the sorting of membrane proteins.

C. Vacuolar Morphology in Wild-Type Cells and Vacuolar Biogenesis Mutants

Several procedures have been developed for visualization of the vacuole using fluorescence microscopy (Pringle *et al.*, 1989; Raymond *et al.*, 1990; Roberts *et al.*, 1990; Weisman *et al.*, 1987). The morphology of the organelle is influenced by the genetic background of the strain, the growth phase of the cells, and perturbations in the growth media (Gomes de Mesquita *et al.*, 1991; Pringle *et al.*, 1989; Raymond *et al.*, 1990; Roberts *et al.*, 1990). Furthermore, the vacuole undergoes dynamic changes as individual cells bud and divide (Raymond *et al.*, 1990; Weisman and Wickner, 1988; Weisman *et al.*, 1987). With regard to this important aspect of vacuolar biogenesis, there are specific mechanisms that ensure that portions of the mother cell vacuole are inherited by newly formed daughter cells (Gomes de Mesquita *et al.*, 1991; Raymond *et al.*, 1990; Shaw and Wickner, 1991; Weisman and Wickner, 1988; Weisman *et al.*, 1987, 1990). Finally, many mutations that disrupt vacuolar function introduce pronounced perturbations in the structure of the vacuole (Banta *et al.*, 1988; Herman and Emr, 1990; Klionsky *et al.*, 1990; Raymond *et al.*, 1990; Raymond, 1990; Rothman and Stevens, 1986; Rothman *et al.*, 1989a; Weisman *et al.*,1990).

Vacuoles in actively proliferating wild-type cells have certain distinctive features. They often appear as several membrane-bound structures clustered in one region of the cytoplasm (Gomes de Mesquita *et al.*, 1991; Pringle *et al.*, 1989; Raymond *et al.*, 1990; Weimken *et al.*, 1970). As cells bud, portions of the maternal vacuole appear to extend into the developing

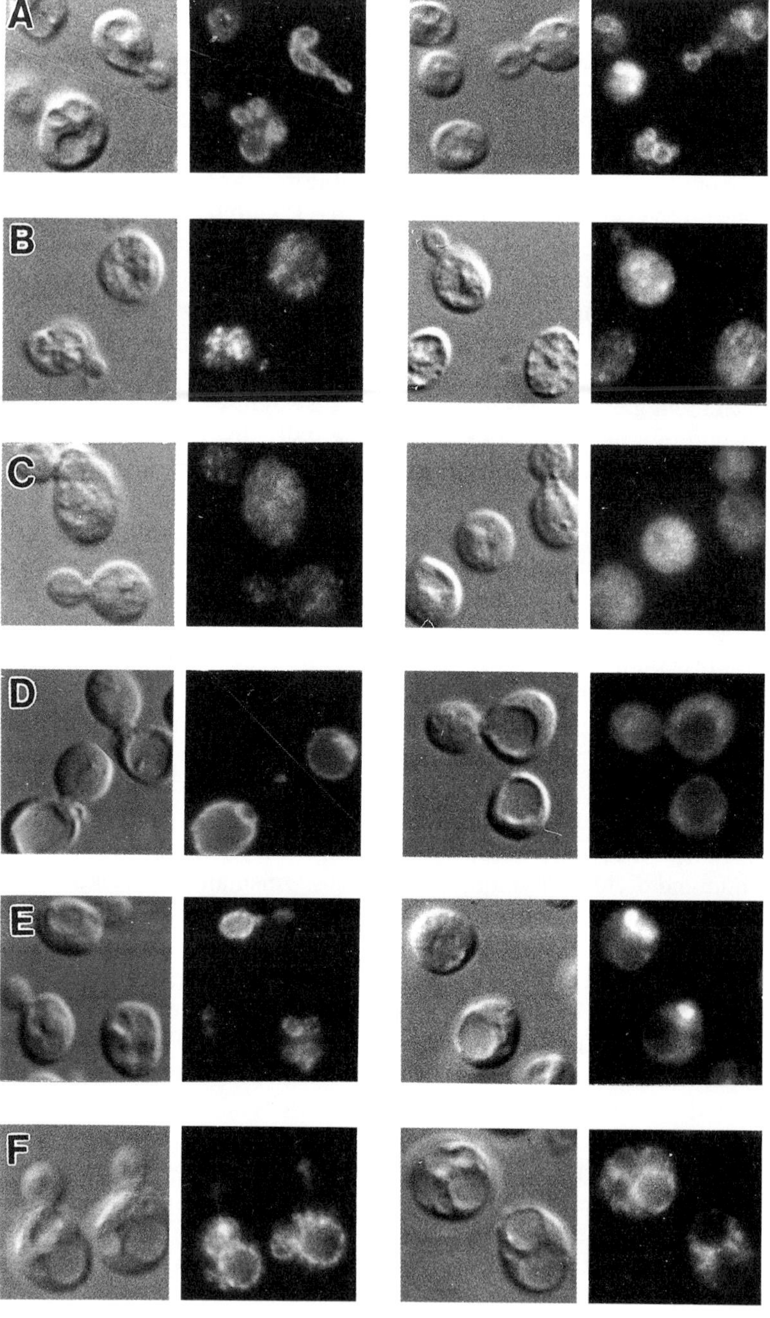
A
B
C
D
E
F

daughter cell. These distinctive vacuolar extensions, which appear as tubules or strings of large, vesicular bodies, have been termed "segregation structures" (Fig. 4,A) (Gomes de Mesquita *et al.*, 1991; Raymond *et al.*, 1990). The segregation of portions of maternal vacuolar material into buds is consistent with previous observations that vacuolar protease activity and an endogenous vacuolar fluorophore (accumulated in *ade2* mutant strains) were passed from one generation of cells to the next (Weisman *et al.*, 1987; Zubenko *et al.*, 1982). Segregation structures are most prevalent when buds have achieved a diameter of at least one fourth that of the mother cell, but the process is generally complete before the onset of nuclear migration to the bud neck that precedes nuclear division (Raymond *et al.*, 1990). Vacuoles segregate into buds at about the same time in bud development as do mitochondria, suggesting a generalized movement of maternal components into the newly forming cell at a specific stage in bud development. Moreover, it is clear from studies of wild-type cells and mutants that specific cellular functions ensure the segregation of vacuoles, demonstrating that vacuolar inheritance is an important aspect of cell growth and vacuolar biogenesis.

The maintenance of complex vacuolar architecture and the establishment of segregation structures seem likely to involve the cytoskeleton. Microtubules probably do not participate in vacuolar inheritance because treatment of yeast cells with the microtubule-disrupting drug nocodazole had no impact on vacuolar segregation structures (Gomes de Mesquita *et Al.*, 1991; Raymond, 1990; C. K. Raymond and T. H. Stevens, unpublished observations). Treatment of cells with nocodazole for several hours was reported to cause vacuole fragmentation in some yeast strains, but the implications of these observations are unclear (Guthrie and Wickner, 1988). Actin filaments may participate in the process of vacuolar inheritance because they have been shown to play a pivotal role in the establishment of cell polarity in *S. cerevisiae* (Novick and Botstein, 1985). Furthermore, actin cables are often spatially coincident with vacuolar segregation structures (C. K. Raymond and T. H. Stevens, unpublished observations), and conditions that perturb actin distribution also affect vacuolar morphol-

---

FIG. 4 Vacuolar morphologies among the *vps* mutants. Nomarski optical images and the corresponding ALP fluorescence signal are shown in the left column, and a similar display of cells stained with antibodies against the vacuolar H^+-ATPase 60 kDa subunit is shown in the right column. (A) Wild-type strain SF838-9D. (B) Class B mutant strain *vps41*. (C) Class C mutant strain *vps11*. (D) Class D mutant strain *vps19*. (E) Class E mutant strain *vps27*. (F) The *vps1* mutant. All mutants are spontaneous Vps^- isolates of SF838-9D except for the *vps1* strain, which is a null mutant constructed by gene substitution.

ogy (Pringle *et al.*, 1989; Raymond *et al.*, 1990; C. K. Raymond and T. H. Stevens, unpublished observations).

There must also be vacuole-specific factors that participate in vacuolar segregation, and there have been several reports of mutations that interfere with vacuolar inheritance (Raymond, 1990, Raymond *et al.*, 1990; Shaw and Wickner, 1991; Weisman *et al.*, 1990). All of the mutants with defects in vacuolar segregation exhibit common phenotypes. The vacuoles are coalesced into a single vacuole that has a simple, spherical shape, segregation structures are rarely observed, and newly formed buds and/or daughter cells possess little if any discernible vacuoles. Interestingly, these mutations have little impact on cell viability, and almost all mother cells that bud contain a vacuole. Taken together, these observations raise the possibility that vacuoles can arise *de novo* from Golgi-derived transport vesicles that are presumably generated in a constitutive manner. Most of the vacuolar segregation mutants have broadly pleiotropic phenotypes that include defects in vacuolar protein sorting, vacuolar acidification, and PrA maturation (see below; see also Raymond, 1990; Raymond *et al.*, 1990; Rothman *et al.*, 1989b; Weisman *et al.*, 1990). The *vac2* mutant isolated by Shaw and Wickner (1991) appears to be an exception. The single allele of *vac2* isolated and characterized has a temperature-conditional defect in vacuolar segregation. At 37° vacuoles fail to segregate; yet the kinetics and efficiency of CPY sorting to the vacuole appear unaffected. These data have two interesting implications. First, they demonstrate that the processes of vacuolar segregation and the sorting of soluble vacuolar proteins can be uncoupled. In further support of this, many of the *vps* mutants show normal vacuolar segregation. Second, the experiments with the *vac2* mutant suggest that the wild-type *VAC2* product may be a specific component of the segregation apparatus. By contrast, the other genes required for segregation may encode products with a less direct influence on inheritance.

Most of the *vps* mutations introduce noticeable changes in vacuolar morphology (Banta *et al.*, 1988; Klionsky *et al.*, 1990; Raymond, 1990; Raymond *et al.*, 1992. Initial characterization of the *vpt* mutant collection (Table II) (Robinson *et al.*, 1988) using vacuolar vital stains showed that there were at least three classes of vacuolar morphology among these strains (Banta *et al.*, 1988; Klionsky *et al.*, 1990). Class A mutants had large vacuoles, class B mutants had multiple fragmented vacuoles, and class C strains had no apparent vacuole (Banta *et al.*, 1988). More recently, mutants with lesions representative of every *VPS* gene (Table II) were subjected to thorough morphological examination using indirect immunofluorescence (Raymond, 1990; Raymond *et al.*, 1992). The antibodies used were specific for the vacuolar integral membrane protein ALP (Section II,E) or for the 60 kDa peripheral membrane subunit of the vacuolar H^+-

ATPase (Section III,E). The class A, B, C nomenclature was extended to include three new groups, which are described next and presented in Fig. 4. Many of the mutants originally designated as having class A vacuoles were found to fall into these new categories (Table III) (Raymond, 1990; Raymond *et al.*, 1992).

The class A *vps* mutants possess vacuoles that are indistinguishable from wild-type cells. Class B cells have multiple fragmented vacuoles that were clearly visible with Nomarski optics and less distinct by antibody staining. Class C null mutants have no apparent vacuoles; yet they accumulated numerous unusual membrane structures when viewed using thin section electron-microscopy techniques (Banta *et al.*, 1988, 1990; Wada *et al.*, 1990; Woolford *et al.*, 1990). The ALP-specific staining pattern is diffuse and punctate, suggesting that the protein is widely distributed throughout the interior of the cell; no cell surface labeling was observed. The class C mutants are described further in Section V,D. Class D *vps* mutants were described in conjunction with mutants defective in vacuolar segregation. They display simultaneous defects in vacuolar segregation and assembly of the 70 and 60 kDa subunits of the vacuolar H^+-ATPase onto the vacuolar membrane (Raymond, 1990; Raymond *et al.*, 1990;

TABLE III

Vacuolar Morphologies of the *vps* Mutants[a]

Class A	Class B	Class C	Class D	Class E	Othe r
vps2	*vps5*	*vps11*	*vps3*	*vps4*	*vps1*
vps8	*vps10*	*vps16*	*vps6*	*vps14*	
vps12	*vps17*	*vps18*	*vps9*	*vps20*	
vps13	*vps41*	*vps33*	*vps15*	*vps22*	
vps30	*vps42*		*vps19*	*vps23*	
vps32	*vps43*		*vps21*	*vps24*	
vps44	*vps46*		*vps26*	*vps25*	
			vps29	*vps27*	
			vps34	*vps28*	
			vps35	*vps31*	
			vps38	*vps36*	
			vps45	*vps37*	
				vps39	

[a] Morphological assignments were based on examination of null mutants, where available (*vps1, vps3, vps8, vps11,* and *vps35*). In all other cases two mutant alleles representing each complementation group were examined, and identical morphological phenotypes were generally observed. Where ambiguities were encountered, several mutant strains of the same complementation group were examined. Exceptions to this are the *vps10, vps26,* and *vps29* mutants, in which only two mutant alleles were available. In these cases one mutant appeared as wild-type and the other appeared as class B (*vps10*) or class D (*vps26* and *vps29*); the more extreme phenotype was used for morphological designation.

Rothman *et al.*, 1989b; Weisman *et al.*, 1990; Raymond *et al.*, 1992). Class E *vps* mutants appear to have wild-type vacuoles as seen by Nomarski optics and by ALP staining (Raymond, 1990; Raymond *et al.*, 1992). However, antibodies directed against the 60 kDa vacuolar H^+-ATPase subunit yielded faint labeling of the vacuolar membrane and intense labeling of a nonvacuolar compartment. The vacuolar H^+-ATPase levels in class E mutants were found to be nearly normal (C. T. Yamashiro *et al.*, unpublished observations). These data suggest that functional vacuolar H^+-ATPase complexes are incorrectly localized in the class E mutants. The *vps1* mutant had a unique morphology in which a central vacuole was surrounded by smaller vacuolar fragments. This vacuolar morphology was similar to that seen in the class B group; thus, *vps1*mutants were originally placed in the class B category (Banta *et al.*, 1988).

Most of the various classes of *vps* mutants have several members because they share certain characteristic features (Table III, Fig. 4). There is variation between *vps* mutants within a group with regard to severity of certain phenotypes, and subdivisions and rearrangements of these groupings are sure to occur as additional information about individual *vps* mutants is acquired. Nonetheless, the morphological classifictions are potentially significant for several reasons. These studies reveal that almost all of the *vps* mutations influence vacuolar assembly. Furthermore, a large collection of sorting mutants can be subdivided into a number of subgroups. Finally, the fact that there are at least six classes of *vps* mutants indicates that there are several discrete molecular processes required for vacuolar biogenesis, and the phenotypic features of mutants within each group may suggest the nature of processes influenced by the wild-type *VPS* products.

V. Molecular Analysis of Genes Required for the Assembly of Vacuoles

A. Gene Products Required for Vacuolar Biogenesis

To understand vacuolar protein sorting and vacuolar biogenesis at the molecular level, several avenues of inquiry will need to converge. More precise characterization of the compartment(s) in which sorting occurs has provided sharper definition of the sorting problem (Section IV,A) (Graham and Emr, 1991). Likewise, cloning and sequencing of *VPS* genes accompanied with characterization of the corresponding gene products is beginning to shed light on the proteins that control vacuolar assembly (Section V,B,C,D). Biochemical reconstitution of vacuolar protein sorting *in vitro* (Section V,E) (Vida *et al.*, 1990) should allow further analysis of the process in a system amenable to external manipulation. Finally, so-called reverse genetic approaches to the protein sorting problem have provided, and should continue to provide, intriguing insights. Specifically, acidification of the vacuolar network in animal cell systems appears to be critical for uncoupling of lysosomal proteins from their cognate mannose-6-phosphate receptors (for reviews see Kornfeld and Mellman, 1989; Mellman *et al.*, 1986; von Figura and Hasilik, 1986). As described in Section III,E, it was therefore of some surprise to discover that acidification of the vacuole (and probably the yeast equivalent of the vacuolar network) by the vacuolar H^+-ATPase in *S. cerevisiae* is only marginally required for efficient CPY sorting (Yamashiro *et al.*, 1990). Similar evidence suggested that proper delivery of lysosomal enzymes to lysosomal compartments occurs via clathrin-coated vesicles (Geuze *et al.*, 1984, 1985; Schulze-Lohoff *et al.*, 1985). Although this may also be true in yeast, clathrin is not required for proper maturation and, presumably, vacuolar localization of CPY (Payne *et al.*, 1988). Hence, the results of reverse genetics underscore a conundrum; the cellular components predicted to participate in vacuolar protein sorting have proven to be dispensible. At the same time, there are many VPS products required for efficient sorting; yet none of their functions are well understood.

In the absence of specific and coherent molecular models for the sorting of soluble vacuolar glycoproteins, research efforts have focused on characterizing specific *VPS* genes and the proteins they encode. The aim is to obtain molecular details about individual components required for protein sorting and vacuolar biogenesis. Information about specific wild-type *VPS* genes can then be integrated with phenotypic analysis of the corresponding *vps* mutant. In addition, the influence of the other *vps* mutations on the expression, function, and localization of specific *VPS* products is being evaluated. Understandably, the *vps* mutants with the most extreme phenotypes and the corresponding wild-type *VPS* genes were chosen for initial study. The following paragraphs describe the fruits of these efforts.

B. *VPS15* Gene Encodes a Novel Protein Kinase

The *vps15* and *vps34* mutants, which possess virtually indistinguishable phenotypes, secrete an exceptionally high percentage of newly synthesized CPY, and they are defective for *PEP4*-dependent maturation of vacuolar proteases that remain within the cell. These mutants cannot grow at elevated temperatures, and they accumulate anomalous intracellular organelles (Banta *et al.*, 1988; Klionsky *et al.*, 1990; Robinson *et al.*, 1988). Both mutants have characteristics of class D *vps* mutants, with corresponding defects in vacuolar segregation and assembly of vacuolar H^+-ATPase subunits onto the vacuolar membrane (Section IV,C). Cloning and sequencing of the *VPS15* gene revealed that the *VPS15* product possesses two intriguing structural motifs (Fig.5) (Herman *et al.*, 1991). The amino-terminal 300 amino acids of this rather large, 1455 residue, hydrophilic protein share significant sequence similarity with the catalytic domains of the serine/threonine family of protein kinases. In addition, a

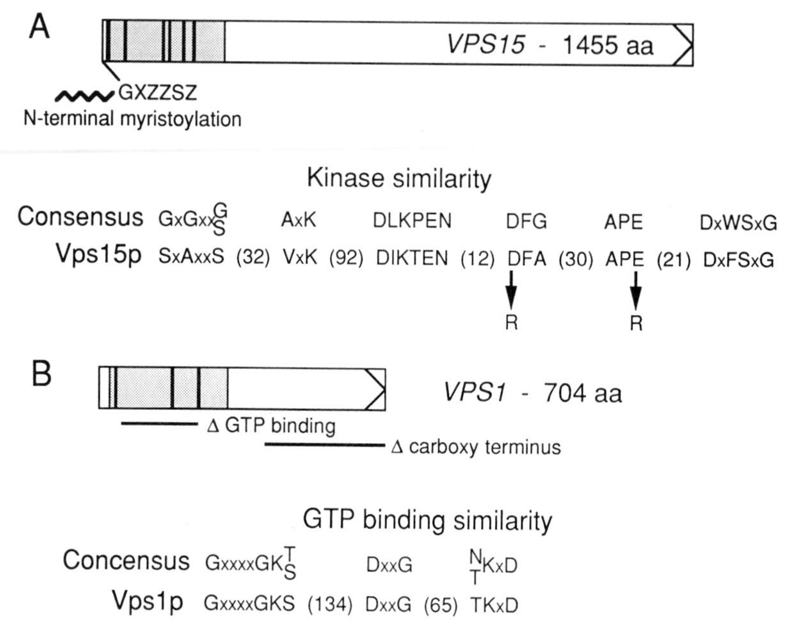

FIG. 5 Functional motifs in Vps15p and Vps1p. (A) The Vps15p kinase homolog (Herman *et al.*, 1991) carries an amino-terminal consensus myristoylation sequence; X, small residue; Z, neutral residues. The shaded region denotes sequences with significant similarity to phosphorylase b kinase from rabbit skeletal muscle (PhK-γ; Reimann *et al.*, 1984) and the product of the *S. pombe wee1*$^+$ gene (Wee1p; Russell and Nurse, 1987). Specific regions of sequence identity are shown as vertical bars in the schematic Vps15p representation, and the corresponding comparisons with regions I, II, VI, VII, VIII, and IX of the consensus kinase subdomains (Hanks *et al.*, 1988) are shown; the numbers in parentheses indicate the number of residues separating the subdomains. Amino acid changes within Vps15p that eliminate function are also displayed (vertical arrows). (B) The Vps1p GTP-binding protein has significant sequence identity (shaded region) with Mx proteins (Rothman *et al.*, 1990), rat dynamin (Obar *et al.*, 1990), and the *Shibire* product from *Drosophilia* (Chen *et al.*, 1991; van der Bliek and Meyerowitz, 1991). The positions of the consensus, tripartite GTP-binding elements are indicated by vertical bars in the Vps1p representation, and comparison with the consensus GTP-binding motifs (Dever *et al.*, 1987) is indicated. The extent of deletion mutations are displayed below the Vps1p schematic.

potential site for the attachment of myristic acid was detected within the amino-terminus of this protein. Subsequent analysis suggests that both regions contribute to the function and/or localization of the *VPS15* protein (Vps15p).

The significance of the myristylation site was not tested by a mutational approach. However, the wild-type Vps15p displays fractionation behavior consistent with a membrane-associated protein (Herman *et al.,* 1991). Furthermore, the protein co-fractionates with Kex2p in differential centrifugation experiments, raising the possibility that it may be localized to a late-Golgi compartment. The importance of the Vps15p kinase domain was evaluated by introducing single point mutations into highly conserved residues of the kinase functional motif (Fig. 5) (Herman *et al.,* 1991). The two specific changes that were made (as separate mutations) were selected because they destroyed the *in vitro* kinase activity and *in vivo* function of other serine/threonine protein kinases. Both mutations eliminated Vps15p function. Although it has not been shown that Vps15p has kinase activity *in vitro,* it was demonstrated that wild-type Vps15p is phosphorylated *in vivo* and that the kinase point mutations dramatically reduce the levels of phosphorylation on the otherwise stable mutant proteins. These data indicate that phosphorylation events mediate vacuolar biogenesis in some manner, and there are likely to be targets and/or regulators of Vps15p-mediated phosphorylation. The *VPS34* protein appears to be one of these (Herman *et al.,*1991). There is evidence that Vps15p and Vps34p associate physically *in vivo*. Furthermore, overproduction of Vps34p partially compensates for the sorting and growth defects of *vps15* kinase domain point mutations. The *VPS34* gene has been cloned and sequenced; its hydrophilic product bore no resemblance to other proteins in sequence data bases, and it had no functional motifs suggestive of function (Herman and Emr, 1990).

At least one other class D *VPS* gene has been analyzed in detail. The *VPS3* gene is predicted to encode a protein of 1011 residues that is hydrophilic in nature and has a predicted molecular weight of 117 kDa; the putative sequence of Vps3p was largely uninformative with regard to sequence identities or functional motifs (Raymond *et al.,* 1990). The 140 kDa VPS3 product, like Vps15p, Vps34p, and some other characterized Vps proteins, was present in yeast cells at very low levels. Studies with a mutant that carried a temperature-conditional *vps3* allele suggested that Vps3p plays a direct role in protein sorting. Unlike Vps15p, overexpression of Vps3p introduced certain perturbations (Herman *et al.,* 1991; Raymond, 1990). Specifically, 100-fold overproduction of Vps3p in wild-type yeast caused secretion of at least 50% of newly synthesized CPY, and it induced a class B-like vacular morphology (Section IV,C; Fig. 4).

C. Vps1p Is a GTP-Binding Protein

The *VPS1* gene was chosen for analysis because *vps1* mutants possess a severe CPY sorting defect, they are unable to grow at high temperatures, and they accumulate unusual intracellular membranes (Fig. 4F) (Rothman and Stevens, 1986). Cloning and sequencing of the gene revealed a 704 amino acid, hydrophilic translation product that had a consensus GTP-binding domain in the amino-terminal region of the protein (Fig. 5) and significant sequence identity with the Mx proteins of vertebrate cells (50% identity over the amino-terminal 250 amino acids of both proteins and about 30% overall; Rothman *et al.,* 1990). Mx proteins are interferon-inducible *in vivo,* and they confer resistance to certain viral infections by an undetermined mechanism (for reviews see Arnheiter and Meier, 1990; Arnheiter *et al.,* 1990; Pavlovic *et al.,* 1990; Staehli, 1990). Subsequently, Vps1p was found to share even greater similarity with the 100 kDa protein dynamin, a microtubule-associated protein purified from bovine brain (Obar *et al.,* 1990; Shpetner and Vallee, 1989; Vallee and Shpetner, 1990). Dynamin and Vps1p share about 70% identity in the amino-terminal 300 amino acids, and an overall identity of about 45% (Obar *et al.,* 1990). Dynamin was proposed to be a microtubule-dependent motor protein (Shpetner and Vallee, 1989; Vallee and Shpetner, 1990), but the *in vivo* function of dynamin remains unclear (Scaife and Margolis, 1990). Finally, the recent sequencing of the *Drosophilia melanogaster Shibire* gene has shown that the Shibire protein is the fruit fly equivalent of dynamin (Chen *et al.,* 1991; van der Bliek and Meyerowitz, 1991). A variety of cells that are endocytically active in wild-type flies appear to be defective for endocytosis in *Shibire* mutants (Poodry, 1990). Thus, both Vps1p and the Shibire protein appear to function in the vacuolar network. In the case of Vps1p, evidence has been presented that it associates with Golgi-like structures *in vivo* (Rothman *et al.,* 1990). The possibility that Vps1p is required for endocytosis in *S. cerevisiae* is under investigation.

The *VPS1* gene is the same as the *SPO15* gene in *S. cerevisiae,* and *SPO15* is required for sporulation of diploid yeast strains (Yeh *et al.,* 1986). Furthermore, *E. coli* expressed Spo15p sedimented with *in vitro* assembled bovine microtubules (Yeh *et al.,* 1991). It was suggested that Spo15p is a bifunctional protein required for vacuolar protein sorting and nuclear division during meiosis. An alternative explanation derives from the fact that vacuolar hydrolases are required for sporulation; *pep4* mutants are sporulation-deficient (Kaneko *et al.,* 1982; Zubenko and Jones, 1981), and diploids that are homozygous mutants in any one of several *vps* loci are unable to sporulate (Jones, 1984; I. Howald-Stevenson and T. H. Stevens, unpublished observations). For example, a mutation in the *VPS11* locus (TableII) was identified previously in a screen for sporulation-defective mutants (as *spo-T7* mutants; Tanaka and Tsuboi, 1985; Tsuboi, 1983; Woolford *et al.*, 1990). The protein sorting function of Vps1p is microtubule-independent. Newly synthesized CPY is delivered to the vacuole and processed with wild-type kinetics and efficiency in yeast cells exposed to the microtubule-disrupting drug nocodazole (Raymond, 1990; C. Vater *et al.*, unpublished observations).

Sequence comparisons among Vps1p, the Mx proteins, dynamin, and the *Shibire* product reveal a common theme. The amino-terminal "half" of these proteins, which includes the GTP-binding domain, is highly conserved, whereas the carboxy-terminal "half" is generally divergent (Arnheiter and Meier, 1990; Chen *et al.*, 1991; Obar *et al.*, 1990; Rothman *et al.*, 1990; van der Bliek and Meyerowitz, 1991). This has led to the proposal that a common GTP-dependent biological function encoded in the first half of these proteins is directed to different, specific cellular targets by the unique carboxy-terminal segments (Arnheiter and Meier, 1990; Raymond, 1990; Rothman *et al.*, 1990). Recent experimental evidence with wild-type and mutant Vps1p proteins supports this hypothesis (Raymond, 1990; C. K. Raymond *et al.*, unpublished observations). The evidence is as follows: (1) the VPS1 protein both binds and hydrolyzes GTP; (2) Vps1p associates with a membrane fraction in wild-type yeast lysates; (3) deletions that remove the carboxy-terminal region of Vps1p largely abolish membrane association and they inactivate the protein; and (4) deletions that remove the GTP-binding region in Vps1p also eliminate Vps1p function, yet the truncated products remain associated with the particulate fraction. These data are consistent with the model that the amino-terminal region of Vps1p uses GTP-binding and hydrolysis to catalyze a function necessary for sorting. This domain is directed to specific intracellular membranes by the carboxy-terminal tail.

D. Characterization of Class C *VPS* Genes

The gene products encoded by the class C *VPS* genes (Section IV,C; Table III) participate in the most fundamental aspects of vacuolar assembly in *S. cerevisiae*. The class C *vps* mutants have no apparent vacuoles; yet they accumulate numerous, aberrant, vesicular bodies (Fig. 4C) (Banta *et al.*, 1988; Klionsky *et al.*, 1990; Wada *et al.*, 1990; Woolford *et al.*, 1990). In addition to missorting soluble vacuolar proteins, there are many additional pleiotropic phenotypes associated with these strains. They are sensitive to high concentrations of salt or calcium, they require fermentable carbon sources for growth, they cannot grow at elevated temperatures, they cannot sporulate, and they lack a storage pool of basic amino acids (Banta *et al.*, 1988; Dulic and Riezman, 1989; Jones, 1983; Kitamoto *et al.*, 1988;

Ohya *et al.*, 1986; Wada *et al.*, 1990). Most significantly, it was recently discovered that class C mutants secrete substantial amounts of unprocessed α-factor precursor (Section III,B,D) (Robinson *et al.*, 1992). This mating pheromone precursor is proteolytically processed in a late-Golgi compartment, the same compartment in which vacuolar protein sorting is likely to occur (Graham and Emr, 1991). Hence, the class C *VPS* genes may be required to maintain the functional integrity of the most distal Golgi compartment(s). Additional support for this hypothesis derives from the observation that the intracellular distribution of another late-Golgi protein, DPAP A (Section III,B), appeared as a few discrete patches in wild-type cells but as a multiple, dispersed pattern of dots in class C mutants (similar to the pattern observed for ALP and shown in Figure 4C) (C. K. Raymond *et al.*, unpublished observations). Partial loss of late-Golgi function in the class C strains may interrupt the vacuolar transport pathway, yielding the extreme vacuolar biogenesis defects in these mutants.

Three class C *VPS* genes have been characterized. The sequence of the *VPS11* gene has been reported twice, once as the *END1* gene (originally proposed to be required for endocytosis; Chvatchko *et al.*, 1986; Dulic and Riezman, 1989) and also as the *PEP5* gene (known to be necessary for synthesis of wild-type levels of CPY; Jones, 1977; Woolford *et al.*, 1990). The product of the *END1/PEP5/VPS11* gene (Table II) was predicted to be a 1030 amino acid, hydrophilic protein, and End1p/Pep5p was found to be a low abundance protein with an apparent molecular weight of 120 kDa (Dulic and Riezman, 1989; Woolford *et al.*, 1990). The gene product appeared to fractionate with vacuolar membranes in wild-type cells (Woolford *et al.*, 1990). Mutants lacking this gene are viable, and, ironically, they retain the capacity to internalize α-factor and are thus endocytosis-proficient (Dulic and Riezman, 1989; Woolford *et al.*, 1990). Studies with the *end1*Δ mutants showed that the inability of this class C mutant to grow on nonfermentable carbon sources was not due to a defect in mitochondrial respiration, and the date are consistent with the suggestion that these mutants are unable to perform gluconeogenesis (Dulic and Riezman, 1989). The cloning and sequencing of *VPS33* were also reported twice (as *VPS33;* Banta *et al.*, 1990; as *SLP1;* Wada *et al.*, 1990). The *VPS33/SLP1* product, which was not essential for cell viability, was predicted to be a hydrophilic protein of 691 amino acids, in good agreement with the observation that Vps33p was a nonabundant protein with an apparent mobility of 75 kDa (Banta *et al.*, 1990). There may be a nucleotide binding site in Vps33p (Banta *et al.*, 1990). Vps33p behaves as a soluble protein in subcellular fractionation experiments. In keeping with tradition, the *VPS18* gene was also characterized by two groups (as *VPS18;* Robinson *et al.*, 1992; as *PEP3;* Preston *et al.*, 1992). *VPS18/PEP3* is a nonessential gene, and the VPS18/PEP3 product is predicted to be a hydrophilic protein of 918 amino acids.

Both Vps11p and Vps18p possess a cysteine-rich, zinc finger-like motif within their carboxy-termini (Robinson *et al.*, 1992; Freemont *et al.*, 1991). They share this feature in common with the 43 kDa protein of rat synaptic nerve termini (Froehner, 1991). The 43 kDa protein was shown to be involved in the clustering of acetylcholine receptors, and it may be vital for maintaining the spatial integrity of neuromuscular synapses. It is not known what role the zinc finger region of the 43 kDa protein plays in this process, but a serine for cysteine substitution within the zinc finger motif of Vps18p compromised the function of the VPS18 protein (Robinson *et al.*, 1992). The sequence similarity between the 43 kDa protein and the two class C products raises the possibility that Vps11p and Vps18p may be involved in the clustering of specific proteins. Coupled with the observation that the function and morphology of the *trans*-Golgi compartment is severly compromised in *vps11* and *vps18* mutants (Robinson *et al.*, 1992), one possibility is that *VPS11* and *VPS18* products assist in maintaining the spatial integrity of the late-Golgi apparatus.

E. Reconstitution of Vacuolar Protein Sorting *in Vitro*

One of the more promising recent developments in research on vacuolar protein sorting in *S. cerevisiae* stems from the successful reconstitution of protein transport to the vacuole *in vitro* (Vida *et al.*, 1990; K. E. Moore and T. H. Stevens, unpublished observations). The assay measures the conversion of Golgi-modified p2 pro-CPY to the mature CPY species (Section II,B). The reaction is performed by combining yeast spheroplasts, which are gently perforated by a combined freeze-thaw, hypotonic lysis procedure, with cytosolic extracts and an exogenous energy source. Several features of this semi-intact cell assay system support the contention that this CPY-processing reaction is an *in vitro* reconstruction of the *in vivo* protein sorting process (Vida *et al.*, 1990; K. E. Moore and T. H. Stevens, unpublished observations): (1) the processing reaction requires the addition of cytosolic extract and ATP for efficient conversion, and only the Golgi-modified CPY (and not the ER-modified p1 proCPY present) is processed; (2) the reaction occurs with relatively rapid kinetics over a narrow temperature range, and the sensitivity of *in vitro* processing to detergent argues that it occurs in sealed compartments; (3) maturation requires the presence of the PEP4 gene product; and (4) several lines of evidence suggest that maturation is coincident with an intercompartmental transfer process (Vida *et al.*, 1990). Interestingly, several of the *vps* mutants were unable to support CPY transport to the vacuole in this semi-intact cell assay.

The *in vitro* transport system may provide a key tool in the molecular dissection of vacuolar biogenesis because it provides the opportunity to identify and purify factors required for *in vitro* CPY maturation. Similar strategies have been used to identify molecular components necessary for interorganellar transport between other compartments of the secretory pathway (Baker *et al.*, 1988; Clary *et al.*, 1990; Wilson *et al.*, 1989). Furthermore, this biochemical system may be used to characterize the *vps* mutants with respect to *in vitro* transport defects. However, before this can occur, several barriers and uncertainties need to be overcome. It remains to be shown that the system reconstitutes protein sorting. Recent data suggest that a pool of p2 pro-CPY passes from a *sec18*-sensitive to a *sec18*-insensitive compartment en route to the vacuole *in vivo,* and the latter material may reside in a postsorting, prevacuolar compartment (Section IV,A) (Graham and Emr, 1991). If the pool of p2 pro-CPY that is processed *in vitro* also resides in such a compartment, then the reaction reconstitutes bulk transport but not protein sorting per se. Furthermore, analysis of the *vps* mutants is complicated by the fact that spheroplasts of many of the *vps* mutants are deficient for processing or intracellular CPY (Robinson *et al.*, 1988). This suggests that perforated *vps* cells cannot be used for *in vitro* analysis and that studies may be limited to analysis of cytosol-specific *vps* defects. Despite these and other potential difficulties, the *in vitro* transport system promises to be a valuable tool for unraveling certain molecular aspects of vacuole biogenesis.

VI. Conclusions and Perspectives

In this review we have explored certain facets of organelle biogenesis in a simple, unicellular, eukaryotic organism. Any effort to understand vacuole assembly in molecular terms presupposes reasonably precise knowledge about the pathways that organelle constituents traverse en route to the vacuole. Much of the research on vacuole biogenesis remains focused on elucidating and clarifying the many ways in which proteins are targetted and transported to this select destination. As we have seen, proteins gain entry into the vacuole through the secretory pathway, via endocytosis from the cell surface, and by direct sequestration from the cytoplasm (Fig. 1). In every case there must be specific molecular assemblies that direct the orderly flow of these constituents. Clearly, one continuing emphasis in research efforts to understand vacuole biogenesis in *S. cerevisiae* will involve more precise descriptions of these trafficking events. For example, most newly synthesized resident vacuolar proteins are sorted from secreted proteins in a late-Golgi compartment; yet it is unclear whether soluble and membrane proteins utilize the same transport apparatus. Similarly, it has yet to be determined whether yeast cells possess a prevacuolar, endosome-like compartment through which newly synthesized vacuolar proteins pass. If such a compartment exists, does it converge with endocytic traffic as is the case in mammalian cells?

Major advances have been made in understanding the sorting of soluble vacuolar glycoproteins. Sorting to the vacuole for soluble proteins has been found to be an active process, whereas cell surface secretion is default. Furthermore, the sorting determinant in the case of CPY appears to be confined to a contiguous tetrapeptide in the propeptide segment of the protein. The apparatus that recognizes this signal has yet to be characterized, and identification of the CPY sorting receptor is one of the next major challenges. Will this receptor recognize multiple hydrolases or will it be specific for CPY? How is the receptor itself targetted to the vacuole? Investigation of vacuolar membrane proteins suggests a simple solution. Data gathered so far are consistent with the model that the vacuole is the default compartment for membrane proteins lacking sorting information (apart from an ER-targetting, hydrophobic, transmembrane segment). If the CPY sorting receptor is a membrane protein, then it could deliver soluble proteins bound to it by a relatively passive mechanism. If, by analogy to the mannose-6-phosphate receptor in mammalian cells, the yeast vacuolar protein receptor recycles for multiple rounds of sorting, then targetting information may not be in the form of entry into the vacuolar network but rather retrieval from a prevacuolar compartment. This is reminiscent of models formulated to explain how proteins retained in the ER nonetheless acquire *cis*-Golgi carbohydrate modifications. The discovery that the default compartment for membrane proteins is probably the vacuole resolves issues about targetting information on vacuolar membrane proteins, but it raises new questions about targetting of membrane proteins to the plasma membrane or retention of proteins in compartments proximal to the vacuole. Although the nature of the sorting process is different from that anticipated, the fundamental questions about membrane protein sorting remain the same.

The emergence of recent reports that proteins can be transported directly from the cytoplasm into the vacuolar lumen demonstrates a complexity in vacuolar protein traffic that is far greater than initially envisioned. Such a process demands a protein translocation mechanism that engages specific, fully synthesized, cytoplasmic proteins. The regulated turnover of FBPase argues that the specificity of such a translocation assembly is subject to regulation, and the *SEC18*-dependence of the process seems to imply that this specificity is regulated at the level of the import machinery and not at the level of FBPase modification. Delivery of specific proteins from the cytoplasm into the vacuole forces us to confront older observations about starvation-induced, *PEP4*-dependent turnover of bulk cellular protein. Do yeast cells possess an autophagic mechanism, and, if so, how is it regulated? These are virtually untapped areas of potential investigation.

The assembly and function of the vacuolar H^+-ATPase highlight several intriguing aspects of vacuolar biogenesis and function. The slow growth rate and impaired homeostasis of mutants lacking vacuolar H^+-ATPase function underscore the importance of this complex in generating the electrochemical gradients necessary for metabolite transport and other processes. Further characterization of these mutants should illuminate and clarify the roles the vacuole plays in overall cellular physiology. It should be pointed out that mutants lacking vacuolar H^+-ATPase function possess nearly normal levels of active vacuolar hydrolases; yet they are exceptionally growth-deficient. By contrast, *pep4* mutants lack many hydrolase activities; yet they exhibit wild-type growth characteristics under normal (laboratory) growth conditions. These data suggest that storage and transport functions of the vacuole are far more central to vegetative cell growth than the digestive/degradative capacity of this organelle. The vacuolar H^+-ATPase also presents a fascinating system to study assembly of a multisubunit enzyme complex. How are the synthesis levels of various subunits, many of which exist in the mature complex in different stoichiometries, coordinated? How are membrane components and peripheral subunits brought together, where is the site of assembly, and what signals assembly to occur? What roles do the many subunits play in assembly and function, and are there factors not associated with the complex that are nonetheless involved in maturation of the vacuolar H^+-ATPase? Does the ATPase acidify other vacuolar network organelles, and, if so, how is its trafficking regulated? These are but some of the questions that await further research into this multisubunit complex.

Vacuolar biogenesis not only involves delivery of proteins to the vacuole but also segregation of the organelle to buds during daughter cell development. There are specific molecular mechanisms that segregate a portion of the maternal vacuole to the bud, but even the most basic aspects of vacuolar inheritance are poorly understood. How are vacuoles transported, which cytoskeletal system, if any, is required for segregation, how is this cytoskeleton involved in the segregation process, and how is organelle movement coordinated with the cell cycle? Morphological studies of vacuoles have also suggested that vacuolar compartments can both separate from one another and that apparently independent vacuoles can fuse. The components that regulate these membrane interactions have yet to be identified. The isolation and charaterization of mutants such as *vac2*

(Section IV,C), which is specifically defective for vacuolar inheritance, will undoubtedly shed light on the dynamic process of organelle segregation.

The sorting of soluble vacuolar glycoproteins has proven to be amenable to genetic analysis with two unanticipated qualifications. First, reverse genetic approaches have shown that molecular components thought to influence sorting in animal cell systems (clathrin and the vacuolar H^+-ATPase) have little impact, if any, on protein sorting in *S. cerevisiae*. Second, selections for sorting mutants have unveiled a wealth of genes involved in the delivery of soluble proteins to the vacuole. To put this apparent complexity into perspective, it is widely estimated that there are approximately 4000 genes in yeast. Given nearly 50 *VPS* complementation groups (a number that is not yet saturated and could easily double), this implies that 1–2% of the yeast genome encodes products that influence vacuolar assembly. Research efforts have naturally focused on *VPS* genes that command distinct roles in vacuolar biogenesis as inferred from unique phenotypes of the corresponding *vps* mutants. Cloning and sequencing of several *VPS* genes have provided some insights into the protein products that control vacuolar assembly; Vps15p, a membrane-anchored protein kinase, and Vps1p, a membrane-associated GTPase, are both required for sorting and organelle assembly. The class C *VPS* genes appear to be involved in establishing and/or maintaining the functional integrity of the late-Golgi apparatus, the compartment from which the vacuole is apparently derived.

The development of new tools for phenotypic analysis of *vps* mutants is beginning to shed light on the diversity of mutants in the *vps* strain collection. For example, indirect immunofluorescence techniques have permitted subclasses of mutants to be identified and have revealed novel membrane-enclosed compartments that accumulate in the mutant strains. The ability to divide the many *vps* mutants into categories, even on a superficial basis, is an important step toward conquering their complexity. It will be of interest to determine whether certain *vps* mutants influence the function and distribution of other *VPS* products. The potential functional association between Vps15p and Vps34p (Section V,B) suggests this kind of interaction. Reverse genetic approaches will continue to be used to assess the role of various protein products in vacuolar biogenesis. This strategy has proven invaluable in assessing the functional significance of polypetides that co-purify with the vacuolar H^+-ATPase (Section III,E). One might expect, by analogy to other portions of the secretory pathway, that small GTP-binding proteins similar to Sec4p, Ypt1p, or the Rab proteins in animal cells (Chavrier *et al.*, 1990a,b) may mediate transport steps between the Golgi compartment and the vacuole. Indeed, small GTP-binding proteins appear to co-purify with the vacuolar membrane (C. K.

Raymond *et al.*, unpublished observations), but potential function for these proteins is unknown. Finally, combining the *in vitro* reconstituted transport systems and *vps* mutant cells will provide a powerful approach to dissecting the role of various *VPS* proteins in vacuolar assembly.

In summary, vacuolar biogenesis in *S. cerevisiae* has proven to be an experimentally tractable system to study the biogenesis of a secretory pathway organelle. Excellent progress has been made in identifying various routes by which proteins are delivered to the vacuole and in understanding vacuolar targetting information on several vacuolar proteins. The fact that many vacuolar functions are dispensable under laboratory growth conditions has permitted the identification and characterization of yeast strains with lesions that profoundly perturb the organelle's biogenesis. The challenges that lie ahead are to gain an appreciation for how the corresponding wild-type products interact with one another in the execution of vacuolar assembly. Lessons learned from this system should provide molecular models for organelle assembly in a variety of diverse organisms.

References

Achstetter, T., and Wolf, D. H. (1985). *Yeast* **1,** 139–157.

Adachi, I., Puopolo, K., Marquez-Sterlin, N., Arai, H., and Forgac, M. (1990). *J. Biol. Chem.* **265,** 967–973.

Ammerer, G., Hunter, C. P., Rothman, J. H., Saari, G. C., Valls, L. A., and Stevens, T. H. (1986). *Mol. Cell. Biol.* **6,** 2490–2499.

Anraku, Y., Umemoto, N., Hirata, R., and Wada, Y. (1989). *J. Bioenerg. Biomembr.* **21,** 589–603.

Arai, H., Terres, G., Pink, S., and Forgac, M. (1988). *J. Biol. Chem.* **263,** 8796–8802.

Arnheiter, H., and Meier, E. (1990). *New Biol.* **2,** 851–857.

Arnheiter, H., Skuntz, S., Noteborn, M., Chang, S., and Meier, E. (1990). *Cell* **62,** 51–61.

Baker, D., Hicke, L., Rexach, M., Schleyer, M., and Sheckman, R. (1988). Cell **54,** 335–344.

Baker, D., Wuestehube, L., Schekman, R., Botstein, D., and Seger, N. (1990). *Proc. Natl. Acad. Sci. U.S.A.* **87,** 355–359.

Balch, W. E., Dunphy, W. G., Braell, W. A., and Rothman, J. E. (1984a). *Cell* **39,** 405–416.

Balch, W. E., Glick, B. S., and Rothman, J. E. (1984b). *Cell* **39,** 525–536.

Bankaitis, V. A., Johnson, L. M., and Emr, S. D. (1986). *Proc. Natl. Acad. Sci. U.S.A.* **83,** 9075–9079.

Banta, L. M., Robinson, J. S., Klionsky, D. J., and Emr, S. D. (1988). *J. Cell Biol.* **107,** 1369–1383.

Banta, L. M., Vida, T. A., Herman, P. K., and Emr, S. D. (1990). *Mol. Cell. Biol.* **10,** 4638–4649.

Beckers, C. J. M., Block, M. R., Glick, B. S., Rothman, J. E., and Balch, W. E. (1989). *Nature (London)* **339,** 397–398.

Blachly-Dyson, E., and Stevens, T. H. (1987). *J. Cell Biol.* **104,** 1183–1191.

Bordallo, C., Schwenke, J., and Suarez Rendueles, M. (1984). *FEBS Lett.* **173,** 199–203.

Bourne, H. R. (1988). *Cell* **53,** 669–671.

Bowman, B. J., and Bowman, E. J. (1986). *J. Membr. Biol.* **94,** 83–97.
Bowman, B. J., Dschida, W. J., Harris, T., and Bowman, E. J. (1989). *J. Biol. Chem.* **264,** 15606–15612.
Bowman, E. J., Siebers, A., and Altendorf, K. (1988). *Proc. Natl. Acad. Sci. U.S.A.* **85,** 7972–7976.
Bowser, R., and Novick, P. (1991). *J. Cell Biol.* **112,** 1117–1131.
Bussey, H. (1988). *Yeast* **4,** 17–26.
Carlson, M., and Botstein, D. (1982). *Cell* **28,**145–154.
Chang, Y.-H., and Smith, J. A. (1989). *J. Biol. Chem.* **264,** 6979–6983.
Chavrier, P., Parton, R. G., Hauri, H. P., Simons, K., and Zerial, M. (1990a). *Cell* **62,** 317–329.
Chavrier, P., Vingron, M., Sander, C., Simons, K., and Zerial, M. (1990b). *Mol. Cell. Biol.* **10,** 6578–6585.
Chen, M. S., Obar, R. A., Schroeder, C. C., Austin, T. W., Poodry, C. A., Wassworth, S. C., and Vallee, R. B. (1991). *Nature (London)* **351,** 583–586.
Chiang, H.-L., and Schekman, R. (1991). *Nature (London)* **350,** 313–318.
Chiang, H.-L., Terlecky, S. R., Plant, C. P., and Dice, J. F. (1989). *Science* **246,** 382–385.
Chvatchko, Y., Howald, I., and Riezman, H. (1986). *Cell* **46,** 355–364.
Clark, D. W., Tkacz, J. S., and Lampen, J. O. (1982). *J. Bacteriol.* **152,** 865–873.
Clary, D. O., Griff, I. C., and Rothman, J. E. (1990). *Cell* **61,** 709–721.
Cleves, A. E., McGee, T. P., Whitters, E. A., Champion, K. M., Aitken, J. R., Dowhan, W., Goebl, M., and Bankaitis, V. A. (1991). *Cell* **64,** 789–800.
Cooper, A. (1990). Ph. D. Thesis, McGill Univ., Montreal.
Cooper, A., and Bussey, H. (1989). *Mol. Cell. Biol.* **9,** 2706–2714.
Cunningham, K. W., and Wickner, W. T. (1989). *Yeast* **5,** 25–33.
Dean, N., and Pelham, R. B. (1990). *J. Cell Biol.* **111,** 369–377.
Deshaies, R. J., and Schekman, R. (1987). *J. Cell Biol.* **105,** 633–645.
Deshaies, R. J., and Schekman, R. (1989). *J. Cell Biol.* **109,** 2653–2664.
Dever, T. E., Glynias, M. J., and Merrick, W. C. (1987). *Proc. Natl. Acad. Sci. U.S.A.* **84,** 1814–1818.
Diaz, R., Mayorga, L. S., Weidman, P. J., Rothman, J. E., and Stahl, P. D. (1989). *Nature (London)* **339,** 398–400.
Dice, J. F. (1990). *Trends Biochem.* **15,** 305–309.
Dmochowska, A., Dignard, D., Henning, D., Thomas, D. Y., and Bussey, H. (1987). *Cell* **50,** 573–584.
Donaldson, J. G., Lippincott-Schwartz, J., Bloom, G. S., Kreis, T. E., and Klausner, R. D. (1990). *J. Cell Biol.* **111,** 2295–2306.
Duden, R., Griffiths, G., Frank, R., Argos, P., and Kreis, T. E. (1991). *Cell* **64,** 649–665.
Dulic, W. G., and Riezman, H. (1989). *EMBO J.* **8,** 1349–1359.
Dunn, W. A. (1990). *J. Cell Biol.* **110,** 1923-1933.
Fok, A. K., and Allen, R. K. (1990). *Int. Rev. Cytol.* **123,** 61–92.
Forgac, M. (1989). *Physiol. Rev.* **69,** 765–796.
Foury, F. (1990). *J. Biol. Chem.* **265,** 18554–18560.
Franzusoff, A., and Schekman, R. (1989). *EMBO J.* **8,** 2695–2702.
Franzusoff, A., Redding, K., Crosby, J., Fuller, R. S., and Schekman, R. (1991). *J. Cell Biol.* **112,** 27–37.
Freemont, P. S., Hanson, I. M., and Trousdale, J. (1991). *Cell* **64,** 483–484.
Froehner, S. C. (1991). *J. Cell Biol.* **114,** 1–7.
Fujiki, Y., Hubbard, A. L., Fowler, S., and Lazarow, P. B. (1982). *J. Cell Biol.* **93,** 97–102.
Fuller, R. S., Sterne, R. E., and Thorner, J. (1988). *Annu. Rev. Physiol.* **50,** 345–362.
Fuller, R. S., Brake, A. J., and Thorner, J. (1989). *Science* **246,** 482–486.

Geuze, H. J., Slot, J. W., Strous, G. J. A. M., Hasilik, A., and von Figura, K. (1984). *J. Cell Biol.* **98,** 2047–2054.

Geuze, H. J., Slot, J. W., Strous, G. J. A. M., Hasilik, A., and von Figura, K. (1985). *J. Cell Biol.* **101,** 2253–2262.

Glickman, J. N., Conibear, E., and Pearse, B. M. F. (1989). *EMBO J.* **8,** 1041–1047.

Gogarten, J. P., Kibak, H., Dittrich, P., Taiz, L., Bowman, E. J., Bowman, B. J., Manolson, M. F., Poole, R. J., Date, T., and Oshima, T. (1989). *Proc. Natl. Acad. Sci. U.S.A.* **86,** 6661–6665.

Goldstein, J. L., Brown, M. S., Anderson, R. G. W., Russell, D. W., and Schneider, W. J. (1985). *Annu. Rev. Cell Biol.* **1,** 1–39.

Gomes de Mesquita, D. S., ten Hoopen, R., and Woldringh, C. L. (1991). *J. Gen. Microbiol.* **137,** 2447–2454.

Gorvel, J.-P., Chavrier, P., Zerial, M., and Gruenberg, J. (1991). *Cell* **64,** 915–925.

Goud, B., Salminen, A., Walworth, N. C., and Novick, P. J. (1988). *Cell* **53,** 753–768.

Graham, T. R., and Emr, S. D. (1991). *J. Cell Biol.* **114,** 207–218.

Groesch, M. E., Ruohola, H., Bacon, R., Rossi, G., and Ferro-Novick, S. (1990). *J. Cell Biol.* **111,** 45–53.

Guerra, R., Valdes-Hervia, J., and Gancedo, J. M. (1988). *FEBS Lett.* **242,** 149–152.

Guthrie, B., and Wickner, W. (1988). *J. Cell Biol.* **107,** 115–120.

Hagen, D. C., McCaffrey, G., and Sprague, G. F., Jr. (1986). *Proc. Natl. Acad. Sci. U.S.A.* **83,** 1418–1422.

Hanks, S. K., Quinn, A. M., and Hunter, T. (1988). *Science* **241,** 42–52.

Hasilik, A., and Tanner, W. (1976). *Biochem. Biophys. Res. Commun.* **72,** 1430–1436.

Hasilik, A., and Tanner, W. (1978). *Eur. J. Biochem.* **85,** 599–608.

Hemmings, B. A., Zubenko, G. S., Hasilik, A., and Jones, E. W. (1981). *Proc. Natl. Acad. Sci. U.S.A.* **78,** 435–439.

Herman, P. K., and Emr, S. D. (1990). *Mol. Cell. Biol.* **10,** 6742–6754.

Herman, P. K., Stack, J. H., DeModena, J. A., and Emr, S. D. (1991). *Cell* **64,** 425–437.

Hirata, R., Ohsumi, Y., and Anraku, Y. (1989). *FEBS Lett.* **244,** 397–401.

Hirsch, J. J., Schiffer, H. H., Muller, H., and Wolf, D. H. (1992). *Eur. J. Biochem.* **203,** 641–653.

Hurtley, S. M. (1991) *Trends Biochem. Sci.* **16,** 165–166.

James, M. N. G., and Sielecki, A. R. (1986). *Nature (London)* **319,** 33–38.

Jenness, D. D., and Spatrick, P. (1986). *Cell* **46,** 345–353.

Jenness, D. D., Burkholder, A. C., and Hartwell, L. H. (1983). *Cell* **35,** 521–529.

Johnson, L. M., Bankaitis, V. A., and Emr, S. D. (1987). *Cell* **48,** 875–885.

Jones, E. W. (1977). *Genetics* **85,** 23–33.

Jones, E. W. (1983). *In* "Yeast Genetics, Fundamental and Applied Aspects" (J. F. T. Spencer, D. M. Spencer, and A. R. W. Smith, eds.), pp. 167–204. Springer-Verlag, New York.

Jones, E. W. (1984). *Annu. Rev. Genet.* **18,** 233–270.

Jones, E. W., Zubenko, G. S., and Parker, R. R. (1982). *Genetics* **102,** 665–677.

Julius, D., Blair, L., Brake, A., Sprague, G. F., Jr., and Thorner, J. (1983). *Cell* **32,** 839–852.

Julius, D., Brake, A., Blair, L., Kunisawa, R., and Thorner, J. (1984a). *Cell* **37,** 1075–1089.

Julius, D., Schekman, R., and Thorner, J. (1984b). *Cell* **36,** 309–318.

Kaiser, C. A., and Schekman, R. (1990). *Cell* **61,** 723–733.

Kaiser, C. A., Preuss, D., Grisafi, P., and Botstein, D. (1988). *Science* **235,** 312–317.

Kane, P. M., Yamashiro, C. T., and Stevens, T. H. (1989a). *J. Biol. Chem.* **264,** 19236–19244.

Kane, P. M., Yamashiro, C. T., Rothman, J. H., and Stevens, T. H. (1989b). *J. Cell Sci., Suppl.* No. **11,** 161–178.

Kane, P. M., Yamashiro, C. T., Wolczyk, D., Neff, N., Goebl, M., and Stevens, T. H. (1990). *Science* **250,** 651–657.

Kane, P. M., Kuehn, M. C., Howald, I., and Stevens, T. H. (1992). *J. Biol. Chem.* **267,** 447–454.

Kaneko, Y., Toh-e, A., and Oshima, Y. (1982). *Mol. Cell. Biol.* **5,** 248–252.

Kaneko, Y., Hayashi, N., Toh-e, A., Banno, I., and Oshima, Y. (1987). *Gene* **58,** 137–148.

Kelly, R. B. (1985). *Science* **230,** 25–32.

Kitamoto, K., Yoshizawa, K., Ohsumi, Y., and Anraku, Y. (1988). *J. Bacteriol.* **170,** 2687–2691.

Klionsky, D. J., and Emr, S. D. (1989). *EMBO J.* **8,** 2241–2250.

Klionsky, D. J., and Emr, S. D. (1990). *J. Biol. Chem.* **265,** 5349–5352.

Klionsky, D. J., Banta, L. M., and Emr, S. D. (1988). *Mol. Cell. Biol.* **8,** 2105–2116.

Klionsky, D. J., Herman, P. K., and Emr, S. D. (1990). *Microbiol. Rev.* **54,** 266–292.

Konopka, J. B., Jenness, D. D., and Hartwell, L. H. (1988). *Cell* **54,** 609–620.

Kornfeld, S., and Mellman, I. (1989). *Annu. Rev. Cell Biol.* **5,** 483–525.

Kuranda, M. J., and Robbins, P. W. (1987). *Proc. Natl. Acad. Sci. U.S.A.* **84,** 2585–2589.

Lippincott-Schwartz, J., Donaldson, J. G., Schweizer, A., Berger, E. G., Hauri, H.-P., Yuan, L. C., and Klausner, R. D. (1990). *Cell* **60,** 821–836.

Makarow, M., and Nevalainen, L. T. (1987). *J. Cell Biol.* **104,** 67–75.

Malhotra, V., Orci, L., Glick, G. S., Block, M. R., and Rothman, J. E. (1988). *Cell* **54,** 221–227.

Manolson, M. F., Rea, P. A., and Poole, R. J. (1985). *J. Biol. Chem.* **260,** 12273–12279.

Mechler, B., Muller, M., Muller, H., Meussdoerffer, F., and Wolf, D. F. (1982). *J. Biol. Chem.* **257,** 11203–11206.

Mechler, B., Muller, H., and Wolf, D. H. (1987). *EMBO J.* **6,** 2157–2163.

Mechler, B., Hirsch, H., Muller, H., and Wolf, D. H. (1988). *EMBO J.* **7,** 1705–1710.

Melancon, P., Glick, B. S., Malhotra, V., Weidman, P. J., Serafini, T., Gleason, M. L., Orci, L., and Rothman, J. E. (1987). *Cell* **51,** 1053–1062.

Mellman, I., Fuchs, R., and Helenius, A. (1986). *Annu. Rev. Biochem.* **55,** 663–700.

Mitchell, J. K., Fonzi, W. A., Wilkerson, J., and Opheim, D. J. (1981). *Biochim. Biophys. Acta* **657,** 482–494.

Moehle, C. M., and Jones, E. W. (1990). *Genetics* **124,** 39–55.

Moehle, C. M., Aynardi, M. W., Kolodny, M. R., Park, F. J., and Jones, E. W. (1987a). *Genetics* **115,** 255–263.

Moehle, C. M., Tizard, R., Lemmon, S. K., Smart, J., and Jones, E. W. (1987b). *Mol. Cell. Biol.* **7,** 4390–4399.

Moehle, C. M., Dixon, C. K., and Jones, E. W. (1989). *J. Cell Biol.* **108,** 309–324.

Mortimere, G. E. (1987). *In* "Lysosomes: Their Role in Protein Breakdown" (H. Glaumann and F. J. Ballard, eds.), pp. 415–444. Academic Press, San Diego.

Nakano, A., and Muramatsu, M. (1989). *J. Cell Biol.* **109,** 2677–2691.

Nakayama, N., Miyajima, A., and Arai, K. (1985). *EMBO J.* **4,** 2643–2648.

Nelson, H., and Nelson, N. (1989). *FEBS Lett.* **247,** 147–153.

Nelson, H., and Nelson, N. (1990). *Proc. Natl. Acad. Sci. U.S.A.* **87,** 3503–3507.

Nelson, H., Mandiyan, S., and Nelson, N. (1989). *J. Biol. Chem.* **264,** 1775–1778

Nelson, N. (1989). *J. Bioenerg. Biomembr.* **21,** 553–571.

Noumi, T., Beltran, C., Nelson, H., and Nelson, N. (1991). *Proc. Natl. Acad. Sci. U.S.A.* **88,** 1938–1942.

Novick, P., and Botstein, D. (1985). *Cell* **40,** 405–416.

Novick, P., Field, C., and Schekman, R. (1980). *Cell* **21,** 205–215.

Novick, P., Ferro, S., and Schekman, R. (1981). *Cell* **25,** 461–469.

Obar, R. A., Collins, C. A., Hammarback, J. A., Shpetner, H. S., and Vallee, R. B. (1990). *Nature (London)* **347,** 256–261.

Ohsumi, Y., and Anraku, Y. (1981). *J. Biol. Chem.* **256,** 2079–2082.

Ohsumi, Y., and Anraku, Y. (1983). *J. Biol. Chem.* **258,** 5614–5617.

Ohya, Y., Ohsumi, Y., and Anraku, Y. (1986). *J. Gen. Microbiol.* **132,** 979–988.

Ohya, Y., Umemoto, N., Tanida, I., Ohta, A., Iida, H., and Anraku, Y. (1991). *J. Biol. Chem.* **266,** 13971–13977.

Opheim, D. J. (1978). *Biochim. Biophys. Acta* **524,** 121–130.

Orci, L., Ravazzola, M., Storch, M.-J., Anderson, R. W., Vassalli, J.-D., and Perrelet, A. (1987). *Cell* **49,** 865–868.

Orci, L., Malhotra, V., Amherdt, M., Serafini, T., and Rothman, J. E. (1989). *Cell* **56,** 357–368.

Pavlovic, J. P., Zurcher, T., Haller, O., and Staehli, P. (1990). *J. Virol.* **64,** 3370–3375.

Payne, G. S., and Schekman, R. (1989). *Science* **245,** 1358–1365.

Payne, G. S., Baker, D., van Tuinen, E., and Schekman, R. (1988). *J. Cell Biol.* **106,** 1453–1461.

Pelham, H. R. B. (1988). *EMBO J.* **7,** 1757–1762.

Pelham, H. R. B. (1989). *Annu. Rev. Cell Biol.* **5,** 1–23.

Perin, M. S., Fried, V. A., Stone, D. K., Xie, X.-S., and Sudhoff, T. C. (1991). *J. Biol. Chem.* **266,** 3877–3881.

Pfanner, N., Glick, B. S., Arden, S. R., and Rathman, J. E. (1990). *J. Cell Biol.* **110,** 955–961.

Plutner, H., Schwaninger, R., Pind, S., and Balch, W. E. (1990). *EMBO J.* **9,** 2375–2383.

Poodry, C. A. (1990). *Dev. Biol.* **138,** 464-472.

Preston, R. A., Murphy, R. F., and Jones, E. W. (1987). *J. Cell Biol.* **105,** 1981–1987.

Preston, R. A., Murphy, R. F., and Jones, E. W. (1989). *Proc. Natl. Acad. Sci. U.S.A.* **86,** 7027–7031.

Preston, R. A., Manolsen, M., Becherer, K., Weidenhammer, E., Kirkpatrick, D., Wright, R., and Jones, E. W. (1992). *Mol. Cell Biol.* **11,** 5801–5812.

Pringle, J. R., Preston, R. A., Adams, A. E. M., Stearns, T., Drubin, D. G., Haarer, B. K., and Jones, E. W. (1989). *Methods Cell Biol.* **31,** 357–435.

Randall, L. L., and Hardy, S. J. S. (1989). *Science* **243,** 1156–1159.

Raymond, C. K. (1990). Ph. D. Thesis, Univ. of Oregon, Eugene.

Raymond, C. K., O'Hara, P. J., Eichinger, G., Rothman, J. H., and Stevens, T. H. (1990). *J. Cell Biol.* **111,** 877–892.

Raymond, C. K., Howald-Stevenson, I., C. Vater, C., and Stevens, T. H. (1992). In preparation.

Redding, K., Holcomb, C., and Fuller, R. S. (1991). *J. Cell Biol.* **113,** 527-538.

Reimann, E. M., Titani, K., Ericsson, L. H., Wade, R. D., Fischer, E. H., and Walsh, K. A. (1984). *Biochemistry* **23,** 4185–4192.

Reneke, J. E., Blumer, K. J., Courchesne, W. E., and Thorner, J. (1988). *Cell* **55,** 221–234.

Rexach, M. F., and Schekman, R. W. (1991). *J. Cell Biol.* **114,** 219–230.

Riezman, H. (1985). *Cell* **40,** 1001–1009.

Rittenhouse, J., Moberly, L., and Marcus, F. (1987). *J. Biol. Chem.* **262,** 10114–10119.

Robbins, A. R. (1988). *In* "Protein Transfer and Organelle Biosgenesis" (R. Das and P. Robbins, eds.), pp. 463–520. Academic Press, San Diego.

Roberts, C. J. (1991). Ph. D. Thesis, Univ. of Oregon, Eugene.

Roberts, C. J., Pohlig, G., Rothman, J. H., and Stevens, T. H. (1989). *J. Cell Biol.* **108,** 1363–1373.

Roberts, C. J., Raymond, C. K., Yamashiro, C. T., and Stevens, T. H. (1990). *Methods Enzymol.* **194,** 644–661.

Roberts, C. J., Nothwehr, S., and Stevens, T. H. (1992). Submitted for publication.

Robinson, J. S., Klionsky, D. J., Banta, L. M., and Emr, S. D. (1988). *Mol. Cell. Biol.* **8,** 4936–4948.

Robinson, J. S., Graham, T. R., and Emr, S. D. (1992). *Mol. Cell. Biol.* **11,** 5813–5824.

Rothblatt, J. A., Deshaies, R. J., Sanders, S. L., Daum, G., and Schekman, R. (1989). *J. Cell Biol.* **109,** 2641–2652.

Rothman, J. E. (1987). *J. Biol. Chem.* **262,** 12502–12510.

Rothman, J. H. (1988). Ph. D. Thesis, Univ. of Oregon, Eugene.

Rothman, J. H., and Stevens, T. H. (1986). *Cell* **47,** 1041–1051.

Rothman, J. H., and Stevens, T. H. (1988). *In* "Protein Transfer and Organelle Biogenesis" (R. Das and P. Robbins, eds.), pp. 159–208. Academic Press, San Diego.

Rothman, J. H., Hunter, C. P., Valls, L. A., and Stevens, T. H. (1986). *Proc. Natl. Acad. Sci. U.S.A.* **83,** 3248–3252.

Rothman, J. H., Howald, I., and Stevens, T. H. (1989a). *EMBO J.* **8,** 2057–2065.

Rothman, J. H., Yamashiro, C. T., Raymond, C. K., Kane, P. M., and Stevens, T. H. (1989b). *J. Cell Biol.* **109,** 93–100.

Rothman, J. H., Raymond, C. K., Gilbert, T., O'Hara, P. J., and Stevens, T. H. (1990). *Cell* **61,** 1063–1074.

Runge, K. W. (1988). *In* "Protein Transfer and Organelle Biosgenesis" (R. Das and P. Robbins, eds.), pp. 317–362. Academic Press, San Diego.

Ruohola, H., Kabcenell, A. K., and Ferro-Novick, S. (1988). *J. Cell Biol.* **107,** 1465–1476.

Russell, P., and Nurse, P. (1987). *Cell* **49,** 559–567.

Salminen, A., and Novick, P. J. (1987). *Cell* **49,** 527–536.

Sato, T., Ohsumi, Y., and Anraku, Y. (1984). *J. Biol. Chem.* **259,** 11509–11511.

Scaife, R., and Margolis, R. L. (1990). *J. Cell Biol.* **111,** 3023–3033.

Schafer, W., Kalisz, H., and Holzer, H. (1987). *Biochim. Biophys. Acta* **925,** 150–155.

Schekman, R. (1985). *Annu. Rev. Cell Biol.* **1,** 115–143.

Schulze-Lohoff, E., Hasilik, A., and von Figura, K. (1985). *J. Cell Biol.* **101,** 824–829.

Schwaiger, H., Hasilik, A., von Figura, K., Weimken, A., and Tanner, W. (1982). *Biochem Biophys. Res. Commun.* **104,** 950–956.

Segev, N., Mulholland, J., and Botstein, D. (1988). *Cell* **52,** 915–924.

Segev, N., (1991) *Science* **252,** 1553–1556.

Serafini, T., Stenbeck, G., Brecht, A., Lottspeich, F., Orci, L., Rothman, J. E., and Weiland, F. T. (1991). *Nature (London)* **349,** 215–220.

Shaw, J. M., and Wickner, W. T. (1991). *EMBO J.* **10,** 1741–1748.

Shih, C.-K., Wagner, R., Feinstein, S., Kanik-Ennulat, C., and Neff, N. (1988). *Mol. Cell. Biol.* **8,** 3094–3103

Shpetner, H. S., and Vallee, R. B. (1989). *Cell* **59,** 421–432.

Singer, B., and Riezman, H. (1990). *J. Cell Biol.* **110,** 1911–1922.

Singer, S. J., Maher, P. A., and Yaffe, M. P. (1987). *Proc. Natl. Acad. Sci. U.S.A.* **84,** 1960–1964.

Sprague, G. F., Jr., Blair, L. C., and Thorner, J. (1983). *Annu. Rev. Microbiol.* **37,** 623–660.

Staehli, P. (1990). *Adv. Virus Res.* **38,** 147–200.

Stevens, T. H., Esmon, B., and Schekman, R. (1982). *Cell* **30,** 439–448.

Stevens, T. H., Rothman, J. H., Payne, G. S., and Schekman, R. (1986). *J Cell Biol.* **102,** 1551–1557.

Suarez-Rendueles, P., and Wolf, D. H. (1987). *J. Bacteriol.* **169,** 4041–4048.

Sun, S.-Z., Xie, X-S., and Stone, D. K. (1987). *J. Biol. Chem.* **262,** 14790–14794.

Taiz, S. L., and Taiz, L. (1991). *Bot. Acta* **104,** 117–121.

Tanaka, H., and Tsuboi, M. (1985). *Mol. Gen. Genet.* **199,** 21–25.

Tang, J. (1979). *Mol. Cell. Biochem.* **26,** 93–109.

Teichert, U., Mechler, B., Muller, H., and Wolf, D. H. (1989). *J. Biol. Chem.* **264,** 16037–16045.

Trumbly, R. J., and Bradley, G. (1983). *J. Bacteriol.* **156,** 36–48.

Tsuboi, M. (1983). *Mol. Gen. Genet.* **191,** 17–21.

Uchida, E., Ohsumi, Y., and Anraku, Y. (1985). *J. Biol. Chem.* **260,** 1090–1095.
Umemoto, N., Yoshihisa, T., Hirata, R., and Anraku, Y. (1990). *J. Biol. Chem.* **265,** 18447–18453.
Umemoto, N., Ohya, Y., and Anraku, Y. (1992). *J. Biol. Chem.* (in press).
Vallee, R. B., and Shpetner, H. S. (1990). *Annu. Rev. Biochem.* **59,** 909–932.
Valls, L. A., Hunter, C. P., Rothman, J. H., and Stevens, T. H. (1987). *Cell* **48,** 887–897.
Valls, L. A., Winther, J. R., and Stevens, T. H. (1990). *J. Cell Biol.* **111,** 361–368.
van der Bliek, A. M., and Meyerowitz, E. M. (1991). *Nature (London)* **351,** 411–414.
Vida, T. A., Graham, T. R., and Emr, S. D. (1990). *J. Cell Biol.* **111,** 2871–2884.
von Figura, K., and Hasilik, A. (1986). *Annu. Rev. Biochem.* **55,** 167–193.
Wada, Y., Ohsumi, Y., and Anraku, Y. (1988). *Cell Struct. Funct.* **13,** 608.
Wada, Y., Kitamoto, K., Kanbe, T., Tanaka, K., and Anraku, Y. (1990). *Mol. Cell. Biol.* **10,** 2214–2223.
Walworth, N. C., Goud, B., Kastan-Kabcenell, A., and Novick, P. J. (1989). *EMBO J.* **8,** 1685–1693.
Waters, M. G. Serafini, T., and Rothman, J. E. (1991). *Nature (London)* **349,** 248–251.
Wattenberg, B. W., Balch, W. E., and Rothman, J. E. (1986). *J. Biol. Chem.* **261,** 2202–2207.
Wattenberg, B. W., Hiebsch, R. H., LeCrureux, L. W., and White, M. P. (1990). *J. Cell Biol.* **110,** 947-954.
Weimken, A., Matile, P., and Moor, H. (1970). *Arch. Mikrobiol.* **70,** 89–103.
Weisman, L. S., and Wickner, W. (1988). *Science* **241,** 589–591.
Weisman, L. S., Bacallao, R., and Wickner, W. (1987). *J. Cell Biol.* **105,** 1539–1547.
Weisman, L. S., Emr, S. D., and Wickner, W. (1990). *Proc. Natl. Acad. Sci. U.S.A.* **87,** 1076–1080.
Wilson, D. W., Wilcox, C. A., Flynn, G. C., Chen, E., Kuang, W.-J., Henzel, W. J., Block, M. R., Ullrich, A., and Rothman, J. E. (1989). *Nature (London)* **339,** 355–359.
Winther, J. R. (1989). Ph. D. Thesis, Carlsberg Lab., Copenhagen.
Winther, J. R., Stevens, T. H., and Kielland-Brandt, M. C. (1991). *Eur. J. Biochem.* **197,** 681–689.
Woolford, C. A., Daniels, L. B., Park, F. J., Jones, E. W., van Arsdell, J. N., and Innis, M. A. (1986). *Mol. Cell. Biol.* **6,** 2500–2510.
Woolford, C. A., Dixon, C. K., Manolson, M. F., Wright, R., and Jones, E. W. (1990). *Genetics* **125,** 739–752.
Xie, X.-S., and Stone, D. K. (1988). *J. Biol. Chem.* **263,** 9859–9867.
Yamashiro, C. T., Kane, P. M., Wolczyk, D. F., Preston, R. A., and Stevens, T. H. (1990). *Mol. Cell. Biol.* **10,** 3737–3749.
Yeh, E., Carbon, J., and Bloom, K. (1986). *Mol. Cell. Biol.* **6,** 158–167.
Yeh, E., Driscoll, R., Coltera, M., Olins, A., and Bloom, K. (1991). *Nature (London)* **349,** 713–714.
Yoshihisa, T., and Anraku, Y. (1989). *Biochem. Biophys. Res. Commun.* **163,** 908–915.
Yoshihisa, T., and Anraku, Y. (1990). *J. Biol. Chem.* **265,** 22418–22425.
Yoshihisa, T., Ohsumi, Y., and Anraku, Y. (1988). *J. Biol. Chem.* **263,** 5158–5163.
Zubenko, G. S., and Jones, E. W. (1981). *Genetics* **97,** 45–64.
Zubenko, G. S., Park, F. J., and Jones, E. W. (1982). *Genetics* **102,** 679–690.
Zubenko, G. S., Park, F. J., and Jones, E. W. (1983). *Proc. Natl. Acad. Sci. U.S.A.* **80,** 510–514.

The Cell Biology of Pattern Formation during *Drosophila* Development

Teresa V. Orenic and Sean B. Carroll
Howard Hughes Medical Institute and Laboratory of Molecular Biology, University of Wisconsin, Madison, Wisconsin 53706

I. Introduction

Biologists have long sought to understand how a single fertilized egg cell develops into a complex animal possessing many different organs and tissues arranged in precise spatial patterns. Decades of study of a wide variety of organisms at many different levels have shown that in many cases the fate of individual cells is determined by their position within a developing field, which may comprise a large or small population of cells that gives rise to particular structures. Positional information within a developing field of cells may be imparted through various mechanisms, including the asymmetric distribution of a diffusible substance [e.g., initial pattern along the anterior–posterior (AP) and dorsal–ventral (DV) axes of *Drosophila* is determined by the graded distribution of morphogens], the differential inheritance of particular substances (cytoplasmic determinants, e.g., polar granules of *Drosophila* specify the germline), or through inductive cell-to-cell interactions.

This review focuses on the role cell-to-cell interactions play during pattern formation in *Drosophila melanogaster*. To understand how cell-to-cell interactions lead to the specification of cell fate and to the formation of specific pattern elements, it has been fruitful to first identify genes and characterize proteins involved in pattern formation. The key advantage to studying pattern formation in *Drosophila* is that it offers the opportunity to carry out exhaustive genetic screens to identify pattern-determining genes. A large number of genetic mutations that alter pattern have been obtained. Genetic and molecular characterization of the genes involved has shown that many of their products are directly involved in cell-to-cell communication, and the signal transduction pathways that are involved in specific pattern-forming processes are beginning to be understood. These cell-to-cell communication mechanisms include direct interactions between transmembrane proteins, interactions at a distance mediated by diffusible factors secreted into the extracellular matrix and bound by cell surface receptors, and direct metabolic coupling through gap junctions, which allow substances to diffuse between adjoining cell cytoplasms. The cellular responses to such signals are also diverse and include activation or repression of transcription of key genes mediating cell phenotype, posttranscriptional regulation of mRNA translation or stability, and posttranslational regulation, such as protein phosphorylation or sequestration of proteins to particular subcellular compartments.

Three models of cell-to-cell interactions involved in *Drosophila* pattern formation will be addressed here in detail. The first concerns the establishment of pattern within the embryonic segments of *Drosophila,* which are fields comprised of epithelial sheets of cells. Cell fate within embryonic segments is specified according to position within the segment and through interactions between adjacent cells. The second model involves the development of the nervous system of *Drosophila,* which requires extensive cell-to-cell interactions. How neuroblasts (NBs) progressively segregate from surrounding epidermal cells will be discussed. The third model concerns the development of the compound eye of *Drosophila,* which is a spectacular example of how positional information is distributed and interpreted. The formation of specific cell types in the ommatidia is determined by the position of each cell in the ommatidial field and the contacts made with other cells. This is not intended to be a comprehensive review of these particular systems; rather, specific examples will be drawn from each that demonstrate the progress that has been attained in understanding global patterning and cell signaling through genetic analysis and how this has paved the way for studying specific developmental prosesses at the cellular and molecular levels.

II. Pattern Formation in the *Drosophila* Epidermis

The epidermis of the *Drosophila* larva and adult is divided into a series of segments, each of which has a unique pattern of denticles, hairs, bristles, and other organs. The genetics, molecular biology, and cell biology of epidermal pattern formation comprise one of the most studied aspects of *Drosophila* development. Extensive genetic analyses to identify mutations that alter the segmental pattern of the larval epidermis have led to the isolation and characterization of a large number of genes required for this process (for reviews see Akam, 1987; Scott and Carroll, 1987; Ingham, 1988a; Nusslein-Volhard, 1991). The focus here is on the establishment of the intrasegmental pattern, and segmentation and embryogenesis will be only briefly reviewed.

A. Embryogenesis and Segmentation

After fertilization synchronous division of the nucleus occurs in the center of the egg. Beginning with the ninth nuclear division (at approximately 70 minutes of development), the nuclei migrate and become arranged in a single layer at the periphery of the embryo to form the syncytial (multinucleate cytoplasm with no individual cells) blastoderm. After the 13th division, cellularization occurs through the invagination of membranes around individual nuclei (2.5 hours). Cellular blastoderm formation is followed by gastrulation and germband extension, a series of morphogenetic events that organize the embryo into its three germ layers and lead to the eventual formation of the larva. During the extension of the germband (between approximately 3.5 and 7.5 hours), which consists of most of the segmented region of the embryo, the length of the ectoderm doubles as the posterior end of the embryo extends anteriorly by folding over on to its dorsal surface. Subsequently, the germband retracts (beginning at 7.5 hours) to restore the normal AP orientation of the embryo (for a more detailed description see Campos-Ortega and Hartenstein, 1985).

Several classes of genes have been found to operate during the early stages of embryogenesis to establish the segmented pattern of the larva (Nusslein-Volhard and Wieschaus, 1980; Nusslein-Volhard *et al.,* 1984; Jurgens *et al.,* 1984; Wieschaus *et al.,* 1984; Nusslein-Volhard, 1991). Some of the products of these genes function principally during the syncytial stages of embryogenesis and act over large distances in the embryo through diffusion, whereas others operate after cellularization and are involved in specifying cell fates and in cell-to-cell interactions. The maternal gene products are provided to the embryo by germline and somatic cells of the mother and function in the syncytium to organize the AP and DV axes of the embryo (Fig. 1) (for review see Nusslein-Volhard, 1991). Three classes of zygotically expressed segmentation genes then act in a hierarchical fashion to divide the embryo into segments and the segments into smaller groups of cells, while the homeotic genes specify individual segmental identity (for reviews see Akam, 1987; Scott and Carroll, 1987; Ingham, 1988a).

Global AP polarity is established through the function of the maternal genes, the products of which are provided to the oocyte by other germline cells. Mutations in many maternal genes cause alterations in the polarity of the embryo. Asymmetric distribution of maternal gene products generates positional information within the embryo and directs position-specific ex-

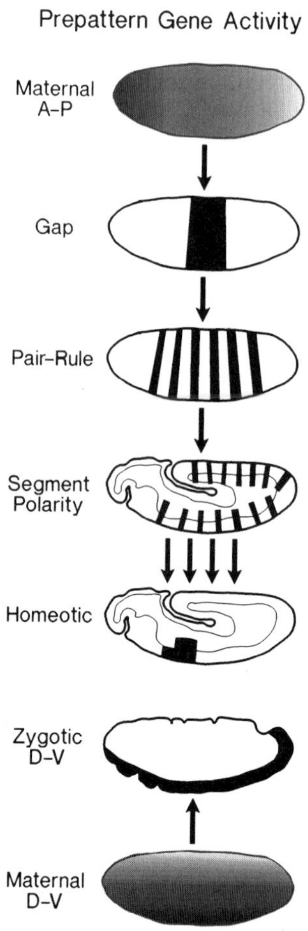

FIG. 1 Schematics of the expression patterns of maternal and zygotic genes that function in pattern formation along the AP and DV axes. The AP maternal genes are expressed in the very early embryo asymmetrically along the AP axis providing global positional information. Responding to the positional information set up by the maternal genes, the gap genes are expressed in blocks of segments that include several segment primordia and begin the subdivision of the embryo. Pair-rule gene expression in alternate parasegments is established by the gap genes and results in the subdivision of the embryo into its segmental repeats and the segmentally repeating pattern of segment polarity gene expression. The segment polarity genes then organize segment polarity and pattern. Segment identity is specified by the homeotic genes. The pattern along the DV axis is determined through the functions of the maternal and zygotic DV genes.

pression of zygotic genes, particularly the segmentation genes (for review see Nusslein-Volhard, 1991). This has been clearly demonstrated by the analysis of the function and expression of the maternal gene *bicoid*. The bicoid protein acts as a morphogen that specifies positional information along the AP axis of the *Drosophila* embryo by controlling early patterns of regulatory gene expression (Driever and Nusslein-Volhard, 1988a). *bicoid* mRNA is deposited at the anterior end of the oocyte and, when translated, the bcd protein diffuses posteriorly to establish an AP protein gradient (Driever and Nusslein-Volhard, 1988b). Different structures along the AP axis are formed in response to the bcd gradient in a concentration-dependent manner.

In response to the bcd gradient and other maternal genes, the first tier of the zygotic segmentation gene hierarchy, the gap genes, is expressed during syncytial blastoderm. The gap genes are expressed in contiguous and partially overlapping domains that include the primordia of several segments (Fig. 1). The level and combination of different gap proteins then regulate the periodic expression of the pair-rule genes, the second tier of segmentation genes. The pair-rule genes are expressed in seven segment-wide stripes in alternate parasegments (Fig. 1); a parasegment is a segment-wide unit that includes the posterior quarter of one segment and the anterior three quarters of the adjacent segment (for reviews see Akam, 1987; Scott and Carroll, 1987; Ingham, 1988a). For example, the pair-rule genes, *ftz* and *eve,* are expressed in complementary patterns of alternating parasegments (Carroll and Scott, 1985; Frasch and Levine, 1987). The other pair-rule genes are similarly expressed but in different registers. Through the concerted actions of the pair-rule genes, the embryo is divided into segments, and the segmentally repeating expression patterns of the segment polarity genes, the third tier of segmentation genes, are established (Fig. 1). The segment polarity gene expression patterns vividly illustrate, at the molecular level, that the genetic program of segmentation has been executed by this stage, even though there is no morphological sign of segmentation until some time later. It is important to emphasize that the genetic regulatory events that have occurred prior to this time have occurred primarily in the context of the syncytial embryo. All the gap and pair-rule genes encode putative transcription factors, and the regulatory cascade of gene expression almost certainly takes place through direct transcriptional regulation among these genes. Because adjacent nuclei share cytoplasm, communication among them can take place directly through the diffusion of gene products within the cytoplasm. However, once the embryo is cellularized, direct diffusion of nuclear transcription factors between nuclei is not possible, and additional molecules are required to establish cell fate within the developing embryo (for reviews see Akam, 1987; Scott and Carroll, 1987; Ingham, 1988a).

B. Function of Segment Polarity Genes in the Establishment of Intrasegmental Pattern

Pattern formation in the epidermis of the embryonic insect segment involves the production of particular structures, such as denticles, and sensory organs, by cells within specific regions of the segment. In addition, it determines cell polarity, which can be detected morphologically by the orientation of the structures produced. Classic grafting experiments in the insects *Rhodnius* and *Oncopeltus* have shown that pattern within segments is determined by a gradient of positional information that is reiterated in every segment (Locke, 1959;Lawrence, 1966). When part of a segment is removed and cells that are not normally adjacent are juxtaposed, regeneration occurs to replace the missing pattern elements. Regeneration is a common developmental phenomenon that takes place when extensive cell death occurs within a development field; it is thought that this occurs through inductive interactions among cells in an effort to regenerate their normal neighbors. Depending on the amount of segmental tissue that is missing, either the original pattern elements will be replaced or mirror image duplications of remaining pattern elements will arise. These phenomena are best explained by a model proposed by Lawrence (1981) in which he assigned values of 0–10 to the segmental gradient. This is reiterated in every segment such that at the segment boundaries the values 0 and 10 are apposed. In *Oncopeltus* when cells of levels 3 and 6 are apposed, pattern elements of levels 4 and 5 are regenerated resulting in the normal pattern. If the cells of the levels 2 and 8 are juxtaposed, the pattern elements that are regenerated are those with the values 1, 0, 10, and 9, resulting in mirror image duplications. From these results it was concluded that regeneration occurs through the route that requires the replacement of the fewest pattern elements (Lawrence, 1981). This is similar to the results of regeneration studies in amphibian limb fields and *Drosophila* imaginal discs from which this rule of intercalation of values through the shortest route was originally derived (French *et al.*, 1976).

The nature of the proposed gradient of positional information within the segments is not known. However, mutations in the segment polarity genes of *Drosophila* cause phenotypes in the embryo that are reminiscent of the duplications and polarity reversals observed in the grafting experiments of *Rhodnius* and *Oncopeltus* (Nusslein-Volhard and Wieschaus, 1980). The pattern within each segment on the ventral surface of the *Drosophila* larva consists of a trapezoidally shaped band of denticles (hair-like structures) in the anterior of the segment followed by clear cuticle in the posterior. The denticles exhibit clear polarity, suggesting that the cells that produce them must be able to sense their position relative to surrounding cells. Mutations in the segment polarity genes cause deletions in specific subsets of every larval segment, and these deletions are often associated with duplication of the remaining pattern elements. In addition, the polarity of the duplicated pattern elements is generally reversed, resulting in mirror image duplications. There is also extensive cell death associated with a number of the segment polarity mutations, suggesting that the mirror image duplications observed in the mutants may be due to regeneration by the remaining cells to replace missing pattern elements. Thus, the segment polarity genes almost certainly function in establishment or interpretation of positional information within the embryonic segment (Nusslein-Volhard and Wieschaus, 1980; for reviews see Martinez Arias, 1989; Wilkins and Gubb, 1991; Ingham, 1991).

When the segment polarity gene products begin to accumulate in the embryo, the animal is completely cellularized (for reviews see Martinez Arias, 1989; Wilkins and Gubb, 1991; Ingham, 1991). Normal segmental pattern and polarity emerge when cells within each segment adopt specific identities and sense their positions relative to other cells. Furthermore, as cell proliferation takes place, identities must be assigned to the new cells, and they must be incorporated into the expanding field. That these processes require communication between cells is underscored by the fact that for most of the segment polarity mutations, their phenotypes do not merely reflect a loss of the specific regions in which they are expressed but include larger regions, suggesting a broad range of influence for these genes. Not surprisingly, then, the segment polarity genes have been found to encode a more diverse class of proteins than the gap and pair-rule genes, including transcription factors (*engrailed, gooseberry,* and *cubitus interruptus Dominant;* Poole *et al.*, 1985; Fjose *et al.,* 1985; Bopp *et al.*, 1986; Baumgartner *et al.,* 1987; Orenic *et al.,* 1990), a transmembrane protein (*patched*; Hooper and Scott, 1989; Nakano *et al.*, 1989), a secreted growth factor (*wingless*; Rijsewijk *et al.,* 1987; Van den Heuvel *et al.*, 1989; Gonzalez *et al.*, 1991), two serine threonine kinases (*fused* and *zeste white-3*; Preat *et al.*, 1990; Siegfried *et al.*, 1990; Bourouis *et al.*, 1990), and a homolog of plakoglobin, a component of desmosomes and gap junctions (*armadillo*; Peifer and Wieschaus; 1990). Some of these proteins clearly function directly in cell-to-cell communication or signal transduction. Analysis of the regulation and function of the segment polarity genes is providing an extensive model for understanding the mechanisms of cell-to-cell signaling and signal transduction involved in pattern formation.

Differential segment polarity gene expression is detectable as early as the onset of gastrulation. This indicates that differences in the stae of specification of cells occurs early and allows one to follow the fate of certain cells from very early stages through later stages. In addition, monitoring segment polarity gene expression allows visualization of pattern in the form of chemical changes in the cell well before morphological differences become detectable. Observation of the expression and regulation of segment polarity genes also shows that determination of cell polarity begins very early. The initial cell states that can be recognized through their unique segment polarity gene expression are not necessarily stable. This is illustrated by the observation that the expression patterns of the segment polarity genes are not static throughout subsequent developmental stages but are actively regulated via cell-to-cell communication and interactions with other segment polarity genes (for reviews see Martinez Arias, 1989; Wilkins and Gubb, 1991; Ingham, 1991). Thus, although the process of specifying cell fate begins very early, the changing patterns of expression of the segment polarity genes, which function in the specification of cell fate, throughout relatively late developmental stages suggests that the final intrasegmental pattern unfolds in a progression. A glimpse of the processes involved in determination of segmental pattern and polarity can be gained through a discussion of the regulation of expression of two segment polarity genes, *wingless* (*wg*) and *engrailed* (*en*), and their interactions with other segment polarity genes.

C. Regulation of *en* Expression and the Progressive Nature of Cell Fate Specification

Each segment of *Drosophila* is composed of two distinct populations of cells, which comprise the anterior and posterior compartments (Garcia-Bellido *et al.*,1973; Lawrence *et al.*, 1979). The *en* gene, which encodes a homeodomain transcription factor (Poole *et al.*,1985; Fjose *et al.*, 1985), functions to specify posterior compartment cell identity (Lawrence and Morata, 1976; Kornberg, 1981a,b; Lawrence and Struhl, 1982). At the time when *en* is first expressed in the embryo, the segmental primordia are each four cells in width. The *en* protein is expressed in the fourth quarter of every segment, defining the posterior compartment (Fig. 2) (DiNardo *et al.*, 1985). The remaining three quarters of the segment comprise the anterior compartment. The *wg* gene is required for appropriate development of the posterior region of the segment, which includes cells from both the anterior and posterior compartments (Nusslein-Volhard *et al.*, 1984; Baker, 1987). The *wg* protein, which is homologous to the mammalian *int-1* proto-oncogene (Rijsewijk *et al.*, 1987), is synthesized in the cells of the third quarter of every segment directly anterior and adjacent to *en*-expressing cells (Baker, 1987). The *wg* protein is actively secreted from these cells and reaches as far as two cells away (Van den Heuvel *et al.*, 1989; Gonzalez *et al.*, 1991).

FIG. 2 Expression pattern of the segment polarity gene *en*. Embryo at the germband extension stage stained with antibodies against *en*, which is expressed within the posterior quarter of every segment.

A detailed study of regulation of *en* expression throughout development suggests that *en*-expressing cells are not stably determined even as late as early germband retraction and provides the best example of the progressive determination of cell fate within segments (Heemskerk *et al.*, 1991). At least four modes of *en* regulation have been identified. First, *en* expression is established by pair-rule gene function. Second, maintenance of *en* expression during and beyond germband extension is dependent on a signal from adjacent *wg*-expressing cells; *wg* expression is thought to provide an activating signal that begins as early as 3 hours of development, suggesting an overlap of pair-rule and *wg* regulation of *en*. Activation by *wg* becomes dispensable at about 5 hours of development (mid-germband extension). Third, continued expression of *en* also requires auto regulation, which can be divided into *wg*- dependent and *wg*- independent phases. Finally, at approximately 7 hours (just before germband retraction begins), an *en*-independent phase begins during which *en* expression is maintained by as yet unidentified factors. It has not yet been determined at what stage the fate of *en*-expressing cells is stably determined, but these data suggest that it does not occur before 7 hours of development.

D. Regulation of *wg* and *en* Expression

For the sake of clarity, we will first briefly outline the current understanding of how *wg* and *en* are expressed and regulated, then present a more detailed description of the data that led to this understanding (Fig. 3). There are several levels of regulation of both *wg* and *en* expression. The position-specific expression of *wg* and *en* is initiated through the function of the pair–rule genes (Ingham *et al.,* 1988). However, the continuous evolution and maintenance of *wg* and *en* expression during later stages of development require interactions between *wg* and *en* and other segment–polarity genes (DiNardo *et al.,* 1988; Martinez Arias *et al.,* 1988; Hidalgo and Ingham, 1990; Ingham *et al.,* 1991; Sampedro and Guerrero, 1991; Ingham, 1991). By germband extension, the segmental primordia are eight cells wide; *en* is expressed in the last two cells and *wg* is expressed in the two cells anterior to these (Fig. 3) (DiNardo *et al.,* 1985; Baker, 1987). At this stage regulation of *wg* and *en* involves maintenance of their expression within their normal domains and repression of ectopic expression in the anterior half of the segment. The segment polarity gene, *nakes* (*nkd*) (Wieschaus *et al.,* 1984), is required to prevent expression of *en* in the cells of the anterior quarter of the segment. Another segment–polarity gene *patched* (*ptc*; Nusslein-Volhard *et al.,* 1984) represses *wg* expression in the second quarter of the segment (DiNardo *et al.,* 1988); *en* and *wg* each require the function of the other gene for each to maintain its wild-type levels of expression beyond germband extension. *wg*-expressing cells provide a positive signal that promotes *en* transcription in adjacent cells. The signal from *wg* cells could likely be the *wg* product itself as it is secreted and is known to be taken up by adjacent cells. However, the factors with which *wg* directly interacts in *en* cells have not been identified. Maintenance of *wg* expression appears to be the result of overriding negative regulation of its transcription by *ptc* (Ingham *et al.,* 1991; Sampedro and Guerrero, 1991; Ingham, 1991). At germband extension *ptc* is expressed in the anterior three quarters of the segment, and its expression overlaps with that of *wg* (Hooper and Scott, 1989; Nakano *et al.,* 1989). Thus, *ptc* appears to selectively repress transcription of *wg* in certain cells of its domain of expression. The differential response of these cells to *ptc* is believed to be mediated by another segment polarity gene, *hedgehog* (*hh;* Ingham *et al.,* 1991; Ingham, 1991), which is produced by the *en*-expressing cells (Mohler, cited in Ingham, 1991). The *ptc* gene encodes a transmembrane protein (Hooper and Scott, 1989; Nakano *et*

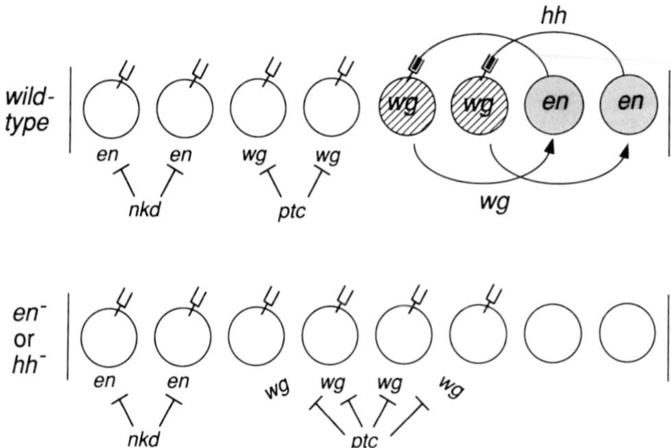

FIG. 3 Regulation of *wg* and *en* expression. This schematic demonstrates the interactions between the segment polarity genes *en, wg, ptc, nkd,* and *hh* within one segment at the germband extension stage (after Ingham *et al.*, 1991). At this stage the segment is approximately eight cells wide. The shaded circles represent *en*-expressing cells; the hatched circles represent *wg*-expressing cells. The two *en*-expressing cells make up the posterior compartment, while the remaining cells make up the anterior compartment. [symbol] represents the membrane *ptc* protein, [symbol] represents *ptc* protein interacting with *hh;* T represents repression; [symbol] represents activation. Regulation of *wg* and *en* expression involves repression of their expression in the anterior half of the segment and maintenance of expression within their normal domains. Although *ptc* is expressed throughout the anterior compartment at this stage, *ptc* represses *wg* expression in the two cells anterior to the normal domain of *wg* expression. The signal transduction pathway through which *ptc* effects this repression has not been deciphered; *nkd* represses *en* expression in the anterior quarter of the segment. Maintenance of *en* expression in its normal domain requires an activating signal from *wg*-expressing cells. The signal could be the *wg* protein itself, which is known to be secreted from cells that transcribe *wg* and has been detected in vesicles within *en*-expressing cells. Other components of this pathway have not been identified. Maintenance of *wg* expression in its normal domain requires communication from *en*-expressing cells. It is thought that *en*-expressing cells send a signal to override *ptc* repression of *wg* transcription in the two most posterior cells of the anterior compartment. It has been proposed that the signal produced by *en* is the product of the *hh* gene, which putatively interferes with *ptc* protein function by directly interacting with it. The nature of this interaction is not known because the structure of the *hh* protein has not been determined. The lower diagram shows the expression of *wg* in *hh* or *en* mutants. In this case *hh* protein is not produced and, therefore, cannot interfere with the function of *ptc* in cells that would normally express *wg*. Thus, *wg* expression is repressed throughout the entire anterior compartment.

al., 1989), and it is conceivable that *hh* protein could act as a ligand that directly interferes with *ptc* function in *wg*-expressing cells (Ingham *et al.*, 1991; Ingham, 1991).

This model of the mechanism of *wg* and *en* regulation was reconstructed primarily from an analysis of the expression patterns of *wg* and *en* in various mutant backgrounds (DiNardo *et al.*, 1988; Martinez Arias *et al.*, 1988; Hidalgo and Ingham, 1990; Ingham *et al.*, 1991; Sampredo and Guerrero, 1991; Ingham, 1991). Some of the details of this analysis are summarized next to demonstrate the utility of the genetic approach employed. That *wg* and *en* each requires the function of the other for maintenance of its expression was demonstrated by analyzing the expression pattern of each in embryos mutant for the other gene. In an *en* mutant *wg* expression is established normally; however, prior to full germband extension the *wg* stripe fades prematurely (Martinez Arias *et al.*, 1988). The same effect is observed on *en* expression in an embryo mutant for *wg* (DiNardo *et al.*, 1988; Martinez Arias *et al.*, 1988). Initiation of expression of *wg* and *en* is not affected by loss of function of either of these genes because pair–rule function is normal.

In *ptc* mutants the *wg* domain expands one to two cells anteriorly; *en* stripes are established normally, but during mid-germband extension, an extra *en* stripe appears immediately adjacent to the cells ectopically expressing *wg*. Ectopic expression of *en* in *ptc* mutants is dependent on *wg* function as in *ptc*; in *wg* double mutants, the ectopic *en* stripe does not appear. Thus, part of the function of *ptc* appears to be suppression of ectopic *wg* and *en* expression in the anterior half of the segment (DiNardo *et al.*, 1988). Expression of *wg* and *en* in *nkd* mutants further demonstrates the interdependent nature of *wg* and *en* expression. In *nkd* mutants *en* is ectopically expressed in the anterior quarter of every segment. During germband extension the *wg* stripes, which are initially established normally, are expanded two or three cells anteriorly. In this case the broadening of the *wg* domain of expression appears to be induced by *en*, which is converse to the case in *ptc* mutants (Martinez Arias *et al.*, 1988).

In the wild-type embryo the effect of *wg* expression on *en* expression is polarized; it only affects transcription in cells posterior to *wg*-expressing cells (Fig. 3). Although one could argue that *wg* is only performing a maintenance function and thereby can only act on cells already expressing *en*, the *wg*-dependent ectopic expression of *en* in anterior compartment cells in *ptc* mutants argues that *wg* can induce *de novo* expression of *en* (DiNardo *et al.*, 1988). Similar agruments based on the expression of *en* and *wg* in *nkd* mutants can be made to suggest that *en* may induce *de novo* expression of *wg*, but in wild-type its effect on *wg* expression is polarized. Thus, it appears that the complex interactions involving *ptc* and *nkd* with *wg* and *en* are partially aimed at maintaining the polarity of *wg* and *en* function. These studies provide insight into the molecular events associated with the determination of polarity, which begins early in development.

Because *ptc* expression is restricted to the anterior compartment during germband extension, coincident with the time during which *en* expression becomes *wg*-dependent, it was initially suggested that *ptc* may function in the anterior of the segment to interfere with the *wg* signal. Thus, the polarity of *en* activation by *wg* would be maintained. This does not appear to be the case, however, because in embryos in which ectopic expression of *ptc* in the posterior compartment is induced, the expression pattern of both *wg* and *en* is normal, and the embryos develop into normal larvae. In addition, *ptc* mutants can be rescued if ubiquitous expression of *ptc* is induced throughout its normal time of function (Ingham *et al.*, 1991; Sampedro and Guerrero, 1991). If *ptc* were acting to suppress induction of *en* by *wg* in anterior cells, one would expect that ectopic *ptc* expression in the posterior compartment would render these cells incapable of responding to the *wg* signal and that *en* expression would fade as it does in *wg* mutants. These data do suggest, however, that *ptc* functions to repress *wg* transcription in the anterior half of the segment and that ectopic *en* expression in *ptc* mutants is a response to the widened domain of *wg* expression.

The signal transduction mechanism through which the *ptc* transmembrane protein exerts its influence on *wg* transcription has not been determined. As discussed earlier, *ptc* expression throughout development overlaps with that of *wg* and yet it selectively represses *wg* expression only in cells anterior to the normal *wg* domain of expression. The fact that *ptc* can be expressed uniformly within the segments without adversely affecting development suggests that *ptc* does not function to specify positional information within segments. However, its apparent selective function within cells that occupy different positions within segments implies that it may function to interpret positional information established by other factors that exhibit regionalized expression. Ingham *et al.* (1991) have suggested that the differential function of *ptc* across the segment may be mediated directly through the function of the *hh* gene. Several lines of evidence suggest that the *hh* protein may be the signal from *en*-expressing cells that antagonizes *ptc* function in *wg*-expressing cells; *hh* has been shown to function in *en*-expressing cells, and it is expressed in posterior compartment cells in the embryo (Mohler, 1988; Mohler, cited in Ingham, 1991). In *hh* mutants *wg* expression decays as it does in *en* mutants (Hidalgo and Ingham, 1990), but in *hh, ptc* double mutants, *wg* expression is maintained in its normal domain and expands anteriorly as occurs in *ptc* mutants. Thus, loss of *ptc* function eliminates the dependence of *wg* expression on *hh* (Fig. 3).

This discussion has focused on the regulation and function of the genes *wg, en, ptc, nkd,* and *hh* because they are involved in the best characterized pathway of determination of segmental pattern and polarity. However, there are a number of other genes that may participate in this or in parallel pathways. For example, many of the segment polarity genes appear to function in patterning of the posterior region of every segment. Mutations in a number of these genes (e. g. *dishevelled* and *armadillo;* Perrimon and Mahowald, 1987; Klingensmith *et al.*, 1989; Peifer *et al.*, 1991) cause a phenotype similar to that caused by the *wg* mutation, and the corresponding gene products may function in transduction of the *wg* signal. Another posterior gene, *gooseberry,* is unique among the segment polarity genes in that its phenotype correlates well with its expression pattern (Baumgartner *et al.*, 1987; Cote *et al.*, 1987). *gooseberry* has been shown to specify positional information within the segment by functioning as a molecular switch to specify the fate of cells in the posterior half of the segment (A. Ungar and R. Holmgren, personal communication).

III. Specification of Cell Fate in the Embryonic and Adult Nervous Systems

A. General

The neurons of the central nervous system (CNS) and the sensory organs of the peripheral nervous system (PNS) of both *Drosophila* larvae and adults are laid out in reproducible patterns. Both the neuroblasts (NBs), which are the CNS precursors, and the sensory mother cells (SMCs), which are the PNS precursors, are derived from ectodermal cells (Poulson, 1950; Hartenstein and Campos-Ortega, 1984; Hartenstein and Posakony, 1989). The NBs are derived from ventral ectodermal tissue known as the neuroectoderm (NR) (Poulson, 1950; Hartenstein and Campos-Ortega, 1984). Cell ablation experiments (Doe and Goodman, 1985) have shown that in insects all cells within the NR have the potential to become NBs; however, only a small fixed number of cells do so while the remaining cells adopt an epidermal fate. During formation of the embryonic nervous system, the NBs delaminate away from the NR and migrate into the interior of the embryo between the ectoderm and mesoderm (Poulson, 1950). The establishment of the pattern of NBs within the neurogenic region was shown to be determined through positional information and cell-to-cell interactions by cell ablation experiments (Doe and Goodman, 1985) in the grasshopper (in which CNS development is homologous to that in *Drosophila;* Thomas *et al.*, 1984) and cell transplantation experiments in *Drosophila* (Technau and Campos-Ortega, 1986). NBs can be distinguished from other cell types by their morphology, e.g., they are enlarged relative to surrounding cells. When an NB is killed by laser ablation just as it is beginning to enlarge, an adjacent cell takes its place. From this experiment it was concluded that more than one cell within a particular region has the capacity to make an NB, but once a cell is selected as an

NB, it inhibits surrounding cells from differentiating as NBs (Doe and Goodman, 1985). Loss of an NB releases the surrounding cells from this inhibition, which has been named lateral inhibition.

Similarly, in the adult PNS, a small number of sensory organs (SO) are selected from a field of epidermal cells, all of which have the capacity to develop as neuronal. This was shown through studies, such as those by Wigglesworth (1940) on bristle (a type of sensory organ) spacing in the insect *Rhodnius*. He determined that there must always be a minimal distance between bristles and that they only arise if the distance between bristles exceeds a maximal limit. This led to models suggesting competition among cells for a neural-promoting factor combined with inhibition imposed by the neural cell on surrounding cells. Genetic analysis has led to the identification of two classes of genes that are required to establish the correct pattern of NBs and SMCs in the CNS and PNS, respectively. The proneural genes are required for cells to differentiate as neuronal (Stern, 1954; Garcia-Bellido and Santamaria, 1978; Garcia-Bellido, 1979; Ghysen and Dambly-Chaudiere, 1988; Jimenez and Campos-Ortega, 1987; for reviews see Ghysen and Dambly-Chaudiere, 1988, 1989; Jan and Jan, 1990), whereas genes of the neurogenic class appear to oppose neurogenesis (Poulson, 1937; Lehmann *et al.*, 1981, 1983; Shannon, 1973; Perrimon *et al.*, 1984; LaBonne and Mahowald, 1985). In embryos mutant for any of the neurogenic genes, cells within the neurogenic region that would normally follow the ectodermal pathway instead follow a neural pathway resulting in a larva with a hyperplasia of neural tissue at the expense of epidermis. This suggests that the neurogenic genes may function in the lateral inhibition process (Lehmann *et al.*, 1983; for reviews see Campos-Ortega, 1988, 1991; Simpson, 1990). Molecular analysis of some of the neurogenic genes indicates that they have structures consistent with a function in cell-to-cell communication and thus could likely act directly in lateral inhibition (Wharton *et al.*, 1985; Kidd *et al.*, 1986; Vassin *et al.*, 1987; Knust *et al.*, 1987a; Kopczynski *et al.*, 1988).

The selection of neural cells is thought to occur progressively. First, it appears that small clusters of cells, called proneural clusters, are selected in specific patterns within ectodermal tissue, and all the cells within one cluster are competent to form a particular NB or SMC. In the embryonic CNS and PNS, establishment of the proneural cluster pattern is controlled by the segmentation genes, which function along the AP axis, the DV genes (Martin-Bermudo *et al.*, 1991; Skeath *et al.*, in preparation), and perhaps the homeotic genes. Second, within each proneural cluster, one or two cells are chosen to become a neural precursor, and the remaining cells follow the epidermal pathway of development. The way in which the NB or SMC is selected is not well understood, nor is the mechanism of lateral inhibition by which the NB or SMC inhibits surrounding cells from becoming neuronal precursors (for reviews see Ghysen and Dambly-Chaudiere, 1989; Jan and Jan, 1990; Simpson, 1990; Campos-Ortega, 1988, 1991; Campos-Ortega and Jan, 1991).

B. Embryonic CNS

Genetic and molecular studies have been used to analyze the functions of the proneural and neurogenic genes in the development of the CNS. Transcripts from the proneural genes of the *achaete-scute complex* (ASC) encode structurally similar proteins that have a basic helix-loop-helix motif (Villares and Cabrera, 1987; Murre *et al.*, 1989). This motif is found in a number of transcription factors, which suggests that the genes of the ASC function in regulation of gene expression. The expression patterns of several members of the ASC have been shown to presage the specification of neural cells in the CNS and PNS and, hence, provide excellent markers with which to follow the events involved in the segregation of NBs (Cabrera *et al.*, 1987; Skeath and Carroll, 1992). For the sake of simplicity, we will describe the expression pattern of only one member of the ASC, *achaete* (*ac*). Other members of the ASC may be expressed in different spatial patterns, but the temporal dynamics of their expression parallel that of the *ac* gene. The *ac* protein is initially expressed in a segmentally repeated pattern of clusters of 5–7 cells within the NR (Figs. 4D and G) (Skeath and Carroll, 1992). These clusters of *ac*-expressing cells most likely represent the equipotential fields of cells that are competent to form particular NBs. A little later in development, *ac* expression becomes restricted to a single cell within each cluster, the NB, which enlarges and delaminates away from the ectoderm (Figs. 4E and H). Thus, it appears that the loss of expression of *ac* and other members of the ASC in the other cells of the cluster correlates with their loss of neuronal competency. Extinction of *ac* expression in epidermal cells is most likely a result of the inhibitory signal sent from the NB; therefore, *ac* can be used as a marker to visualize the consequences of lateral inhibition at the molecular level.

As mentioned, the neurogenic genes are believed to function in the lateral inhibition process. The functions of the neurogenic genes in the CNS can be observed through their effect on the expression of the ASC genes during neural specification. Expression of *ac* has been found to be altered in embryos mutant for the neurogenic genes (Skeath and Carroll, 1992). In neurogenic mutants the initial pattern of *ac* expression is normal, but its later restriction to the neuroblast does not occur; rather all cells of the cluster continue to express *ac*, enlarge, and delaminate (Fig. 4F). Thus, mutations in the neurogenic genes short-circuit the process of lateral inhibition, resulting in the continued expression of *ac* (and presumably the other ASC genes) within all cells of the cluster, which correlates directly with their adoption of a neural fate.

C. Adult PNS

The cuticle of the *Drosophila* adult has a fixed pattern of sensory organs. The ASC has been shown to be required for the formation of most external sense organs found on the cuticle of the adult. Mutations in *ac* or *sc*, for example, cause the loss of specific SOs, and animals doubly mutant for *ac* and *sc* lack most SOs (Stern, 1954; Garcia-Bellido and Santamaria, 1978; Garcia-Bellido, 1979; Dambly-Chaudiere and Ghysen, 1987; Ghysen and Dambly-Chaudiere, 1988, 1989; Jan and Jan, 1990). The primordia of the adult structures, such as wings, eyes, thoraces, and legs, are derived from 10–50 embryonic cells, which in the larva give rise to eipthelial pouches known as imaginal discs. The different imaginal discs are distinctive in size and shape, and during metamorphosis each undergoes extensive morphogenesis to produce a particular adult body part. The imaginal discs can be isolated during larval development and stained with probes that detect specific developmental gene products to visualize their changing expression patterns. The expression patterns of the *ac* and *sc* genes in the wing discs, which give rise to the wings and thorax, correlate to the regions from which specific SOs arise, as deduced from wing disc fate maps and the use of specific SO markers (Romani *et al.*, 1989; Skeath and Carroll, 1991; Cubas *et al.*, 1991). Because of cross-regulatory interactions between *ac* and *sc*, the two genes are expressed in identical patterns within the developing wing disc (Martinez and Modollel; 1991; Skeath and Carroll, 1991). How the patterns of *ac* and *sc* expression in imaginal discs are initially established is not known.

The temporal pattern of expression of *ac* and *sc* during imaginal development parallels that observed within the embryonic CNS. However, unlike the situation in the CNS, there are no specialized neural regions in the imaginal discs; rather, the sensory organs arise from specific sites scattered throughout the ectoderm. Similar to what has been observed in the embryonic CNS, *ac* and *sc* are initially expressed in clusters of cells in regions from which SOs arise (Figs. 4A and B) (Skeath and Carroll, 1991; Cubas *et al.*, 1991). These clusters of *ac*/*sc*-expressing cells most likely represent fields of equipotential cells that are competent to form particular SMCs. In the imaginal discs, as in the embryo, *ac*/*sc* expression is quickly restricted from its initial expression in cell clusters to one or two cells, the SMCs (Fig. 4C) (Skeath and Carroll, 1991; Cubas *et al.*, 1991). Only the neural precursor, the SMC, retains high levels of *ac* and *sc*, and loss of *ac* and *sc* expression in the other cells of the cluster is correlated with a loss

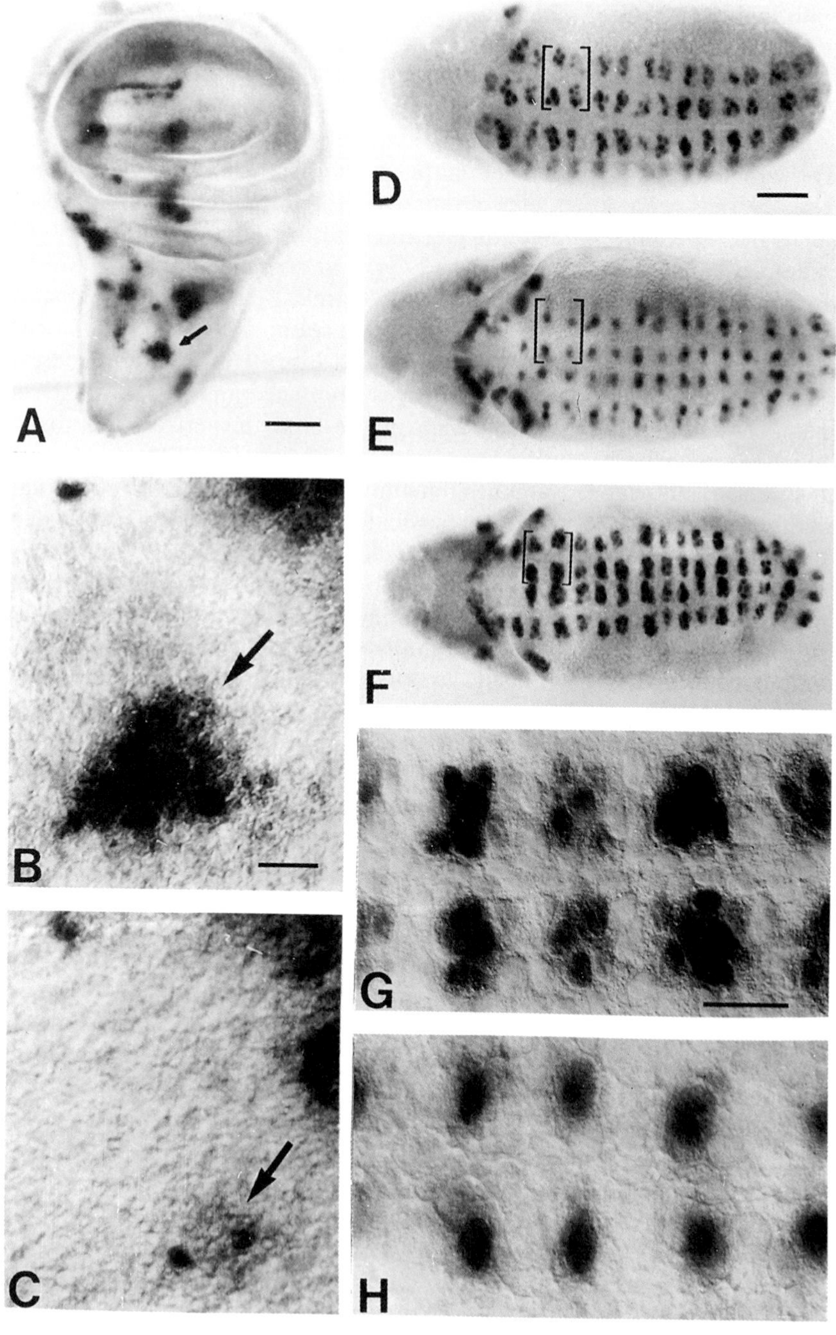
A
B
C
D
E
F
G
H

of neural competency. Thus, as in the CNS, the patterns of *ac* and *sc* expression reflect the molecular events that foreshadow the morphological changes associated with neurogenesis.

The loss of neural competency from the cells in proneural clusters within the adult promordia is thought to be mediated by a lateral inhibitory process triggered by the SMC (for reviews see Ghysen and Dambly-Chaudiere, 1989; Jan and Jan, 1990; Simpson, 1990). The neurogenic genes that function in this process in the CNS are also thought to function similarly in the adult epidermis. This has been demonstrated through the use of a technique called clonal analysis, which is used to assay the function of genes, such as the neurogenic genes, that function more than once in development. The embryonic lethality of the neurogenic genes precludes the use of standard genetics to examine mutant phenotypes within the adult cuticle. This method involves generation of a patch of genetically mutant cells, i.e., a mosaic, in otherwise wild-type tissue. This allows one to determine the focus of function of the gene within a developmental field. A widely used method to generate mosaics is mitotic recombination. Mitotic recombination is induced by X-irradiating animals at various developmental stages. At a low frequency, this will induce exchange between the chromatids of homologous chromosomes in a mitotic cell. During chromosomal segregation, one of the daughter cells may receive two copies of the mutant chromosome, making it a homozygous

---

FIG. 4 *ac* expression in imaginal discs and the embryonic CNS. (A) *ac* expression in a wing disc of the third instar larva. The clusters of *ac*-expressing cells represent equipotential fields of cells that are competent to form SMCs. The arrow points to a cluster called the dorsal central cluster from which two SMCs are derived, the anterior and posterior dorsal central bristles. (B) Magnification of the dorsal central cluster shown in A. (C) By puparium formation, the cluster of *ac*-expressing cells shown in A and B has refined to two cells, the presumptive SMCs. (D) *ac* expression in an embryo undergoing germband extension (ventral view). There are eight clusters of 5–7 *ac*-expressing cells in every segment. The bracketed region is a hemisegment. The two pairs of posterior clusters in every segment are in the posterior compartment, within the *en* region of expression. (E) The clusters of *ac*-expressing cells are refined to one cell in this embryo, which is almost at full germband extension (ventral view). (F) *ac* expression in a neurogenic mutant, *neuralized* (most of the neurogenic mutants have identical phenotypes and the expression of *ac* in all neurogenic mutants at this stage of development is essentially identical), at the same stage as that shown in F (ventral view). The clusters fail to refine to one cell as occurs in wild-type embryos, (G) High magnification view of the *ac*-expressing clusters in two hemisegments (four clusters per hemisegment) in the mid-germband extended embryo. (H) High magnification view of most of two hemisegments in an embryo, which is almost at full germband extension showing that the proneural clusters have refined to a single *ac*-expressing cell. In all figures anterior is to the left. A, bar, 50 μm. B,C, bar, 10 μm. D,E,F, bar, 50 μm. G,H, bar, 20 μm. (Courtesy of Jim Skeath.)

mutant. Genotypically mutant cells can be unambiguously identified by using a chromosome that has a cell autonomous marker linked to the mutant gene of interest.

Mosaic analysis with the neurogenic mutations, [specifically, *Notch* (*N*) and *Delta* (*Dl*)], shows that their functions are required in the cells that express *ac* and *sc* (Heitzler and Simpson, 1991). Clones of cells mutant for these genes on the thorax form extopic bristles within the regions where single wild-type bristles normally occur. Formation of ectopic bristles is dependent upon *ac/sc* function. This is consistent with the model that clusters of ASC-expressing cells are competent to form neuronal structures and that as one cell is selected it inhibits the others from following suit. Loss of neurogenic gene function causes a breakdown in this process, and cells of the cluster that normally would become epidermal adopt neuronal fates.

D. Possible Modes of the Function of Neurogenic Genes

A number of the neurogenic genes have been cloned and some of them encode proteins, the structures of which are consistent with a direct function in signal transduction during lateral inhibition. The genes *N* and *Dl* both encode transmembrane proteins with multiple epidermal growth factor (EGF) repeats in their extracellular domains (Wharton *et al.*, 1985; Kidd *et al.*, 1986; Vassin *et al.*, 1987; Knust *et al.*, 1987a; Kopczynski *et al.*, 1988). Mosaic analysis has also been used to determine whether the functions of *N* and *Dl* are cell autonomous, i.e., whether their domains of function are confined to their region of expression (cell autonomous function) or whether it can exert an influence outside this region (cell nonautonomous function). Autonomous function would be expected of a gene that encodes a protein that receives or transduces a signal within a cell, for example, a transmembrane protein that functions as a receptor or an intracellular kinase, the activity of which is triggered by a reaction at the cell surface. In contrast, a gene that is involved in the transmission of a signal to another cell should be nonautonomous, for example, a transmembrane protein that transmits a signal by direct interaction with the membrane component of an adjacent cell or a diffusible factor released by one cell and that is taken up by another cell. Mosaics made by one method, mitotic recombination (Section III,C), were used to assay the autonomy of *N* and *Dl* in adults (Heitzler and Simpson, 1991). In this experiment *N* was determined to be cell autonomous. In patches of cells that are homozygous mutant for *N*, all the cells exhibit the *N* phenotype (i.e., they make bristles); thus, these cells could not be rescued by surrounding cells that produce wild type *N*. In addition, no cells outside the mutant patches of tissue exhibited the mutant phenotype. This result suggests that *N* is cell autonomous and functions to receive or transduce signals from adjacent cells. In *Dl* clones, however, cells in the center made bristles, but cells mutant for *Dl* bordered by wild-type cells were epidermal, indicating that they were rescued by surrounding wild-type cells, and that *Dl* function is, therefore, nonautonomous. Thus, *Dl* probably functions in transmission of signals to adjacent cells.

A second method to test for cell autonomy involves the generation of mosaics in the embryo by cellular transplantation (Technau and Campos-Ortega, 1986). Cells from the neurogenic region of an embryo mutant for one of the neurogenic genes were transplanted into the neurogenic region of a wild-type embryo. If a gene functions autonomously, one would expect that transfer of a mutant cell into wild-type tissue would not cause it to differentiate differently than it would have in the mutant context. One would not expect its phenotype to be rescued by the surrounding wild-type tissue, nor would one expect it to induce aberrant development of the surrounding wild-type tissue. In these experiments when single *N* or *Dl* cells were transplanted, cells of both mutant genotypes were often able to differentiate into epidermal cells. This suggests that *N* and *Dl* function cell nonautonomously because mutant cells can be rescued by surrounding wild-type cells. For *N* this result is the converse of that obtained by mitotic recombination.

A possible interpretation for these contradicting results has been suggested by Heitzler and Simpson (1991). As mentioned, genetic analysis, cell transplantation, and laser ablations suggest that all cells within the neurogenic region are competent to form NBs. This raises the question as to whether at some point early in development, the neurogenic region is comprised of tightly packed proneural clusters (defined by expression of the proneural genes such as *ac*). Simpson (1991) suggested that this may not be the case but, instead, that the proneural clusters arise at different times during neurogenesis and that all cells are not competent to form NBs at the same time. Thus, if a neurogenic mutant cell were transplanted into a wild-type region at a time when it was no longer competent to form NBs, it would differentiate into an epidermoblast.

Subsequent mosaic analysis has also addressed the possibility that cells are sensitive to the dosage level of *N* and *Dl* (Heitzler and Simpson, 1991). Mosaics were generated in which the clonal tissue was wild-type for *N* but differed in the number of copies of the N^+ gene from the adjacent tissue. At the interface between tissues that had different copy numbers of the N^+ gene, bristles were always produced from cells that had lower gene dosages of *N*. Similar experiments with *Dl* showed that cells with a lower copy number of the *Dl* gene always make epidermal cells.

These results led Heitzler and Simpson to propose a model in which *N* and *Dl* are involved in the process of singling out the neural precursors as well as functioning later in lateral inhibition (Fig. 5). In this model the

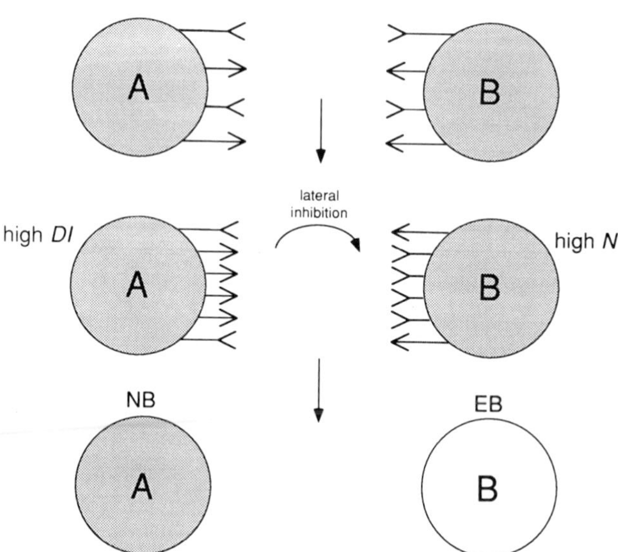

FIG. 5 Diagram of Heitzler and Simpson model for selection of neural cells and lateral inhibition (after Heitzler and Simpson, 1991). ⅄ represents the *N* protein; ↗ represents the *Dl* protein. The shaded circles are cells that express *ac*. All cells in a proneural cluster express *N* and *ac;* however, random fluctuations can result in one cell expressing slightly lower levels of *N* (cell A). This leads to higher levels of the putative neural inhibitor, *Dl,* in cell A. Cells can sense the levels of *N* and *Dl* in surrounding cells, and this leads to amplification of the differences in the levels of *N* and *Dl* in nieghboring cells. Higher levels of *N* in cell B cause it to be effectively inhibited from following the neural pathway of development by cell A. Loss of *ac* expression by cell B is evidence at the molecular level that lateral inhibition has taken place.

neural precursor, which can sense the level of *N* in adjacent cells, is selected through competition between cells (Fig. 5). The model suggests that although all cells in a proneural cluster express *N,* slight fluctuations can occur, resulting in one cell having lower levels of *N* than adjacent cells. This lower level of *N* expression would lead the cell to produce more inhibitory signal relative to surrounding cells. *N* could function directly to receive the inhibitory signal, and this would occur most effectively in cells that express higher levels of *N* relative to surrounding cells. Amplification of these differences over time would allow a particular cell to adopt a neural fate and to most effectively inhibit other cells from adopting the neural fate. The nonautonomous function of *Dl* and the fact that cells that express more *Dl* than surrounding cells develop as neuronal are consistent with a function for *Dl* in acting as or mediating the inhibitory signal. Biochemical data support the idea of direct interactions between *N* and

Dl; it has been shown in tissue culture that cells that express *N* and *Dl* can bind each other specifically (Fehon *et al.*, 1990).

This model is attractive in that it suggests a mechanism for both the initial selection of the neural precursor and lateral inhibition. However, it does not address the possible roles of the other neurogenic genes, such as *neuralized, mastermind, enhancer of split,* and *big brain* (Lehmann *et al.*, 1981, 1983). Genetic analysis of the interactions between the neurogenic genes suggests that they all function in a common pathway with the exception of *big brain* (Campos-Ortega *et al.*, 1984; Vassin *et al.*, 1985; de la Concha *et al.*, 1988). These studies further suggest that the *neuralized* and *mastermind* genes, both of which encode nuclear proteins (Boulianne *et al.*, 1991; Smoller *et al.*, 1990), function before *N* and *Dl* and that *Enhancer of split,* a gene complex that encodes multiple proteins with helix-loop-helix motifs (Knust *et al.*, 1987b; Klambt *et al.*, 1989), functions at the end of the pathway. To understand the molecular basis for cellular events of neural precursor selection and lateral inhibition, the functions of the neurogenic gene products and interactions between them will have to be studied at the cellular and biochemical levels.

IV. Specification of Cell Fate in the *Drosophila* Eye

A. General

The compound eye of *Drosophila* is made up of approximately 800 identical units called ommatidia (Fig. 6). Each ommatidium consists of eight photoreceptor cells (R1–R8) and 12 accessory cells (Waddington and Perry, 1960; Ready *et al.*, 1976). The eye is derived from the eye portion of the eye–antennal imaginal disc, which is comprised of a single epithelial cell layer. Differentiation of this epithelial tissue into highly organized ommatidial arrays begins during the third larval instar. Morphogenesis and differentiation are closely associated with the morphogenetic furrow, a depression in the eye disc that traverses the disc from the posterior margin to the anterior margin. Behind the furrow cells are arranged in a precise pattern of ommatidial units (Fig. 6B). Anterior to the furrow, relatively little organization is observed (Ready *et al.*, 1976; Tomlinson, 1985; Tomlinson and Ready, 1987a,b).

The organization of the eye is reflected not only in the arrangement of cells within each ommatidial unit but also in the precise spatial arrangement of the ommatidia themselves (Fig. 6B). Several genes have been identified that are required for the spacing of the ommatidia (Baker *et al.*, 1990; Baker and Rubin, 1989; Cagan and Ready, 1989; Renfranz and

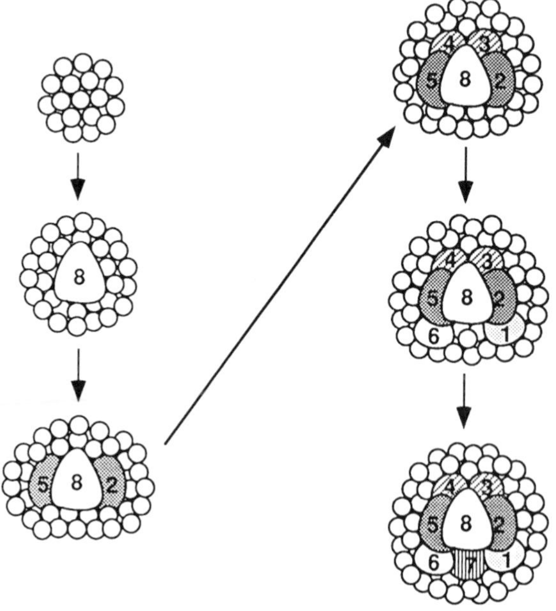
8
5
2
4
3
6
1
7
A

B

Benzer, 1989; Karpilov *et al.*, 1989). Mutations in these genes increase or decrease the distances between the ommatidia and disrupt ommatidial structure. The photoreceptor cell, R8, is the first cell to differentiate, and its specification is closely tied to ommatidial spacing (Tomlinson and Ready, 1987a,b). Antibodies that detect R8-specific antigens show that at very early stages of ommatidial assembly (before other photoreceptor cells can be identified), the presumptive R8 cells are precisely spaced relative to one another. It has been suggested that lateral inhibition by R8 on surrounding cells is involved in establishing and maintaining appropriate spacing (Baker *et al.*, 1990; Mlodzik *et al.*, 1990). Interestingly, one of the genes that has been shown to be required for appropriate spacing is the *Notch* gene (Baker and Rubin, 1989), which is required for lateral inhibition in the CNS and PNS.

Through morphological studies of developing ommatidia combined with the use of neuronal-specific antibodies, it was determined that the fates of photoreceptor cells R1–R8 are specified sequentially (Fig. 6A) (Tomlinson and Ready, 1987a,b). After R8 the next cells to be specified are R2 and R5, then R3 and R4, followed by R1 and R6, and finally R7. R2 and R5 contact R8; R3 and R4 contact both R8 and R2 or R5, and R7 contacts R8, R1, and R6. It was initially thought that the ommatidial pattern was lineage-dependent; however, it has been shown that the specification of each photoreceptor cell depends upon its position within the ommatidial field and its contacts with other R cells (Ready *et al.*, 1976; Lawrence and Green, 1979). This was determined partly through the use of mosaic analysis in the eye to mark the precursors of the photoreceptor cells. The mosaic tissue in these experiments was marked with a cell autonomous marker that did not alter eye development. If the establishment of ommatidial pattern was dependent on lineage rather than position, one would expect that the observation of many different mosaic ommatidia would reveal a precise pattern of marked cells, showing that specific R cells are always derived from a particular ancestor. This was not the case, however; ommatidial cells were always randomly marked. This study combined with the morphological studies that showed that the R cells always arise from a specific position in the ommatidium and that as development proceeds

---

FIG. 6 (A) Diagram showing the progressive differentiation of photoreceptors R1–R8 in developing ommatidium. (B) Eye imaginal disc stained with 22C10, a neuronal specific antibody. (Courtesy of Nadean Brown and Steve Paddock.) This shows the organization of the ommatidia posterior to morphogenetic furrow. Anterior to the furrow, little organization is observed. The insert is a closeup showing individual ommatidia. Anterior is up. B, bar, 50 μm.

new cells added to the ommatidium make specific contacts with existing differentiating cells led to the proposal that the fate of R cells is established by their position within the ommatidium and through the specific cell-to-cell contacts they make (Tomlinson and Ready, 1987a,b). Unfortunately, it is not possible to perform cell ablation studies in the *Drosophila* eye discs as in the grasshopper embryonic CNS or transplantation studies as in the *Drosophila* embryonic CNS to directly show that cell contacts are crucial to the establishment of ommatidial pattern. This was demonstrated through the use of genetic screens to find mutations that disrupted cell-to-cell communication and thereby altered the fate of specific ommatidial cells (for reviews see Tomlinson, 1988; Basler and Hafen, 1988b; Ready, 1989; Banerjee and Zipursky, 1990; Hafen and Basler, 1991; Hafen, 1991). This discussion will focus on the specification of the fate of one photoreceptor cell, R7, because it is an elegant example of a case in which a genetic approach has been used to dissect a complex signal transduction pathway.

B. Specification of the Fate of R7

Through genetic screens to identify genes that function in specification of R7 cell fate and analysis of the function of the corresponding gene products, it has been determined that the specification of the R7 fate occurs through an inductive signal from the R8 photoreceptor cell, one of the cells R7 contacts (Fig. 7). Thus, the model suggesting that position and combination of cell contacts specify ommatidial cell fate is at least partially correct for the specification of R7, which requires at least one inductive signal from another ommatidial cell, R8. In addition to R8, R7 contacts both R1 and R6, although it has not been determined whether these two cells play a role in induction of R7 fate. Several mutations have been identified that specifically affect the development of the R7 cell (*sevenless, bride of sevenless, seven in absentia,* and *Son of sevenless*; Harris *et al.*, 1976; Reinke and Zipursky, 1988; Carthew and Rubin, 1990; Rogge *et al.*, 1991).

The *sevenless (sev)* mutation causes the cell that would normally develop as an R7 cell to adopt the fate of an accessory cell (Tomlinson and Ready, 1986). Mosaic analysis has shown that *sev* functions autonomously within the R7 cell, suggesting that it may be involved in the reception or transduction of a signal within the R7 cell (Campos-Ortega *et al.*, 1979; Tomlinson and Ready, 1987a,b). Consistent with this idea, *sev* encodes a transmembrane protein with a cytoplasmic C-terminal domain that has strong similarity to tyrosine kinases and a large extracellular domain (Hafen *et al.*, 1987; Basler and Hafen, 1988a; Bowtell *et al.*, 1988). The *sev*

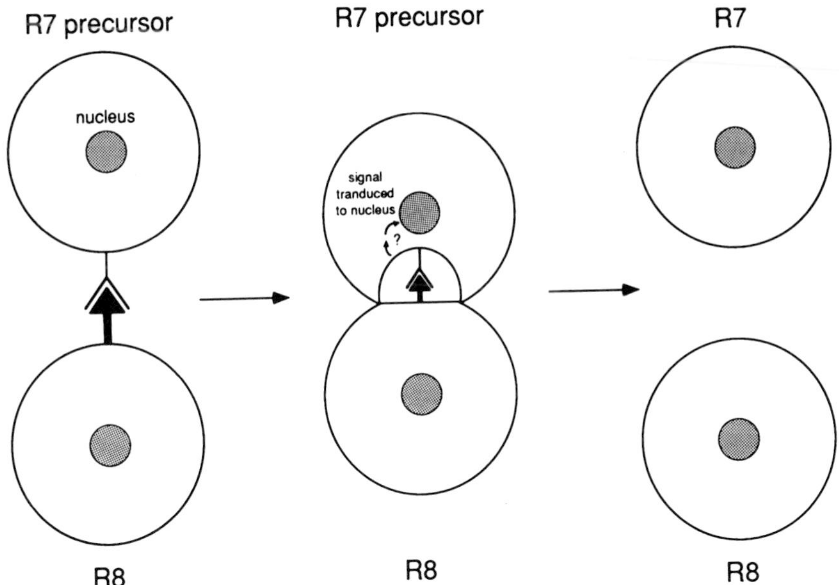

FIG. 7 Diagram showing the interactions between R8 and the R7 precursor that lead to specification of R7 cell fate. ⊬ represents the *sev* transmembrane tyrosine kinase, ➚ represents the boss transmembrane protein. The boss protein on the surface of the R8 cell interacts with *sev*, its receptor on the membrane of the R7 precursor. This activates a signal transduction pathway that leads to the specification of the R7 cell fate. Antibodies against boss show that it is internalized within R7. It is not known whether internalization of R8 is an essential part of the signal transduction pathway.

protein is associated with the cell membranes, but it is not expressed exclusively within R7, as might be expected. Rather it is expressed in R1, R3, R6, and R7. It is expressed at high levels in R3 and R4, then in R7, and later in certain accessory cells (Banerjee *et al.*, 1987; Tomlinson *et al.*, 1987). Function of the tyrosine kinase domain is sufficient to direct an ommatidial cell to assume an R7 fate. This was shown by overexpression of an N-terminal truncated *sev (sev-S11)* protein, in which most of the extracellular domain is missing, within those cells that normally express *sev* (Basler *et al.*, 1991). This results in a constitutively active protein that causes the transformation of certain non-R7 cells into R7. This transformation appears to be independent of ligand because when *sev-S11* is expressed in a *bride of sevenless (boss,* the putative ligand for *sev,* discussed next) mutant background, the same effect is observed and because most of the extracellular domain is missing in *sev-S11*. The fact that the R7 fate can be induced by *sev-S11* protein alone in cells that do not contact R1 and R6 suggests that if R1 and R6 sent inductive signals to R7, they must go through the *sev* pathway.

A mutation in the *boss* gene also causes loss of the R7 cell (Reinke and Zipursky, 1988). Mosaic analysis with *boss* mutations shows that it functions nonautonomously; if *boss* is mutant in R8 within a particular ommatidium, R7 will be missing. If the *boss* gene is mutant in R1–R7, the R7 cell will be wild-type (Reinke and Zipursky, 1988). Thus, boss is required in R8 for appropriate specification of the fate of R7, which is consistent with a function for *boss* in transmitting an inductive signal from R8 to R7. Correspondingly, *boss* has been shown to encode a transmembrane protein with a large extracellular domain (Hart *et al.*, 1990). This implies that *boss* might be the ligand for the *sev* receptor. Recent data suggest that this may indeed be the case. Staining of eye imaginal discs with antibodies specific for the *boss* protein shows that *boss* expression is restricted to R8 and is found on the apical surface of R8 (Kramer *et al.*, 1991). In addition, *boss* protein can be detected within large vesicles in R7. Internalization of *boss* into R7 is not observed in a *sev* background but in eyes homozygous for *sev* alleles that affect the tyrosine kinase domain, but not the extracellular domain, internalization is normal. This implies that the *sev* extracellular domain is specifically required for internalization of *boss*. Consistent with these data, it has been demonstrated that tissue culture cells expressing *boss* and *sev* aggregate to each other specifically. It has not been determined whether internalization of *boss* is an essential feature of the signal transduction pathway that activates R7 cell fate; however, the transformation of non-R7 cells into R7 by the ligand-independent activity of the constitutively active *sev-S11* protein suggests that internalization of *boss* is not required.

V. Conclusion

We have discussed three model systems of *Drosophila* in which cell-to-cell interactions are required for differentiation of specific cell types. The elucidation of the mechanisms by which cell-to-cell interactions lead to specification of cell fate requires that the circuitry of the signal transduction pathways involved be deciphered. Through genetic and molecular analyses, which have paved the way for biochemical and cell-biological studies, some of the molecules that directly mediate cell-to-cell interaction have been identified. For example, development biologists have long sought to identify the molecules that participate in the process of induction of specific tissue or cell types. The genetic analysis of the specification of cell fate in the eye has led to the identification of the R8-specific transmembrane protein *boss,* which interacts with its receptor, *sev,* in the R7 precursor cell and sets off a cascade of events that culminate in the induction of the R7 cell fate (Fig. 7) (Kramer *et al.,* 1991). The *sev* gene encodes a tyrosine kinase (Hafen *et al.,* 1987; Basler and Hafen, 1988a; Bowtell *et al.,* 1988). A number of vertebrate receptor tyrosine kinases are involved in signal transduction pathways important for normal cell proliferation and, in some cases, cell differentiation. Although the *sev* protein differs from other receptor tyrosine kinases in its extracellular domain, its intracellular kinase domain shares substantial similarity to the kinase domains of other receptor tyrosine kinases. Thus, it is interesting to speculate that they may participate in similar signal transduction pathways. In addition, there may be other as yet unidentified vertebrate receptor tyrosine kinases similar to *sev* that function in pattern-forming processes.

Identifying the components of the normal and oncogenic pathways of vertebrate receptor tyrosine kinases has proven difficult partially because these proteins may participate in multiple processes. The discovery of the involvement of a receptor tyrosine kinase in a genetically accessible pathway, which when disrupted causes a specific phenotype, should greatly simplify the identification of other participants. Indeed, two possible downstream components of the *sev* pathway have been identified through genetic screens. One of these genes, *seven in absentia,* encodes a nuclear protein and mutations in it also produce the *sev* phenotype (Carthew and Rubin, 1990). This gene may function at the end of the *sev* pathway to regulate gene expression that leads to R7 determination. Another of these genes, *son-of-sevenless (sos),* was identified through a screen for dominant mutations that suppress the *sev* phenotype; further genetic analysis of this gene suggests that it may interact directly with *sev* (Rogge *et al.,* 1991). Interestingly, null mutations in *sos* cause recessive embryonic lethality, indicating that it probably functions in multiple signal transduction pathways. The approach used to identify *sos,* suppressor–enhancer genetics, led to the isolation of an allele of *sos* that appears to specifically alter its function in the *sev* pathway. This dominant *sos* allele suppresses the phenotype of only one *sev* allele. This approach may be generally useful in identifying components of specific signal transduction pathways that may be widely distributed or that participate in multiple processes.

Suppressor–enhancer genetics may also be useful in the identification of components of other pathways that function in pattern formation, such as the establishment of segmental pattern. Cloning of a number of the segment–polarity genes has led to the identification of a number of genes that encode proteins, which could function in multiple signaling pathways, such as the maternal gene products of *zeste white-3* and *fused,* both of which encode serine threonine kinases (Siegfried *et al.*, 1990; Bourouis *et al.*, 1990; Preat *et al.*, 1990) and *armadillo,* which is homologous to plakoglobin, a component of desmosomes and gap junctions (Peifer and Wieschaus, 1990). Consistent with this idea is the fact that *zeste white-3* also functions in the selection of sensory organs of the PNS (Simpson *et al.*, 1988; Bourouis *et al.*, 1989). It may be fortuitous that mutations in these genes cause specific pattern defects; thus, there may be many other as yet unidentified genes that function in the establishment of segmental pattern.

There is compelling evidence that these pathways of *Drosophila* development may provide general models for pathways of cell-to-cell interactions involved in pattern formation. For example, the model proposed by Heitzler and Simpson (1991) to explain how one cell from a proneural cluster is selected to become a neuroblast is similar to that proposed for the function of *lin-12,* a gene that is required for cell-to-cell interactions between the anchor cell (AC) and the ventral uterine cell (VU) of the *Caenorhabditis elegans* gonad (Seydoux and Greenwald, 1989; Greenwald *et al.*, 1983; Ferguson and Horvitz, 1985). Lineage studies indicate that there are two cells that can develop with equal probability into either an AC or a VU cell (Kimble and Hirsh, 1979). Laser ablation experiments have shown that specification of the fate of a cell such as VU requires an inducing signal from the AC (Kimble, 1981), and genetic analysis shows that *lin-12* is required autonomously in the VU cell to receive the signal from the AC (Seydoux and Greenwald, 1989). The model suggests that although both cells express *lin-12, lin-12* does not function until it is activated by a signal in the adjacent cell, and only one cell will produce enough signal to activate *lin-12* function in the other cell (Seydoux and Greenwald, 1989). The cell in which *lin-12* has been activated loses the capacity to produce a signal, and this cell adopts the VU fate. Consistent with the proposed similarities in their mechanisms of function, the *lin-12* and *N* gene products are structurally similar with multiple EGF repeats in their extracellular domains (Greenwald, 1985; Yochem *et al.*, 1988; Wharton *et al.*, 1985; Kidd *et al.*, 1986).

In addition, the segment polarity genes, many of which are involved in cell-to-cell interactions required for the specification cell fate within the embryonic epidermis, also function in pattern formation within the primordia of adult structures, the imaginal discs. This distinguishes them from the other classes of segmentation genes, most of which function only within the syncytial embryo. As in the segmental primordia, the segment–polarity genes are expressed in subsets of cells within imaginal discs (for review see Wilkins and Gubb, 1991). Study of the role of these genes in pattern formation within both the embryonic segment and the imaginal discs should provide an excellent opportunity to compare the function of a class of genes in setting up pattern within two outwardly dissimilar developmental fields and may lead to the formulation of general models of pattern formation.

As discussed, the genetic analysis of specification of cell fate has led to the identification of a number of molecules involved in signal transduction. Although the complete pathways have not been determined, some of the responses elicited have been identified. In many cases the responses are activation, or repression, of transcription. For example, secretion of the *wg* protein and its incorporation into adjacent *en*-expressing cells results in activation of *en* expression (DiNardo *et al.*, 1988; Martinez Arias *et al.*, 1988), and one of the results of lateral inhibition by the NB of the CNS on surrounding cells is inactivation of *ac* expression (Skeath and Carroll, 1992) (Fig. 4F). The *wg* protein also mediates posttranscriptional regulation on another member of the segment–polarity class, *armadillo* (Riggleman *et al.*, 1990; Peifer *et al.*, 1991). The *armadillo* mRNA is transcribed uniformly in all cells of the segment (Riggleman *et al.*, 1989); however, the accumulation of the *armadillo* protein appears to be directly related to the levels of *wg* protein present and is dependent on *wg* function (Riggleman *et al.*, 1990).

For all the models discussed, much has yet to be learned about the pathways involved in specifying pattern. However, the combination of genetic and molecular approaches to identifying the components of these processes should greatly facilitate their analysis at the cellular and biochemical levels.

Acknowledgments

The authors gratefully acknowledge Jim Skeath, Nadean Brown, and Steve Paddock for providing photographs for figures. The authors also thank Nadean Brown, Steve Paddock, Grace Panganiban, Jim Skeath, and Jim Williams for their helpful comments on the manuscript and Leanne Olds for help in producing figures. T. Orenic is supported by an NIH Postdoctoral Fellowship (GM-13899-01A1), and work in the Carroll Laboratory is supported by the National Science Foundation, the Shaw Scholars' Program of the Milwaukee Foundation, and the Howard Hughes Medical Institute.

References

Akam, M. (1987). *Development* **101,** 1–22.

Baker, N. (1987). *EMBO J.***6,** 1765–1773.

Baker, N. E., and Rubin, G. M. (1989). *Nature* (*London*) **340,** 150–153.

Baker, N. E., Mlodzik, M., and Rubin, G. M. (1990). *Science* **250,** 1370–1375.

Banerjee, U., and Zipursky, L. (1990). *Neuron* **4,** 177–187.

Banerjee, U., Renfranz, P. J., Pollock, J. A., and Benzer, S. (1987). *Cell* **49,** 281–291.
Basler, K., and Hafen, E. (1988a). *Cell* **54,** 299–311.
Basler, K., and Hafen, E. (1988b). *Trends Genet.* **4,** 74–79.
Basler, K., Christen, B., and Hafen, E. (1991). *Cell* **64,** 1069–1081.
Baumgartner, S., Bopp, D., Burri, M., and Noll, M. (1987). *Genes Dev.* **1,** 1247–1267.
Bopp, D., Burri, M., Baumgartner, S., Frigerio, G., and Noll, M. (1986). *Cell* **47,** 1033–1040.
Boulianne, G., de la Concha, A., Campos-Ortega, J. A., and Jan, L. Y., and Jan, Y. N. (1991). *EMBO J.* **10,** 2975–2983.
Bourouis, M., Heitzler, P., El Messal, M., and Simpson, P. (1989). *EMBO J.* **7,** 3899–3906.
Bourouis, M., Moore, P., Ruel, L., Grau, Y., Heitzler, P., and Simpson, P. (1990). *EMBO J.* **9,** 2877–2884.
Bowtell, D. D. L., Simon, M. A., and Rubin, G. M. (1988). *Genes Dev.* **2,** 620–634.
Cabrera, C. V., Martinez Arias, A., and Bate, M. (1987). *Cell* **50,** 425–433.
Cagan R., and Ready, D. F. (1989). *Genes Dev.* **3,** 1099–1112.
Campos-Ortega, J. A. (1988). *Trends Neurosci.* **11,** 400–405.
Campos-Ortega, J. A. (1991). *Int. Rev. Cyto.* **124,** 1–41.
Campos-Ortega, J. A., and Hartenstein, V. (1985). "The Embryonic Development of *Drosophila melanogaster.*" Springer-Verlag, Berlin.
Campos-Ortega, J. A., and Jan, Y. N. (1991). *Annu. Rev. Neurosci.* **14,** 399–420.
Campos-Ortega, J. A., Jurgens, G., and Hofbauer, A. (1979). *Wilhelm Roux's Arch. Dev. Biol.* **186,** 27–50.
Campos-Ortega, J. A., Lehmann, R., Jimenez, F., and Dietrich, U. (1984). *In* "Organizing Principles of Neural Development" (S. C. Sharma, ed.), pp. 129–144. Plenum, New York.
Carroll, S. B., and Scott, M. P. (1985). *Cell* **43,** 47–57.
Carthew, R. W., and Rubin, G. M. (1990). *Cell* **63,** 561–577.
Cote, S., Preiss, A., Haller, J., Schuh, R., Kienlin, A., Seifert, E., and Jackle, H. (1987). *EMBO J.* **6,** 2793–2801.
Cubas, P., de Celis, J.-F., Campuzano, S., and Modolell, J. (1991). *Genes Dev.* **5,** 996–1008.
Dambly-Chaudiere, C., and Ghysen, A. (1987). *Genes Dev.* **1,** 297–306.
de la Concha, A., Dietrich, U., Weigel, D., and Campos-Ortega, J. A. (1988). *Genetics* **118,** 499–508.
DiNardo, S., Kiner, J. M., Theis, J., and O'Farrell, P. (1985). *Cell* **43,** 59–69.
DiNardo, S., Sher, E., Heemskerk-Jongens, J., Kassis, J., and O'Farrell, P. H. (1988). *Nature (London)* **332,** 604–609.
Doe, C. Q., and Goodman, C. S. (1985). *Dev. Biol.* **111,** 206–219.
Driever, W., and Nusslein-Volhard, C. (1988a). *Cell* **54,** 95–104.
Driever, W., and Nusslein-Volhard, C. (1988b). *Cell* **54,** 83–93.
Fehon, R. G., Kooh, P. J., Rebay, I., Regan, C. L., Xu, T., Muskavitch, M. A. T., and Artavanis-Tsakonas, S. (1990). *Cell* **61,** 523–534.
Ferguson, E. L., and Horvitz, H. R. (1985). *Genetics* **110,** 17–72.
Fjose, A., McGinnis, W. J., and Gehring, W. J. (1985). *Nature (London)* **313,** 284–289.
Frasch, M., and Levine, M. (1987). *Genes Dev.* **1,** 981–995.
French, A., Bryant, P. J., and Bryant, S. V. (1976). *Science* **193,** 969–981.
Garcia-Bellido, A. (1979). *Genetics* **91,** 491–520.
Garcia-Bellido, A., and Santamaria, P. (1978). *Genetics* **88,** 469–486.
Garcia-Bellido, A., Ripoll, P., and Morata, G. (1973). *Nature (London), New Biol.* **245,** 251–253.
Ghysen, A., and Dambly-Chaudiere, C. (1988). *Genes Dev.* **2,** 495–501.
Ghysen, A., and Dambly-Chaudiere, C. (1989). *Trends Genet.* **5,** 251–255.
Gonzalez, F., Swales, L., Bejsovec, A., Skaer, H., and Martinez Arias, A. (1991). *Mech. Dev.* **35,** 43–54.

Greenwald, I. S. (1985). *Cell* **43,** 583–590.
Greenwald, I. S., Sternberg, P. W., and Horvitz, H. R. (1983). *Cell* **34,** 435–444.
Hafen, E. (1991). *Curr. Opinion Genet. Dev.* **1,** 263–274.
Hafen, E., and Basler, K. (1991). *Development, Suppl.* 1, 123–130.
Haven, E., Basler, K., Edstroem, J.-E., and Rubin, G. M. (1987). *Science* **236,** 55–63.
Harris, W. A., Stark, W. S., and Walker, J. A. (1976). *J. Physiol.* (*London*) **256,** 415–439.
Hart, A. C., Kramer, H., Van Vactor, D. L., Paidhungat, M., and Zipursky, L. (1990). *Genes Dev.* **4,** 1835–1847.
Hartenstein, V., and Campos-Ortega, J. A. (1984). *Wilhelm Roux's Arch, Dev. Biol.* **193,** 308–325.
Hartenstein, V., and Posakony, J. W. (1989). *Development* **107,** 389–405.
Heemskerk, J., DiNardo, S., Kostriken, R., and O'Farrell, P. H. (1991). *Nature* (*London*) **352,** 404–410.
Heitzler, P., and Simpson, P. (1991). *Cell* **64,** 1083–1092.
Hidalgo, A., and Ingham, P. A. (1990). *Development* **110,** 291–301.
Hooper, J. E., and Scott, M. P. (1989). *Cell* **59,** 751–765.
Ingham, P. W. (1988). *Nature* (*London*) **335,** 25–33.
Ingham, P. W. (1991). *Curr. Opinion Genet. Dev.* **1,** 261–267.
Ingham, P. W., Baker, N. E., and Martinez-Arias, A. (1988). *Nature* (*London*) **331,** 73–75.
Ingham, P. W., Taylor, A. M., and Nakano, Y. (1991). *Nature* (*London*) **353,** 184–187.
Jan, Y. U., and Jan, L. Y. (1990). *Trends Neurosci.* **13,** 493–498.
Jimenez, F., and Campos-Ortega, J. A. (1987). *J. Neurogenet.* **4,** 179–200.
Jurgens, G., Wieschaus, E., Nusslein-Volhard, C., and Kluding, H. (1984). *Wilhelm Roux's Arch. Dev. Biol.* **193,** 283–295.
Karpilov, J., Kolodkin, A., Bork, T., and Venkatesh, T. (1989). *Genes Dev.* **3,** 1834–1844.
Kidd, S., Kelley, M. R., and Young, M. W. (1986). *Mol. Cell. Biol.* **6,** 3094–3108.
Kimble, J. (1981). *Dev. Biol.* **87,** 286–300.
Kimble, J., and Hirsh, D. (1979). *Dev. Biol.* **70,** 396–417.
Klambt, C., Knust, E., Tietze, K., and Campos-Ortega, J. A. (1989). *Embo J.* **8,** 203–210.
Klingensmith J., Noll, E., and Perrimon, N. (1989). *Dev. Biol.* **134,** 130–145.
Knust, E., Dietrich, U., Tepass, U., Bremer, K. A., Weifel D., Vassin, H., and Campos-Ortega, J. A. (1987a). *EMBO J.* **6,** 761–766.
Knust, E., Tietze, K., and Campos-Ortega, J. A. (1987b). *EMBO J.* **6,** 4113–4123.
Kopczynski, C. C., Alton, A. K., Fechte., K., Kooh, P. J., and Muskavitch, M. A. T. (1988). *Genes Dev.* **2,** 1723–1735.
Kornberg, T. (1981a). *Proc. Natl. Acad. Sci. U.S.A.* **78,** 1095–1098.
Kornberg, T. (1981b). *Dev. Biol.* **86,** 363–381.
Kramer, H., Cagan, R. L., and Zipursky, L. (1991). *Nature* (*London*) **352,** 207–212.
LaBonne, S. G., and Mahowald, A. P. (1985). *Dev. Biol.* **110,** 264–267.
Lawrence, P. A. (1966). *J. Theor. Biol.* **25,** 1–47.
Lawrence, P. A. (1981). *Cell* **26,** 30–10.
Lawrence, P. A., and Green, S. M. (1979). *Dev. Biol.* **71,** 142–152.
Lawrence, P. A., and Morata, G. (1976). *Dev. Biol.* **50,** 321–337.
Lawrence, P. A., and Struhl, G. (1982). *EMBO J.* **1,** 827–833.
Lawrence, P. A., Struhl, G., and Morata, G. (1979). *J. Embryol. Exp. Mophol.* **51,** 195–208.
Lehmann, R., Dietrich, U., Jimenez, F., and Campos-Ortega, J. A. (1981). *Wilhelm Roux's Arch. Dev. Biol.* **190,** 226–229.
Lehman, R. Jimenez, F., Dietrich, U., and Campos-Ortega, J. A. (1983). *Wilhelm Roux's Arch. Dev. Biol.* **192,** 62–74.
Locke, M. (1959). *J. Exp. Biol.* **36,** 459–477.
Martin-Bermudo, M. D., Martinez, C., Rodriguez, A., and Jimenez, F. (1991) *Development* **113,** 445–454.

Martinez, C., and Modollel, J. (1991). *Science* **251,** 1485–1487.
Martinez Arias, A. (1989). *Trends Genet.* **5,** 262–267.
Martinez Arias, A., Baker, N., and Ingham, P. (1988). *Development* **103,** 157–170.
Mlodzik, M., Baker, N. E., and Rubin, G. M. (1990). *Genes Dev.* **4,** 1848–1861.
Mohler, J. (1988). *Genetics* **120,** 1061–1072.
Murre, C., McCaw, P. S., Vaessin, H., Caudy, M., Jan, L. Y., Jan, Y. N., Cabrera, C. V., Buskin, J. N., Hauschka, S. D., Lassar, A. B., Weintraub, W., and Baltimore, D. (1989). *Cell* **58,** 537–544.
Nakano, Y., Guerrero, I., Hidalgo, A., Taylor, A., Whittle, J., and Ingham, P. (1989). *Nature (London)* **341,** 508–513.
Nusslein-Volhard, C. (1991). *Development, Suppl.* 1, 1–10.
Nusslein-Volhard, C., and Wieschaus, E. (1980). *Nature (London)* **287,** 795–801.
Nusslein-Volhard, C., Wieschaus, E., and Kluding, H. (1984). *Wilhem Roux's Arch. Dev. Biol.* **193,** 267–282.
Orenic, T. V., Slusarski, D. C., Kroll, K. L., and Holmgren, R. A. (1990). *Genes Dev.* **4,** 1053–1067.
Peifer, M., and Wieschaus, E. (1990). *Cell* **63,** 1167–1178.
Peifer, M., Rauskolb. C., Williams, M., Riggleman, B., and Wieschaus, E. (1991). *Development* **111,** 1029–1043.
Perrimon, N., and Mahowald, A. P. (1987). *Dev. Biol.* **119,** 587–600.
Perrimon, N., Engstrom, L., and Mahowald, A. P. (1984). *Dev. Biol.* **105,** 404–414.
Poole, S., Kauvner, L., Drees B., and Kornberg, T. (1985). *Cell* **40,** 37–43.
Poulson, D. F. (1937). *Proc. Natl. Acad. Sci. U.S.A.* **23,** 133–137.
Poulson, D. F. (1950). "Biology of *Drosophila,*" (M. Demerec, ed.) pp. 268–274. John Wiley & Sons, New York.
Preat, T., Therond, P., Lamour-Isnard, C., Limbourg-Bouchon, B., Tricoire, H., Erk, I., Mariol, M. C., and Busson, D. (1990). *Nature (London)* **347,** 87–89.
Ready, D. F. (1989. *Trends Neurosci.* **12,** 102–110.
Ready, D. F., Hanson, T. E., and Benzer, S. (1976). *Dev. Biol.* **53,** 217–240.
Reinke, R., and Zipursky, S. L. (1988). *Cell* **55,** 321–330.
Renfranz, P. J., and Benzer, S. (1989). *Dev. Biol.* **136,** 411–429.
Riggleman, B., Wieschaus, E., and Schedl, P. (1989). *Genes Dev.* **3,** 96–113.
Riggleman, B., Schedl, P., and Wieschaus, E. (1990). *Cell* **63,** 549–560.
Rijsewijk, F., Schuermann, M., Wagenaar, E., Parren, P., Weigel, D., and Nusse, R. (1987). *Cell* **50,** 649–657.
Rogge, R. D., Karlovitch, C. A., and Banerjee, U. (1991). *Cell* **64,** 39–48.
Romani, S., Campuzano, S., Macogno, E., and Modollel, J. (1989). *Genes Dev.* **3,** 997–1007.
Sampedro, J., and Guerrero, I. (1991). *Nature (London)* **353,** 187–190.
Scott, M. P., and Carroll, S. B. (1987). *Cell* **51,** 689–698.
Seydoux, G., and Greenwald, I. (1989). *Cell* **57,** 1237–1245.
Shannon, M. (1973). *J. Exp. Zool.* **183,** 383–400.
Siegfried, E., Perkins, L., Capaci, T., and Perrimon, N. (1990). *Nature (London)* **345,** 825–829.
Simpson, P. (1990). *Development* **109,** 509–519.
Simpson, P., El Messal, M., Moscoso del Prado, J., and Ripoll, P. (1988). *Development* **103,** 391–401.
Skeath, J. B., and Carroll, S. B. (1991). *Genes Dev.* **5,** 984–995.
Skeath, J. B., and Carroll, S. B. (1992). *Development* **114,** 939–946.
Skeath, J. B., Panganiban, G., Selegue, J., and Carroll, S. (1992). Submitted for publication.
Smoller, D., Friedel, C., Schmid, A., Bettler, D., Lam, L., and Yedvobnick, B. (1990). *Genes Dev.* **4,** 1688–1700.

Stern, C. (1954). *Am. Sci.* **42,** 213–247.
Technau, G. M., and Campos-Ortega, J. A. (1986). *Wilhelm Roux's Arch. Dev. Biol.* **195,** 445–454.
Thomas, J. B., Bastiani, M. J., Bate, M., and Goodman, C. S. (1984). *Nature* (*London*) **310,** 203–207.
Tomlinson, A. (1985). *J. Embryol. Exp. Morphol.* **89,** 313–331.
Tomlinson, A. (1988). *Development* **104,** 183–193.
Tomlinson, A., and Ready, D. F. (1986). *Science* **231,** 400–402.
Tomlinson, A., and Ready, D. F. (1987a). *Dev. Biol.* **120,** 366–376.
Tomlinson, A., and Ready, D. F. (1987b). *Dev. Biol.* **123,** 264–275.
Tomlinson, A., Bowtell, D. D. L., Hafen, E., and Rubin, G. M. (1987). *Cell* **51,** 143–150.
Van den Heuvel, M., Nusse R., Johnston, P., and Lawrence, P. (1989). *Cell* **59,** 739–749.
Vassin, H., Vielmetter, J., and Campos-Ortega, J. A. (1985). *J. Neurogenet.* **2,** 291–308.
Vassin, H., Bremer, K. A., Knust, E., and Campos-Ortega, J. A. (1987). *EMBO J.* **11,** 3431–3440.
Villares, R., and Cabrera, V. C. (1987). *Cell* **50,** 415–424.
Waddington, C. H., and Perry, M. M. (1960). *Proc. R. Soc. London, Ser. B* **153,** 155–178.
Wharton, K. A., Johansen, K. M., Xu, T., and Artavanis-Tsakonas, S. (1985). *Cell* **43,** 567–581.
Wieschaus, E., Nusslein-Volhard, C., and Jurgens, G. (1984). *Wilhelm Roux's Arch. Dev. Biol.* **193,** 296–307.
Wigglesworth, V. B. (1940). *J. Exp. Biol.* **17,** 180–200.
Wilkins, A. S. and Gubb, D. (1991). *Dev. Biol.* **145,** 1–12.
Yochem, J., Weston, K., and Greenwald, I. (1988). *Nature* (*London*) **355,** 547–550.

Assays of Random Motility of Polymorphonuclear Leukocytes *in Vitro*

Leon P. Bignold
Department of Pathology, University of Adelaide, Adelaide, South Australia, Australia

I. Introduction

Random motility is a basic function of polymorphonuclear leukocytes (PMNs), which underlies the movement of these cells in various infective and noninfective lesions of tissues. This movement of PMNs is ameboid in type and is not a single cell-biological phenomenon but rather has several components. The major components of ameboid movement are the motor of the cell, which generates the motile forces, the transmission of these forces to the periphery (exterior) of the cell, and the traction of the periphery on the substratum on which the cell crawls. By analogy with an automobile proceeding along a road, forces are generated by an engine and transmitted by gear box, shafts, and axles to the wheels. Where the tires of the wheels make contact with the road, adequate traction must exist for the vehicle to advance. In studies of amboid cells, the mechanisms of the presumed motor (de Bruyn, 1944a; Allen, 1961, 1973; Abercrombie, 1980; Stossell, 1988, 1989, 1990) and traction (adhesion; Weiss, 1960; Curtis, 1967; Culp, 1978; Grinnell, 1978; Kishimoto *et al.*, 1989) have received special attention.

In any *in vitro* assay of random motility of cells, changes may result from injury to the cells during separation from their milieu *in vivo*, inappropriate conditions of incubation, or alterations to one or more of these cell-biological components of ameboid movement. Furthermore, in particular assays the degree of alteration of each component of ameboid movement of PMNs could vary according to the individual features of the assay system.

Supporting these notions of the importance of cell-biological issues to assays of PMN movements are the conflicting results of previous *in vitro* studies of both the maximal rates of PMN motility and the effects of agents such as drugs. Such discrepancies between results of various studies have occurred not only according to the type of assay employed but also according to technical details, such as the composition of the incubation medium and the nature of the substratum on which the cells crawl (Table I). Thus, reports of average maximal rates of motility have been as high as 34 μm/min in whole blood clot preparations (McCutcheon, 1923) and as low as 1–2 μm/min in filter membrane/Boyden chamber assays (Boyden, 1962; Keller and Sorkin, 1966; Wilkinson, 1982). Similarly, many agents, such as anti-inflammatory drugs, have been reported to enhance,

TABLE I

Rates of Random Motility of PMN *in Vitro* According to Assay

Assay conditions	Rate (μm/min)	Reference
In whole blood clots (not attached to artificial substrata)	10–50 (av 34) 36.6 (av)	McCutcheon (1923) Henderson (1928)
In plasma–fibrin clots (not attached to artificial substrata)[a]	19.4 (av)	Lewis (1934)
Sedimented from blood to glass,		
incubated in plasma clots	10–15 (est[b])	Harris (1953)
incubated in BSS + SA or serum	10.0 ± 1.0[c] 7.2–8.0	Ramsey (1972a) Howard (1982)
Sedimented from separated suspension containing protein to glass or plastic,	10–13 2.5–17.3 (av 7.4)	Grinnell (1982) Maher *et al.* (1984)
incubated in BSS + SA or FCS	6.2 ± 1.0[d] in 0.05% SA, and 8.2 ± 1.5[d] in 2% SA	Howard (1986)
Sedimented from separated suspension containing no protein, incubated in BSS with protein	0.0 if attached at 37°C but 0.8–1.3 if attached at 22°C 0.3–5.5 (temperatures not stated)	Lackie (1982) Keller *et al.* (1983)
Sedimented from separated suspension containing no protein to glass or plastic,		
incubated in BSS alone	0–0.3	Keller *et al.* (1983)
In collagen gels	5–8	Grinnell (1982)

[a] Rat neutrophils.
[b] Estimated from illustrations.
[c] Mean ± 95% confidence limits.
[d] Mean ± 1 SD. BSS, balanced salt solution; SA, serum albumin; FCS, fetal calf serum.

decrease, or have no effect on the random motility of PMNs (Giroud and Roch-Arvellier, 1982; Keller, 1985).

The purpose of the present review is to consider how alterations of the components of ameboid movement (especially motor, transmission, and traction) might occur in commonly used assays of motility of these cells. Additional phenomena that may be related to motility [cell viability, polarization, spreading, zeiosis (blebbing), membrane fluidity, and streaming] are discussed. For each component the relevant literature is reviewed to ascertain whether any useful assay of its activity or involvement in motility of PMNs has been described.

II. Early Concepts of Random Motility of PMNs

The peculiar movements involving frontal projections of free-living, single-celled organisms found in brackish water (later named amebas) were apparently first described by Rösel von Rosenhof in 1755 (Lorch, 1973). Later authors showed that the active motility of these cells is continual under favorable conditions but tend to cease when sublethal cell injury is induced by physical or chemical insults (Bovee and Jahn, 1973). Amebas rendered nonmotile by such adverse conditions can regain normal motility by restoration of favorable environmental conditions.

The motility of PMNs was discovered during the investigation of the accumulation of leukocytes in acute inflammatory foci (for reviews see Rolleston, 1934; Rebuck and Crowley, 1955; Buckley, 1963; Rather, 1972; Movat, 1985; Colditz, 1985). The similarity of PMN movements to those of amebas by virtue of pseudopodia was emphasized by Lieberkühn who in 1870 apparently introduced the term "ameboid motion"(Rebuck and Crowley, 1955).

In the first half of the twentieth century, the dominant concept of the nature of PMN motility seems to have been that PMNs are unceasingly active, independent, ameba-like cells operating in the particular interior milieu of complex multicellular animals. Early *in vitro* techniques to study the motile phenomena of PMNs, including those of Ponder (1908), Jolly (1913), Comandon (1917), and the ingenious test tube and chamber experiments of Wright and Colebrook (1921) were unsatisfactory because physiological conditions were poorly preserved in the incubation chambers and the cells were short-lived. However, more reliable techniques were developed, such as those using a sealed, polished glass chamber in which films of freshly drawn blood were allowed to clot (Sabin, 1923). The polishing of the surfaces of the slide and coverslip prevented the PMNs adhering to the glass, and ceaseless activity of PMNs for many hours was observed.

Sabin (1923) therefore considered that "the normal neutrophil is never round and never still." The same technique was used by McCutcheon (1923, reviewed in 1946) who found that PMNs were capable of speeds ranging from 10 to 50 μm/min, with an average of 34 μm/min. These rates were later confirmed by Henderson (1928) and von Philipsborn in the 1950s (Dittrich, 1962) using similar techniques. Applying the notions of sublethal cell injury and recoverable motility (derived from studies of amebas) to the neutrophil, it seemed likely that cells in blood clot and without contact with a foreign surface were in optimal conditions, especially because spontaneous motility was highest under these circumstances (McCutcheon, 1923).

III. Methodology and Concepts of PMN Motility Since 1960

A. Use of Separated PMNs and Plasma-Free Media

A major change in methodology of neutrophil motility and chemotaxis since 1960 has been the use of PMNs separated from whole blood and suspended in plasma-free media. This has followed the recognition of the numerous and diverse effects that products of activation of enzyme systems, such as of complement (Hahn, 1969; Goldstein, 1988; Müller-Eberhard, 1988) and of the contact activation and kinin systems (Roche e Silva, 1964; Wilhelm, 1971; Cochrane and Griffin, 1982) in plasma, may have in biological experiments. In particular, fragments of the complement system were shown to be chemotactic for PMNs (Ward *et al.*, 1965; Ward, 1968). From this work it seems that a factor included in plasma with all leukocytes might (1) activate a plasma enzyme system (such as complement, or the contact-activated system) and generate active molecules that have biological effects (stimulatory or inhibitory) on PMNs, (2) degrade (by enzymatic action, binding, precipitation, or other means) a component of plasma that is essential for functioning of PMNs, (3) be itself converted by a normal plasma enzyme (such as a protease) to an unsuspected secondary factor that acts on PMNs, (4) be itself degraded by a normal plasma enzyme or rendered inactive by binding to a plasma protein, including antibody, (5) act on the PMNs, or other leukocytes (especially monocytes and lymphocytes) to induce them to activate or degrade a component of plasma or the factor itself, or (6) act on monocytes and lymphocytes to induce them to secrete factors, such as cytokines, which act on neutrophils.

Nevertheless, separation of PMNs from whole blood may be injurious to the cell, especially if erythrocyte-lysing techniques are used (Cutts,

1970). Furthermore, functions of PMNs deteriorate more rapidly in plasma-free media than in plasma-containing solutions (Wilkinson, 1982). Few detailed studies of PMN function in relation to separation procedures and conditions of incubation have been published (Ferrante *et al.*, 1980; Bignold, 1987b), although duration of survival (to the end-point of nonviability as assessed by dye exclusion) of PMNs in simple incubation may be a suitable measure of sublethal injury because the more severe the injury, the shorter period of time the cell is likely to survive.

B. Chemokinesis

The use of separated cells and plasma-free media, such as Hanks solution, is essential for many investigations, but such PMNs are characterized by spherical shape and low motility (i.e., no longer ceaselessly active). Nevertheless, under these conditions a wide variety of factors, including formyl (synthetic bacterial) peptides, induce polarized shape and frequently increased motility. The term "chemokinesis" was reintroduced into the literature of leukocyte locomotion to describe increases of spontaneous motility under these circumstances (Keller *et al.*, 1977a; see also Bignold, 1988a,b; Keller *et al.*, 1977b, 1982; Wilkinson, 1988a,b). Characteristically, such increases occur in filter membrane assays when a protein with no chemotactic effect, such as serum albumin, is included in the incubation medium (Wilkinson, 1976; Keller *et al.*, 1977a, 1983). The rates of movement achieved by chemokinetic effects have never been reported to exceed the random motility of neutrophils in the optimal conditions reported by McCutcheon (1923).

Although useful in describing certain simple experimental results, chemokinesis does not specify which of several significantly different groups of causes are responsible for the increased motility. First, the term does not allow for the possibility that the increased motility is that induced by restoration of physiological conditions of incubation and hence reversal of sublethal injury. Thus, PMNs separated from whole blood and held in simple solutions (e.g., Hanks) can be considered in unfavorable conditions. Restoration of some important components of serum to the medium allows for some improvement of motility but may amount to no more than removal of an injurious factor (absence of serum) from the environment of the cells. Second, chemokinesis does not indicate which cell-biological component of ameboid movement is being affected. The early literature of chemokinesis of PMNs (Keller *et al.*, 1977a) referred to enhanced intrinsic locomotor capacity without giving its meaning in more specific cell-biological terms. It was implied that the motor of the cell was being affected, but later work on the effects of serum albumin indicated that adhesion (traction) can be the major mechanism of chemokinesis (Keller *et al.*, 1977b, 1983; Keller and Cottier, 1982). In the case of *N*-formylmethionyl-leucyl-phenylalanine (fMLP) the chemokinetic effects of this factor could represent a stimulus to the motor of the cell, but effects on adhesiveness, which might account for increased motility in certain assays, have been described (O'Flaherty *et al.*, 1977, 1978; Fehr and Dahinden, 1979; Smith *et al.*, 1979). Yet other components of ameboid movement might be involved.

C. Significance of Attachment to Artificial Substrata

The use of cells adherent to artifical substrata, such as glass or plastic, was common both before 1960 (Harris, 1954) and since. However, evidence has accumulated that such attachment of cells to substrata can itself cause cell-biological events within the cell. In cells that show cytoplasmic streaming (Section V,E), adhesion to substratum causes the rearward moving ectoplasm (fountain streaming) to become stationary in relation to the cell membrane (Allen, 1961, 1973). PMNs that are spherical in suspension can become polar and locomote when they attach to glass without any other factor being operative (Keller *et al.*, 1983). Under certain conditions attachment to glass can cause polymerization of actin in PMNs (Southwick *et al.*, 1989) and loss of the contents of cytoplasmic granules to the exterior (frustrated phagocytosis; for review see Klebanoff and Clark, 1978). The extremely flattened shape of the cell referred to as spreading (Section V,B) results from attachment and cannot be induced in neutrophils in suspension (L. P. Bignold, unpublished observations). Systematic studies of the mechanisms of these effects or of effects of attachment on the responses of PMNs to chemical agents, such as inflammatory mediators, have not been reported. Furthermore, it is not known what particular properties (physical or chemical) of the substratum are responsible for the effects.

IV. Cell-Biological Components of Ameboid Movement

A. The Motor of the Ameboid Cell

The motor of the ameboid cell is considered to reside in the cytoplasm because isolated cytoplasm from other ameboid types are capable of contraction independently of both cell membrane and nucleus (Allen *et al.*, 1960; Wolpert *et al.*, 1964; Wolpert, 1971; Kuroda, 1979; for review see

Taylor and Condeelis, 1979). Proposed motors involve either modifications of the actin–myosin mechanism of skeletal muscle mechanism to accommodate the low content of myosin in these cells (Huxley, 1979; Taylor, 1986; Sheterline, 1983; Adams and Pollard, 1989; Higashi-Fujime, 1991), or purely actin-based theories (Stossell, 1982, 1988, 1989, 1990; Southwick and Stossell, 1983; Hartwig *et al.,* 1985). The latter are most popular, especially because mutant *Dictyostelium,* which are myosin-deficient, retain their ability to move in ameboid fashion (Fukui *et al.,* 1990).

The primary observation that has been used to support a role of actin as the motor of PMNs is that factors, such as fMLP, induce the polymerization of monomeric actin to the filamentous form (F-actin) (Fechheimer and Zigmond, 1983; Wallace *et al.,* 1984). Polymerization of actin under the influence of fMLP is unlike that induced by adhesion to solid substrata because it has a different time–course and is not dependent on extracellular Ca^{2+} (Southwick *et al.,* 1989). Whether or not either the occurrence or the rates of polymerization of actin in PMNs can be used as an assay of the motor of the cell is unclear at the present time.

B. Transmission of Forces from Cytoplasm to the Periphery

Each of the general theories of ameboid movement, as reviewed by de Bruyn (1944a), Allen (1961, 1973), and Abercrombie (1980), requires different theoretical mechanism of transmission of cytoplasmic forces to the substratum (Fig. 1).

1. Rolling Balloon Models

These include the sol-gel theories of Mast (1931), Goldacre (1961), Allen (1961), Bovee and Jahn (1973), Taylor and Fechheimer (1982), and Stossell (1990), and have in common the notion that cytoplasmic contractility can move the cell in particular directions with little or no involvement of the cell periphery. The outer coat is seen as a bag that is rolled over and over by the movements of cytoplasm in a manner analogous to the continuous tread of a military tank (Allen and Kamiya, 1964; see also Harris, 1973a; Grebecki, 1984, 1986; Bignold, 1987a). For the rolling balloon model to work, interaction between the cytoplasmic contractile apparatus and the inner side of the cell membrane is necessary. Without interaction of these components, the cytoplasmic contractile apparatus would cause the cytoplasm to spin inside the cell without advancing the whole cell at all (Fig. 1a). According to the military tank analogy, the corresponding structure is the cog wheel on the end of the axle, which interdigitates with the inner side of the tread to force the tank along. In the absence of the cog wheel,

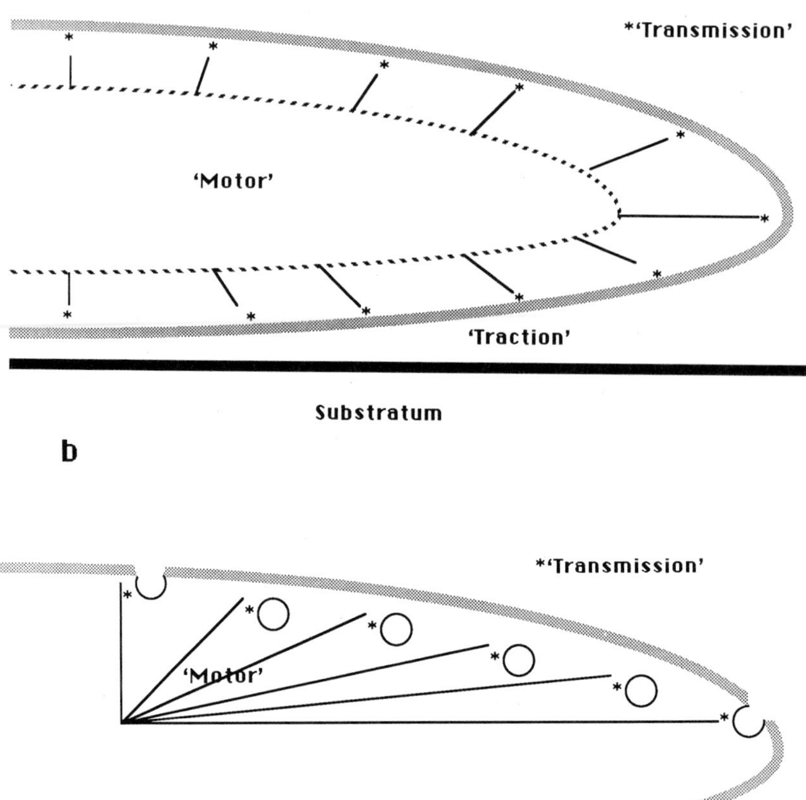

FIG. 1 Diagrams indicating by asterisks the locations of necessary transmission mechanisms of forces from the motor of the cell to the periphery according to the rolling balloon (a), membrane recycling (b), surface wave (c), extension–contraction (d), and membrane ratchet (e) theories of ameboid movement.

c

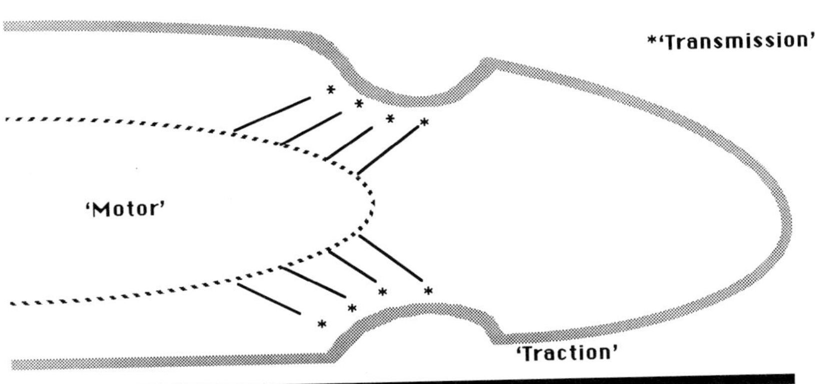

d

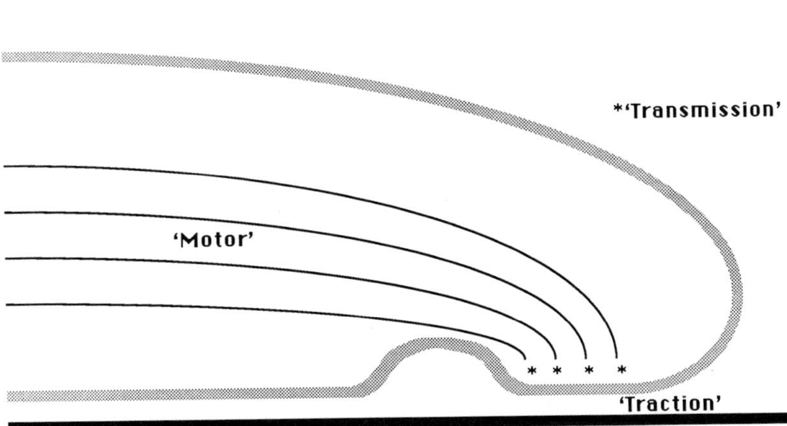

FIG. 1 (*continued*)

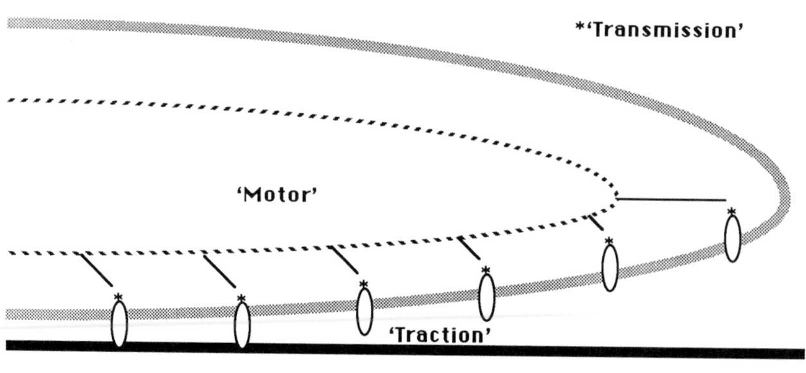

FIG. 1 *(continued)*

the axle spins uselessly without advancing the tank at all. The mechanism of the transmission could be a specific cytoplasm–membrane interaction, as has been envisaged between the cytoskeleton and the membrane (Weatherbee, 1981) and in particular between actin and membrane myosin (Adams and Pollard, 1989), or a looser arrangement of nonspecific viscous coupling of macromolecules of the two structures (Lackie, 1986).

2. Membrane Recycling Models

These theories propose that cell membrane is formed at the front of the moving cell and resorbed at the rear. This idea was proposed for amebas by Goldacre (1961), and later for fibroblasts by Abercrombie *et al.* (1970, 1972; see also Harris, 1983a; Dunn, 1980). The structure by which the pieces of membrane might be moved through the cytoplasm has been considered the vesicle (Dunn, 1980; Bretscher, 1984, 1987, 1988; Singer and Kupfer, 1986), akin to the endocytotic vesicle. Although endocytotic vesicles are considered to form from and reinsert into the adjacent cell membrane, in these endocytotic vesicle models of ameboid movement, the vesicles are said to move forwards in the submembranous zone of the cytoplasm and not to insert into the membrane until they reach the front of the cell. The vesicles are possibly pushed to the front of the cell by microtubular mechanisms (Sheetz *et al.,* 1986). For membrane recycling to work, transmission mechanisms (Fig. 1b) would be required for the engagement and release of cytoplasmic forces selectively from the vesicles (rather than the inner plasma membrane as a whole) to effect their transference to the front of the cell.

3. Surface Wave Theories

These notions are similar to the method of movement of mollusks, which move by waves of contractility of the gastropod (Trueman and Jones, 1977; Elder, 1980). For their application to PMNs, these theories rely on transverse waves that have been described on the surfaces of these cells (Lewis, 1934; de Bruyn, 1944a,b, 1946; Senda *et al.*, 1975; Shields and Haston, 1985). For surface wave theories, transmission would be required for engagement of cytoplasmic forces with the concentric band of inner plasma membrane corresponding to the trough of the wave (Fig. 1c).

4. Cyclical Extension/Attachment/Contraction/Detachment

These models envisage active forward protrusions of the cell body, which attach to substratum and then act as forward anchors by which the contractile apparatus of the cell pulls the cell body along. The additional features of the model require that the tips of the processes be nonadhesive at other times (i.e., while the processes are being pushed forwards and after the cell body has caught up with the site of attachment) and that contractility of the cytoplasm be coordinated with the adherent phase of the the tip of the process (because contractility could not occur during extension). Such notions have been proposed for amebas (Bovee and Jahn, 1973), growing neurites in culture (Wessels *et al.*, 1973), metazoan cells in culture (Huxley, 1973), and PMNs (Sullivan and Mandell, 1983). For these theories the transmission is required to enable engagement of forces with the relevant patches of membrane in the forward anchors (Fig. 1d) with the appropriate timing.

5. Membrane Ratchet

This model was proposed by the author (Bignold, 1987a) and is based on the fluid-mosiac model of the cell periphery (Singer and Nicholson, 1972). It is speculated that there are laterally mobile transmembranous structures in the cell membrane that are adhesive for the substratum externally and attached to the contractile apparatus internally. Both adhesion and attachment are strong when the structure is at the front of the cell and are lost when the transmembranous structure is displaced to the rear of the cell. This adhesive/attachment activity is postulated to be under the control of an intracellular front-to-rear gradient of a hypothetical third factor, which could possibly be intracellular water. Such a mechanism could provide explanations of formation of the pseudopod as well as chemotaxis (Bignold, 1987a). In the membrane ratchet theory the transmission mechanism is the hypothetical laterally moveable structure of variable adhesiveness (Fig. 1e).

C. Traction

Theories of traction are essentially those of the mechanisms of adhesion of cells to surfaces. These involve analysis of possible physiochemical forces of intermolecular attraction (for reviews see Pethica, 1961; Curtis, 1967, 1973, 1988; Dembo and Bell, 1987; Bongrand, 1988), identification of possible receptors on cell surfaces and the corresponding ligands on substrata (for reviews see Gallin, 1985; Kishimoto *et al.*, 1989; Albelda and Buck, 1990; Patarroyo, 1991), as well as studies of the influences of electrical charge, wettability, and roughness (Harris, 1973b; for review see Grinnell, 1978). An additional factor that may be relevant *in vivo* but not in relation to foreign solid substrata is viscous coupling between complex macromolecules on the surfaces of cells (Lackie, 1986).

A morphological correlate of cell adhesion is the focal adhesive contact that certain cells, especially cultured tissue cells, exhibit in relation to solid substrata to which the cell is attached *in vitro* (Birchmeier, 1981; Burridge *et al.*, 1987, 1988). PMNs exhibit weaker focal contacts than do fibroblasts (Wilkinson, 1978; Wilkinson and Lackie, 1979), but when adhesion of these cells is tested under conditions of flow, PMNs are more adhesive than fibroblasts and many other cell types (Forrester and Lackie, 1984). The importance of focal contacts to adhesion of PMNs is therefore unclear.

It seems likely that all of these mechanisms of adhesion of cells occur but act in differing combinations of intensities under various circumstances.

Traction is important to PMN motility *in vitro* because excess adhesiveness of PMNs to a surface can reduce measured motility by immobilizing the cells, whereas insufficient adhesiveness prevents cells being included in the assay, thus reducing apparent motility in populations of cells (Wilkinson *et al.*, 1982; Keller *et al.*, 1983). Factors influencing adhesion of PMNs to the substrata, which are used in assays of random motility and chemotaxis (including glass, cellulose ester, and plastics), are not well understood but appear to include the following.

1. Viability of the Cells

Only living PMNs adhere to substrata in the presence of plasma proteins, and do so by a mechanism that requires extracellular Ca^{2+}. However, dead cells can adhere to substrata such as glass, provided proteins are absent. Adhesion in the latter circumstances does not seem to be dependent on extracellular Ca^{2+}. Adhesion of these types has been referred to as active and passive, respectively (Taylor, 1961; Grinnell, 1978; Schreiner and Hopen, 1979) or physiological and nonphysiological (Wilkinson, 1982; Curtis, 1988). The reasons why viability of the cell is essential for adhesion under physiological conditions (and especially whether or not protein-dependent active adhesion is related to protein-supported cell viability; Section IV,A) as well as the mechanisms of passive adhesion have been little discussed.

2. Membranous Structures

The ultimate structures responsible for adhesion of cells to a surface must be in the cell periphery. According to present understanding, the periphery of the cell includes the lipid bilayer forming the plasma membrane to which is attached glycoproteins of various sizes. The gylcoproteins are known to show various relationships to the lipid bilayer, including hydrophobic transmembranous segments and glycosyl-phosphatidylinositol anchoring (Ferguson and Williams, 1988; Low, 1989; Cross, 1990; Singer, 1990). The extramembranous components (on the outer side of the cell) of the glycoproteins carry the carbohydrate moieties, which in many cells are sufficiently concentrated membranes to be visualized microscopically as a glycocalyx.

In respect of PMNs, particular membrane proteins with roles in adhesion to endothelium have been identified and include integrins (for reviews see Gallin, 1985; Hogg, 1989; Kishimoto *et al.*, 1989) as well as other families of glycoproteins. There is some doubt, however, that integrins are responsible for adhesion of cells to artificial substrata, which are used in assays of chemotaxis of PMNs (Forsyth and Levinsky, 1989). Nevertheless, all such glycoproteins may have nonspecific roles in binding of cells to artificial surfaces.

3. Intermediaries

It has long been recognized that the adhesion of cells to substratum can be markedly affected by extraneous substances, such as proteins and cations in the medium (Weiss, 1960; Culp, 1978; Grinnell, 1978). Of greatest relevance to adhesion of PMNs are plasma proteins, which have considerable effects on such adhesion (Dalgren, 1979; Lackie, 1982; Wilkinson, 1982; Keller *et al.*, 1983; Valerius, 1983). Nevertheless, in relation to substrata used for assays of chemotaxis, there have been few studies of the relative importance of the various proteins in plasma. For polycarbonate filtration membrane (Costar-Nuclepore, Cambridge, Massachusetts), gamma globulins are the most strongly pro-adhesive protein, whereas albumin, fibrinogen, and fibronectin are antiadhesive (Bignold *et al.*, 1990). The mechanisms of effects of these proteins may include bridges between undefined structures on the cell membrane and the substratum, and possibly stabilization of membrane structures, enhancing viability of the cell (Section IV,C,1).

Divalent cations, especially Ca^{2+}, have been frequently studied for roles in adhesion (Weiss, 1960; Curtis, 1973; Grinnell, 1978) and are generally recognized to be essential for the adhesion of living cells (i.e., active adhesion) to artificial substrata. Early notions of the mechanism of action of Ca^{2+} included stabilization of the cell coat (Rinaldini, 1958) and forming ionic bridges between membrane and substratum (Weiss, 1960; Bangham, 1964; Wilkinson, 1978). There has been great interest in the role of intracytoplasmic free Ca^{2+} in various cell processes, and transmembrane fluxes of the cation have been considered (for reviews see Westwick and Poll, 1986; Campbell, 1987; Farese, 1988; Sadler and Badwey, 1988). The significance of external membrane Ca^{2+} for PMN adhesion, however, is currently a relatively neglected area of study.

4. Properties of Substratum Relevant to Adhesion

Although the substratum would seem to be a critical factor in the adhesion and hence motility of PMNs in assays of motility and chemotaxis, the properties of the most common of these substrata (glass and plastic in direct microscopy, and cellulose ester and polycarbonate in Boyden chamber assays) have been largely ignored. There have been occasional studies of the effects treatment of glass and plastic on the adhesiveness of cultured cells, such as macrophages (Rich and Harris, 1981) and fibroblasts (Ramsey *et al.*, 1984) but few corresponding studies involving PMNs. Moreover, there have been few comparisons of adhesiveness of PMNs to the substrata of the different types of assays.

V. Other Cell-Biological Phenomena Related to Ameboid Movement

A. Polarization

Formation of an extension of the cell body at the front of the cell is a necessary prerequisite of ameboid movement and occurs independently of gradients of chemotactic factors. The simple morphology of the pseudopod does not seem to be related to function because in amebas size, shape, and number of pseudopods can vary from species to species without being related to the rates of movement of which members of the species are capable (Bovee, 1964). Similarly, the size and shape of pseudopods do not seem to be related to comparative rates of movement among leukocytes (Robineaux, 1964).

In attempting to quantify polarization of PMNs, formation of a pseudopod by populations of cells can be studied when the cells are in suspension so that the morphological effects of adhesion and spreading are avoided. The relevant index of polarization in suspension can be either the proportion of cells that become polar under the influence of a factor in a given period or the rate at which the cells become polar (Bignold, 1987b). Both large and small extensions of the cell bodies of PMNs occur in suspension under the influence of a factor, such as *N*-formyl peptide. Cells exhibiting these extensions have been classified as well and poorly polarized (Shields and Haston, 1985), bipolar and amorphous (Jadwin *et al.,* 1981), and long polarized and short polarized (Bignold and Ferrante, 1988). Roos *et al.* (1987) provided a more detailed classification of PMN shapes as spherical smooth cells, polarized with and without tail, spherical smooth cells with unifocal surface projections, spherical rough cells, and nonpolar cells with surface projections. In this laboratory (Harkin and Bignold, unpublished), a modification of the classification of Roos *et al.* (1987) has been used, comprising spherical cells and cells with extensions as follows: Type I (spherical with multiple, randomly directed extensions, Fig. 2a), Type II (spherical with one or two processes restricted to one side of the cell, Fig. 2b), Type III (elongation of the cell body with the pseudopod being relatively short and rounded, Fig. 2c), and Type IV (larger, broader pseudopod with the cell body being narrower toward the tail, where a uropod may be present, Fig. 2d). Time–courses of the relative proportions of these types in a preparation of PMNs in suspension indicate that Types I, II, and III appear to be transitional forms between spherical and the Type IV form because their numbers peak in sequence (Fig. 3).

Another feature of the pseudopods of amebas (Allen, 1961, 1973) and leukocytes (Robineaux, 1964; Zigmond, 1978) is a clear (granule-free) zone of cytoplasm (the hyaline tip or hyaloplasmic veil) in the submembranous region of its frontal extremity. The width of this zone does not appear to be related to speeds of locomotion of cells, although no precise studies of this point appear to have been made.

In relation to studies of motility and chemotaxis of PMN *in vitro,* it has been suggested that polarization in suspension could be used as an index of the motor of the cell in random motility because it correlates with locomotion of PMNs on glass (Keller, 1983). However, polarization of neutrophils in plasma was shown not to correlate with motility of these

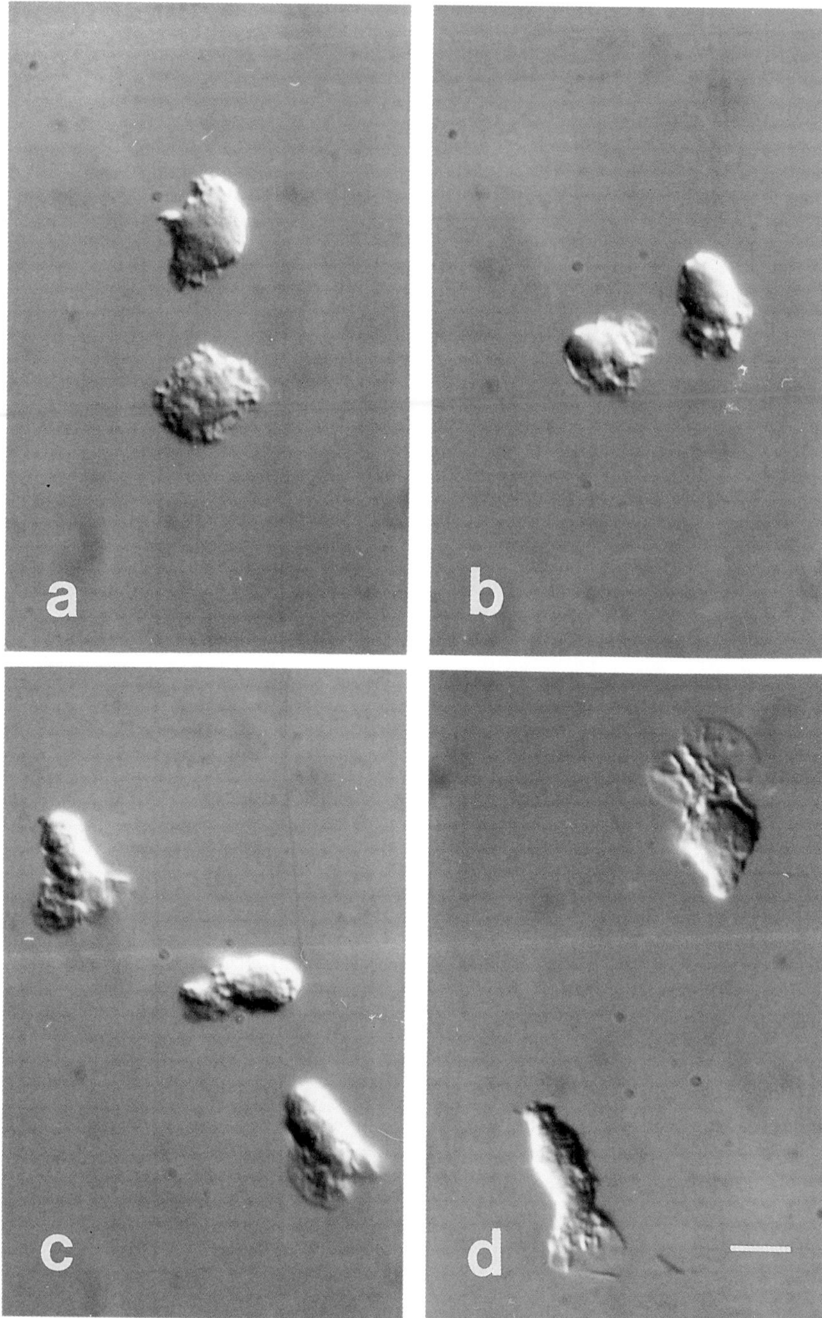
a
b
c
d

cells in cellulose ester membrane (Bignold, 1986), but factors particular to the membrane may have caused this result. Alternatively, because the chemotactic factor fMLP causes polarization of PMNs in suspension, this shape change has been considered a manifestation of chemotaxis (Newman and Wilkinson, 1989). If either of these suggestions is correct, defective polarization of PMNs could be a basis either of reduced random motility or of chemotaxis of PMNs in *in vitro* assays, but whether or not this should be assessed when the cells are on a substratum (and if so which substratum) or when the cells are in suspension is unclear.

Few theories of formation of pseudopods by ameboid cells have been advanced (Bignold, 1987a). For amebas a common view is that the pseudopod forms by herniation of liquid endoplasm through a weak point in the gelled ectoplasm (Bovee and Jahn, 1973; Stossel, 1990). Odell and Frisch (1975) proposed that the pseudopod produces substances that direct the contractility of the cytoplasm toward itself. Vasiliev (1991) suggested that for *Dictyostelium* and cultured mammalian fibroblasts, reorganization of the actin–myosin cortex as well as of microtubules and intermediate filaments may be involved. For PMNs (which do not show prominent endoplasm and ectoplasm) polarization has been suggested to be due to chemotactic factors acting on one end of the cell and inducing transmembranous flows into the cell at that site. Local intracellular increased Ca^{2+} may in turn cause reorientation of microtubules to polarize the cell (Gallin *et al.*, 1978; see also Allison, 1973). PMNs, however, contain few microtubules to support this change, and such a mechanism does not easily explain polarization of cells when suspended in solutions (i.e., nongradient presence) of chemotactic factors (Shields and Haston, 1985). Stossell (1988, 1989) and Wyman *et al.* (1990) have suggested that regional actin gelation in the cytoplasm forms the pseudopod, but this would not account for the long persistence of polarization after actin polymerization in response to the chemotactic factor fMLP has subsided (Southwick *et al.*, 1989). Actin polymerization is probably related to levels of intracellular free Ca^{2+}, and fMLP has been found to activate the phosphoinositide–phosphokinase (second messenger) system that controls these levels (for reviews see Berridge, 1986; Snyderman *et al.*, 1986; Naccache, 1987; Farese, 1988; Gill, 1989; Bansal and Majerus, 1990; Tsien and Tsien, 1990). A role for intracellular fluctuations of Ca^{2+} in polarization of neutrophils therefore

---

FIG. 2 A classification of extensions of PMNs in suspension. (a) Type I (spherical with multiple, randomly directed extensions). (b) Type II (spherical with one or two processes restricted to one side of the cell). (c) Type III (elongation of the cell body with the pseudopod being relatively short and rounded). (d) Type IV (larger, broader pseudopod with the cell body being narrower toward the tail, where a uropod may be present). Nomarski optics. Bar, 10 μm. (From D. G. Harkin and L. P. Bignold, unpublished observations.)

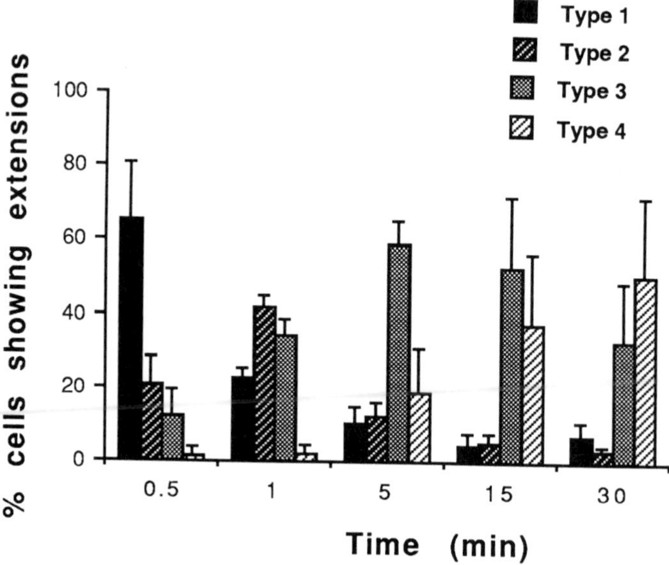

FIG. 3 Time–courses of relative proportions of types of shapes of PMNs in suspension when exposed to *N*-formyl-methionyl-leucyl-phenylalanine (fMLP). Type I is predominant at 0.5 minute, Type II at 1 minute, Type III at 15 minutes, and Type IV at 30 minutes, suggesting that Types I–III may be transitional forms between spherical and Type IV under the influence of this agent. (From D. G. Harkin and L. P. Bignold, unpublished observations.)

remains possible (Marks and Maxfield, 1990; Bengtsson, 1990; Eberle *et al.*, 1990).

Another possible mechanism of polarization of PMNs is that the mechanical properties of cell membrane may have a major role in determining cell polarity (Svetina and Zeks, 1990).

B. Spreading

Spreading is the degree of flattening a cell exhibits when attached to a solid substratum (Fig. 4a), believed to be an active cell-biological process closely related to adhesion (Bessis and de Boisfleury, 1976; Vasiliev, 1982, 1985). In PMNs, spreading is more marked under conditions of passive adhesion compared with conditions of active adhesion (Keller *et al.*, 1983). Spreading of neutrophils on polycarbonate membrane in gamma globulin exceeds that occurring in Hanks solution alone (passive adhesion) and

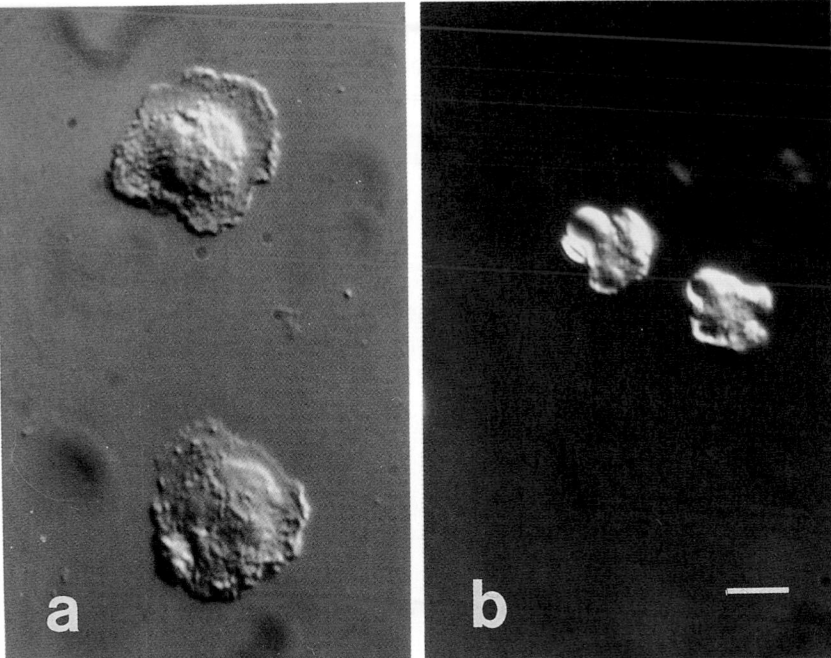

FIG. 4 (a) Spreading of PMNs on glass when sedimented from separated suspension in the presence of 0.1% human gamma globulin. (b) Blebbing (zeiosis) of PMNs induced by cytochalasin B (10 μg/ml) with *N*-formyl peptide (10^{-6} *M*) in Hanks solution. Nomarski optics. Bar, 10 μm.

correlates more closely with reduced random motility than does measured adhesiveness (Bignold *et al.*, 1990).

C. Zeiosis

The formation of small, spherical surface projections of cells including neutrophils under the influence of toxins, such as cytochalasin B, has long been recognized and referred to as zeiosis (Godman and Miranda, 1978) or blebbing (Bessis and de Boisfleury, 1986; Shroeder, 1978) (Fig. 4b). Blebs of PMNs have been found to have a high actin content and to be capable of independent movement (Malawista and de Boisfleury-Chevance, 1982; Dyett *et al.*, 1986; Stossell, 1990; Malawista and de Boisfleury-Chevance, 1991). However, whole PMNs that have undergone zeiosis are less capable of motility than are cells that have not (Bignold and Ferrante, 1988). The mechanisms of the formation of these structures have not been widely discussed (Harris, 1973a) but could possibly represent a loss of intracellular coordination of the motor and transmission of the cell.

D. Membrane Fluidity

This term refers to the lateral moveability of constituents of cell membranes in relation to the fluid-mosiac model of cell membrane (de Petris and Raff, 1973) and, in particular, to the motions of phospholipid acyl side chains (Edidin, 1987). Its detection depends on visualization of cells while under the influence of causative agents. Increased fluidity of the membranes of PMNs have been reported as an effect of binding of chemotactic peptide, and the fluidity in turn affects the affinity and numbers of binding site on cell membrane (Collins *et al.*, 1990). Speculatively, it would seem that if the transmission of the cell involves the cell membrane (Bignold, 1987a), then membrane fluidity could be an index of the efficiency of this system. However, whether or not membrane fluidity is associated with increased movement, chemotaxis, polarizaion, or spreading has apparently not been investigated.

E. Streaming

Streaming refers to the forward movement of a central column of cytoplasm (endoplasm) through a well-defined cylinder of stationary outer cytoplasm (ectoplasm) and is distinct from simple forward flow of cytoplasm into pseudopods. It can be observed in isolated cytoplasm (Allen, 1973). The ectoplasm is stationary only when the cell is attached to a solid substratum. In suspended cells the ectoplasm moves backwards in relation to the cell (fountain streaming; Allen, 1961, 1973). Streaming is prominent in certain plant cells (Kuroda, 1990), and in *Chaos chaos* but absent in *Hyalodiscus* (Hülsmann and Habery, 1973). Kuroda (1990) suggested that the mechanism of streaming may involve interactions between actin in the endoplasm and myosin in the ectoplasm.

Although mentioned in early studies of PMN movements (Sabin, 1923; Mudd *et al.*, 1934), streaming has not subsequently been noted (Robineaux, 1964; Wilkinson, 1978; Wilkinson and Lackie, 1979). The possible significance of streaming to random motility of PMNs therefore seems limited.

VI. Current Assays of PMN Motility in Relation to Ameboid Movement

The general features of the various types of assays of random motility and chemotaxis of neutrophils that are in current use have been previously described (Bignold, 1988a, 1989). The following discussion deals with the ways in which the outcomes of these assays can be influenced by the component cell-biological phenomena related to ameboid movement.

A. Direct Microscopic Assays

Since 1970 assays involving direct microscopy of living PMNs have included those using cells being sedimented onto glass or plastic from either whole blood (Ramsey, 1972a,b; Howard, 1982, 1986; Elgefors and Olling, 1984) or from suspensions of PMNs in artificial media, such as Hanks solution (Zigmond and Hirsch, 1973; Cheung *et al.*, 1982; Maher *et al.*, 1984). The potential significance of source of the adherent cells is that during sedimentation of cells from whole blood, components of plasma, especially proteins, adhere to the substratum, so that the resultant PMNs are subsequently allowed to migrate on substratum that has been coated to a greater or lesser extent with protein and possible other intermediaries (Section IV,C; Fig. 5a and Table I). In experiments using cells sedimented from simple media, such as Hanks solution alone, the same problem may still be present because plasma proteins are not necessarily completely removed from the surfaces of PMNs by washing. Therefore, even in the preparations in which exclusion of plasma is intended, some may remain attached to the cell surface and act as intermediaries of unknown concentration in the subsequent assay of motility.

In general, the cell-biological problems related to random motility that arise in these methods have not been addressed. In early methods using clots of whole blood and polished chambers (Sabin 1923; McCutcheon, 1923, Section 2) the cells were subjected to no techniques of separation from whole blood, and the substratum therefore was fibrin strands in the presence of plasma proteins as possible intermediaries. The rates of migration obtained by this method ranged from 10 to 50 μm/min, with an average speed of approximately 34 μm/min. In contrast, when PMNs are used after they have been allowed to sediment onto and become adherent to glass, their migration rates are much reduced. If the sedimentation has been from a fresh blood clot with subsequent rinsing in physiological salt solutions, the substratum is effectively glass, which is coated with an unknown amount of plasma proteins and possibly other constituents. The

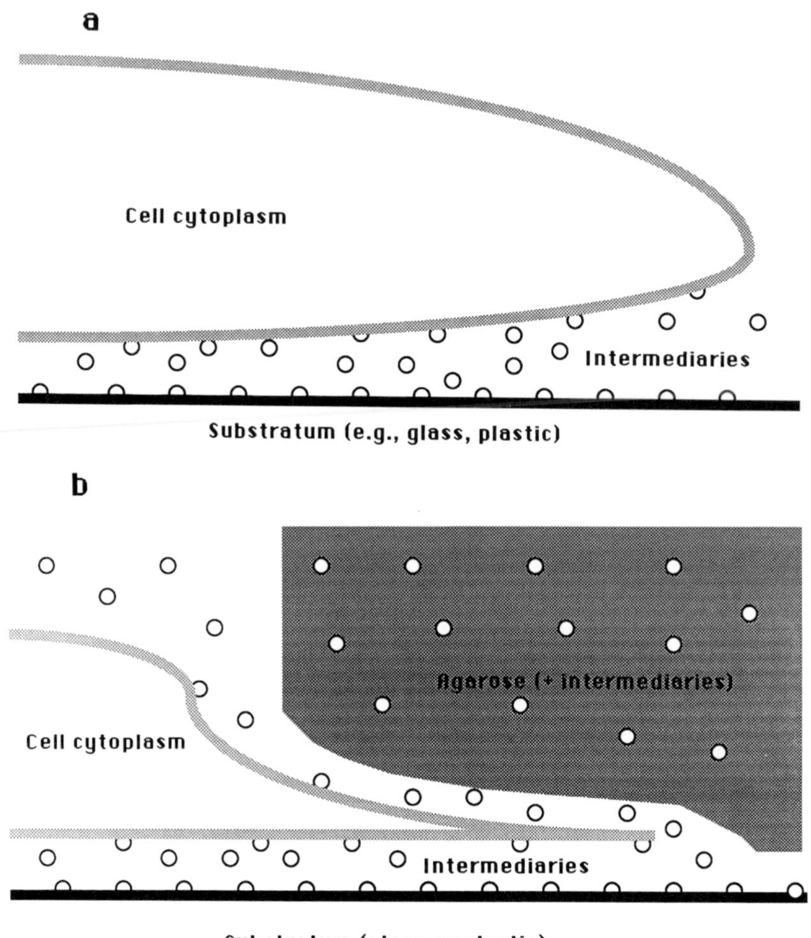

FIG. 5 Diagrams of direct microscopic (a), under agarose (b), and filtration membrane (c) assays of motility of PMNs.

rates of migration under these circumstances is 7–10 μm/min (Ramsey, 1972a,b). In techniques in which the neutrophils have been first separated from blood and then allowed to sediment onto the glass from simple media such as Hanks solution, without protein, the rates of migration are even lower than for cells directly separated from blood (Zigmond and Hirsch, 1973; Zigmond, 1977; Lackie, 1982; Keller *et al.*, 1983) (Table I). The mechanism of the higher migration rates that occur in plasma clots in comparison with all other assays of chemotaxis has not been established. The mechanism(s) of the differences in rates of motility of PMN in blood

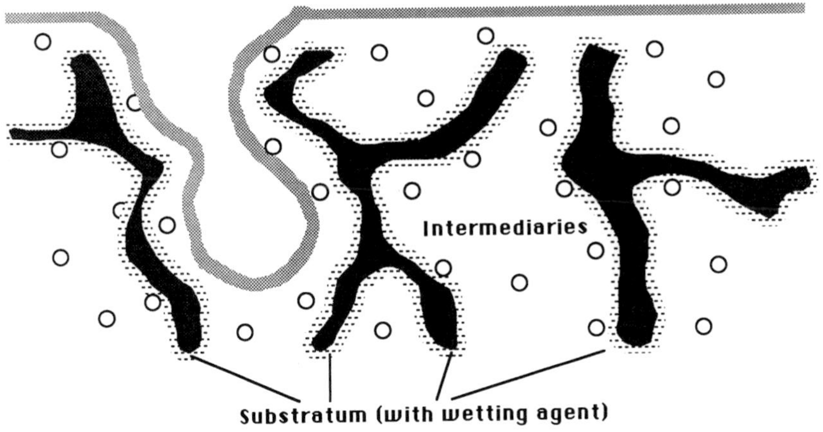

FIG. 5 *(continued)*

clots compared with separated cells on solid substrata with simple media could be the following: (1) that the cells are more viable (not being subjected to any technique of separation from whole blood); (2) that the (unknown) intermediaries in plasma (in relation to the fibrin substratum) are critical for optimal migration of cells; (3) that adhesion to fibrin is more suitable for motility of neutrophils perhaps being either greater or lesser than adhesion to solid substrata; or (4) that substances released in blood during clotting under the conditions of the assay enhance random motility over that occurring under the other conditions (i.e., induce chemokinesis).

Other modifications of direct microscopic methods of measurement of random motility and chemotaxis have included attempts to avoid solid substrata by using purified gels, such as collagen (Allan and Wilkinson, 1978; Brown, 1982; Grinnell, 1982; Reid and Newman, 1991). These techniques raise similar questions of cell viability and interaction with substratum.

Ease of estimation of other cell-biological phenomena associated with random motility varies with the details of the direct microscopic assay. In assays involving cells adherent to glass, rates and directions of random motility can be assessed without difficulty, as can spreading and adhesion to the transparent substratum (Keller *et al.*, 1979, 1983; Keller and Zimmermann, 1985). However, when the cells are tested during migration in plasma clot or collagen gels, the shapes of the cells tend to adapt to the fibers of the gel, so that assessment of spreading is not possible. Assessment of adhesion is also difficult because applying a distraction force to a cell attached to nonrigid material (the gel) is likely to disrupt the latter structure.

B. Asymmetric Migration Assays

The principle of asymmetric migration of neutrophils from a starting locus has been used for assays of chemotaxis at least since 1925 (Harris, 1954). In its usual current form (see, e.g., Nelson *et al.*, 1975, 1976, 1978), three equally spaced wells are cut in a row in solid agarose in a plastic or glass Petri dish, and cells are placed in the middle well. When incubated at 37°C, cells in the well migrate outward by insinuating themselves between the agarose and the floor of the dish. Characteristically, rates of migration of cells from the well are less than 5 μm/min, indicating that the environment of the cells must be the major influence on their rates of movement.

Cell-biological problems of the "under agarose" assay methods include the following (Fig. 5b). The lower surfaces of the cells move on the glass or plastic of the floor of the plate, but the upper surfaces of the cells are in contact with the agarose above. Whether or not traction (adhesiveness) is optimal for either or both materials in the assays is usually not established. Protein additives to the agarose have been recognized to be important for allowing random motility in "under agarose" assays (Chenoweth *et al.*, 1979), but whether or not these proteins act by reducing adhesion of the agarose to the floor of the plate (allowing simply easier insinuation of the cells) or act on the cell directly to allow either appropriate traction or preservation of cell viability or have another effect is unknown. Polarization spreading, and adhesion have been little studied in relation to agarose and protein additives, so that the contribution of these cell-biological factors to random motility in this type of assay is difficult to assess. In unpublished studies by the present author, neutrophils were found to be more adherent to agarose containing either gelatin or serum albumin than to glass, but the influence of this factor on cell motility between two such surfaces was not established. Certain preparations of agarose may contain toxic impurities (Dahlgren, 1986), but this source of possible effects on the viability of the cells, and hence on component mechanisms of random motility including adhesion, was not investigated.

C. Filter Membrane (Boyden Chamber) Assays

The third major method of measuring random motility of PMNs involves filtration membranes and was introduced by Boyden (1962). The method was subsequently popularized by Wilkinson (1982), Keller and Sorkin

(1965a,b, 1966), and Ward *et al.* (1965). The technique involves a chamber consisting of an upper and a lower compartment separated by the microporous membrane. Cells in suspension are placed in the upper compartment and after a suitable period of incubation, either the distance the cells have migrated into the filter (Zigmond and Hirsch, 1973) or the proportion of cells that have migrated completely through the membrane (Boyden, 1962; Keller *et al.*, 1972) is assessed. Characteristically, rates of random migration are 1–2 μm/min.

Aspects of random motility and related cell-biological phenomena are important in these assays (Fig. 5c). First, with cellulose ester membranes the cells negotiate a dense mesh-like substratum, which must, at least in part, account for the low rates of detected motility. Second, the presence of protein additives affecting adhesion in the incubation medium appears to be of particular importance (Bignold, 1986; see also Section III).

Other cell-biological features, especially polarization and spreading of PMNs in relation to cellulose ester membrane, cannot be assessed in Boyden chamber-type assays because the cells cannot be microscopically observed during this period of the experiment (the filtration membrane is not optically clear).

In addition to these problems, consideration of the properties of filtration membranes as substrata is complicated by the different wetting agents that the various manufacturers include in their preparations of filtration membranes (Millipore: Triton X; Sartorius: methylcellulose and Nuclepore: polyvinylpyrollidone) (Cahn, 1967). Wettability of membrane and hence concentrations of wetting agent appear to vary between batches of each manufacturer's product. In the case of Nuclepore (polycarbonate) membrane, removal of the wetting agent can be effected and is known to be essential for adhesion of neutrophils (Bignold, 1987c). However, with cellulose ester membranes no suitable wetting agent-free membrane of suitable pore size is marketed. Attempts to extract the wetting agent prior to use in chemotactic assays has, in the author's laboratory, resulted in membranes that do not permit the ingress of neutrophils. Whether this was due to the membrane being rendered hyperadhesive or disorganized has not been established. Little is therefore known of the possible relative significance of the wetting agents or the cellulose ester substratum, or their interaction with protein intermediaries for cell viability, adhesion, polarization, spreading, or other cell-biological phenomena for assays of chemotaxis using such membranes.

As an alternative to cellulose ester, filtration membranes composed of nuclear track-etched polycarbonate (Nuclepore, Pleasanton, California) are available (Bignold, 1988a). These are translucent and permit studies of adhesion, spreading, random motility, and chemotaxis of PMNs on the same substratum (Bignold *et al.*, 1990). Such membranes can be incorporated into transparent chambers (Sykes and Moore, 1959), so that photographic movements of the cells on the same membrane, as is used for the chemotaxis assay, is feasible (L. P. Bignold, unpublished observations). Without clearer understanding of the cell biology of random motility, however, it is not possible to establish whether or not this technique is ideal.

VII. Conclusions

Alterations to random motility, including increases (chemokinesis) of PMNs in an *in vitro* assay can theoretically result from (1) induction of or recovery from sublethal cell injury to the cells during separation or incubation, (2) changes to the motor of the cell, (3) changes to traction (adhesion) of the cells for the particular substratum, and (4) possibly variation in the transmission of the cells' motile mechanism(s).

There is as yet no well-known assay of sublethal injury to PMNs, although zeiosis (blebbing) may be an indication of certain injuries having occurred. Duration of survival (to the end-point of nonviability as assessed by dye exclusion) of the cells in simple incubation may be a suitable measure of sublethal injury but has not been investigated.

The activity of the motor of PMNs is difficult to assay, especially in view of the uncertainty concerning its nature. Either polarization or rapidity of actin polymerization may be suitable methods. Cytoplasmic streaming is not seen in PMNs according to recent reports and therefore seems to be of no value in assays of the motor of these cells. No assay is yet available for the functioning of the transmission of cytoplasmic forces through the membrane of the cell to the substratum, although speculatively membrane fluidity could be involved.

Assays of adhesion are well recognized, but attention is drawn to groups of factors (cell viability, membrane structures, intermediaries, and substratum) that are involved in this phenomenon.

Of the various assay types, direct microscopy allows for concurrent documentation of adhesion and additional features, such as polarization and spreading. Under agarose methods are complicated by the presence of two substrata between which the cells move, and to one of which (agarose) adhesion is difficult to measure. Boyden chamber methods using cellulose ester filter membranes can allow assays of adhesion but not of polarization or spreading. Polycarbonate membranes in Boyden chambers allow such assessment of polarization and spreading.

Acknowledgments

This work was supported by the Australian Research Council.

References

Abercrombie, M. (1980). *Proc.R. Soc. London, Ser. B* **207,** 129–147.

Abercrombie, M., Heaysman, J. E. M., and Pegrum, S. M. (1970). *Exp. Cell Res.* **62,** 389–398.

Abercrombie, M., Heaysman, J. E. M., and Pegrum, S. M. (1972). *Exp. Cell Res.* **73,** 536–539.

Adams, R. J., and Pollard, T. D. (1989). *Cell Motil. Cytoskel.* **14,** 178–182.

Albelda, S. M., and Buck, C. A. (1990). *FASEB J.* **4,** 2868–2880.

Allan, R. B., and Wilkinson, P. C. (1978). *Exp. Cell Res.* **111,** 191–203.

Allen, R. D. (1961). *In* "The Cell: Biochemistry, Physiology, Morphology" (J. Brachet and A. E. Mirsky, eds.), Vol. 2, pp. 135–216. Academic Press, New York.

Allen, R. D. (1973). *In* "The Biology of the Amoebae" (K. W. Jeon, ed.), pp. 201–248. Academic Press, New York.

Allen, R. D., and Kamiya, N., eds.) (1964). "Primitive Motile Systems in Biology." Academic Press, New York.

Allen, R. D., Cooledge, J. W., and Hall, P. J. (1960). *Nature* (*London*) **187,** 896–899.

Allison, A. C. (1973). *Ciba Found. Symp.* **14,** 109–142.

Bangham, A. D. (1964). *Ann. N.Y. Acad. Sci.* **116,** 945–949.

Bansal, V. S., and Majerus, P. W. (1990). *Annu. Rev. Cell Biol.* **6,** 41–67.

Bengtsson, T. (1990). *Exp. Cell Res.* **191,** 57–63.

Berridge, M. J. (1986). *J. Cell Sci., Suppl.* No. 4, 137–154.

Bessis, M., and de Boisfleury, A. (1976). *Blood Cells* **2,** 365–410.

Bignold, L. P. (1986). *Cell Biol. Int. Rep.* **10,** 535–543.

Bignold, L. P. (1987a). *Experientia* **43,** 680–686.

Bignold, L. P. (1987b). *Cell Biol. Int. Rep.* **11,** 19–25.

Bignold, L. P. (1987c). *J. Immunol. Methods* **105,** 275–280.

Bignold, L. P. (1988a). *J. Immunol. Methods* **108,** 1–18.

Bignold, L. P. (1988b). *J. Immunol. Methods* **110,** 145–148.

Bignold, L. P. (1989). *J. Immunol. Methods* **118,** 217–225.

Bignold, L. P., and Ferrante, A. (1988). *Cell Biol. Int. Rep.* **12,** 195–203.

Bignold, L. P., Rogers, S. D., and Harkin, D. G. (1990) *Eur. J. Cell Biol.* **53,** 27–34.

Birchmeier, W. (1981). *Trends Biochem. Sci.* **6,** 234–237.

Bongrand, P., ed. (1988). "Physical Basis of Cell–Cell Adhesion." CRC Press, Boca Raton, Florida.

Bovee, E. C. (1964). *In* "Primitive Motile Systems in Cell Biology" (R. D. Allen and N. Kayima, eds.), pp. 189–220. Academic Press, New York.

Bovee, E. C., and Jahn, T. L. (1973). *In* "The Biology of the Amoebae" (K. W. Jeon, ed.), pp. 249–290. Academic Press, New York.

Boyden, S. V. (1962). *J. Exp. Med.* **115,** 453–466.

Bretscher, M. S. (1984). *Science* **224,** 681–686.

Bretscher, M. S. (1987). *Sci. Am.* **257,** 44–50.

Bretscher, M. S. (1988). *J. Cell Biol.* **106,** 235–237.

Brown, A. F. (1982). *J. Cell Sci.* **58,** 455–467.
Buckley, I. K. (1963). *Int. Rev. Exp. Pathol.* **2,** 241–356.
Burridge, K., Molony, L., and Kelly, T. (1987). *J. Cell Sci., Suppl.* No. 8, 211–230.
Burridge, K., Fath, K., Kelly, T., Nuckolls, G., and Turner, C. (1988). *Annu. Rev. Cell Biol.* **4,** 487–525.
Cahn, R. D. (1967). *Science* **155,** 195–196.
Campbell, A. K. (1987). *Clin. Sci.* **72,** 1–10.
Chenoweth, D. E., Rowe, J. G., and Hughli, T. E. (1979). *J. Immunol. Methods* **25,** 337–353.
Cheung, A. T. W., Miller, M. E., and Keller, S. R. (1982). *J. Reticuloendothel. Soc.* **31,** 193–205.
Cochrane, C. G., and Griffin, J. H. (1982). *Adv. Immunol.* **33,** 241–306.
Colditz, I. G. (1985). *Surv. Synth. Pathol. Res.* **4,** 44–68.
Collins, J. M., Scott, R. B., and Grogan, W. M. (1990). *J. Cell Physiol.* **144,** 42–51.
Comandon, J. (1917). *C. R. Seances Soc. Biol. Ses Fil.* **80,** 314–316.
Cross, G. A. M. (1990). *Annu. Rev. Cell Biol.* **6,** 1–39.
Culp, L. A. (1978). *Curr. Top. Membr. Transp.* **11,** 327–396.
Curtis, A. S. G. (1967). "The Cell Surface." Logos Press and Academic Press, New York.
Curtis, A. S. G. (1973). *Prog. Biophys. Mol. Biol.* **27,** 315–386.
Curtis, A. S. G. (1988). *In* "Physical Basis of Cell–Cell Adhesion" (P. Bongrand, ed.), pp. 207–226. CRC Press, Boca Raton, Florida.
Cutts, J. H. (1970). "Cell Separation." Academic Press, New York.
Dahlgren, C. (1979). *Linköping Univ. Med. Diss.* **66.** Cited in Lackie (1982).
Dahlgren, C. (1986). *J. Immunol. Methods* **88,** 287–288.
de Bruyn, P. P. H. (1944a). *Anat. Rec.* **89,** 43–63.
de Bruyn, P. P. H. (1944b). *Q. Rev. Biol.* **22,** 1–24.
de Bruyn, P. P. H. (1946). *Anat. Rec.* **95,** 177–192.
Dembo, M., and Bell, G. I. (1987). *Curr. Top. Membr. Transp.* **19,** 71–89.
de Petris, S., and Raff, M. C. (1973). *Ciba Found. Symp.* **14,** 27–40.
Dittrich, H. (1962). *In* "The Physiology and Pathology of Leukocytes" (H. Braunsteiner and D. Zucker-Franklin, eds.), pp. 130–151. Grune & Stratton, New York.
Dunn, G. A. (1980). *In* "Cell Adhesion and Motility" (A. S. G. Curtis and J. D. Pitts, eds.), pp. 409–424. Cambridge Univ. Press, Cambridge.
Dyett, D. E., Malawista, S. E., Naccache, P. H., and Sha'afi, R. I. (1986). *J. Clin. Invest.* **77,** 34–37.
Eberle, M., Traynor-Kaplan, A. E., Sklar, L. A., and Norgauer, J. (1990). *J. Biol. Chem.* **265,** 16725–16728.
Edidin, M. (1987). *Curr. Top. Membr. Transp.* **29,** 91–127.
Elder, H. Y. (1980) *In* "Aspects of Animal Movement" (H. Y. Elder and E. R. Trueman, eds.), pp. 71–92. Cambridge Univ. Press, Cambridge.
Elgefors, B., and Olling, S. (1984). *Acta Pathol. Microbiol. Scand., Sect. C* **92C,** 113–119.
Farese, R. V. (1988). *Am. J. Med. Sci.* **296,** 223–230.
Fechheimer, M., and Zigmond, S. H. (1983). *Cell Motil.* **3,** 349–361.
Fehr, J., and Dahinden, C. (1979). *J. Clin. Invest.* **64,** 8–16.
Ferguson, M. A. J., and Williams, A. F. (1988). *Annu. Rev. Biochem.* **57,** 285–320.
Ferrante, A., Beard, J. L., and Thong, Y. H. (1980). *Clin. Exp. Immunol.* **39,** 532–537.
Forrester, J. V., and Lackie, J. M. (1984). *J. Cell Sci.* **70,** 93–110.
Forsyth, K. D., and Levinsky, R. J. (1989). *Clin. Exp. Immunol.* **75,** 265–268.
Fukui, Y., De Lozanne, A., and Spudich, J. A. (1990). *J. Cell Biol.* **110,** 367–378.
Gallin J. I. (1985). *J. Infect. Dis.* **152,** 661–664.
Gallin J. I., Gallin, E. K., Malech, H. L., and Cramer, E. B. (1978). *In* "Leukocyte Chemotaxis: Methods, Physiology and Clinical Implications" (J. I. Gallin and P. G. Quie, eds.), pp. 123–142. Raven, New York.

Gill, D. L. (1989). *Nature (London)* **342,** 16–18.
Giroud, J. P., and Roch-Arvellier, M. (1982). *Trends Pharmacol. Sci.* **3,** 447–450.
Godman, G. C., and Miranda, A. F. (1978). *Front. Biol.* **46,** 277–429.
Goldacre, R. J. (1961). *Exp. Cell Res., Suppl.* No. 8, 1–16.
Goldstein, I. M. (1988). *In* "Inflammation: Basic Principles and Clinical Correlates" (J. I. Gallin, I. M. Goldstein, and R. Snyderman, eds.), pp. 55–74. Raven, New York.
Grebecki, A. (1984). *Protoplasma* **123,** 116–124.
Grebecki, A. (1986). *J. Cell Sci.* **83,** 23–36.
Grinnell, F. (1978). *Int. Rev. Cytol.* **53,** 65.
Grinnell, F. (1982). *J. Cell Sci.* **58,** 95–144.
Hahn, F. (1969). *In* "Polypeptides which Affect Smooth Muscles and Blood Vessels" (M. Schacter, ed.), pp. 275–292. Pergamon, New York.
Harris, A. K. (1973a). *Ciba Found. Symp.* **14,** 3–26.
Harris, A. K. (1973b). *Exp. Cell Res.* **77,** 285–297.
Harris, H. (1953). *J. Pathol. Bacteriol.* **64,** 135–146.
Harris, H. (1954). *Physiol.Rev.* **34,** 529–562.
Hartwig, J. H., Niederman, R., and Lind, S. E. (1985). *Subcell. Biochem.* **11,** 1–49.
Henderson, M. (1928). *Anat. Rec.* **38,** 71–95.
Higashi-Fujime, S. (1991). *Int. Rev. Cytol.* **125,** 95–138.
Hogg, N. (1989). *Immunol. Today* **10,** 111–114.
Howard, T. H. (1982). *Blood* **59,** 946–951.
Howard, T. H. (1986). *Blood* **67,** 1036–1042.
Hülsmann, N., and Habery, M. (1973). *Acta Protozool.* **12,** 71–82.
Huxley, H. E. (1973). *Nature (London)* **243,** 445–449.
Huxley, H. E. (1979). *In* "Cell Motility: Molecules and Organisation" (S. Hatano, H. Ishikawa, and H. Sato, eds.), pp. 3–9. Univ. Park Press, Baltimore.
Jadwin, D. F., Smith, C. W., and Meadows, T. R. (1981). *Am. J. Clin. Pathol.* **76,** 395–402.
Jolly, J. (1913). *C. R. Seances Soc. Biol. Ses Fil.* **64,** 504–506.
Keller, H. U. (1983). *Cell Motil.* **3,** 47–60.
Keller, H. U. (1985). *In* "Handbook of Inflammation. Vol. 5: The Pharmacology of Inflammation" (I. L. Bonta, M. A. Bray, and M. J. Parnham eds.), pp. 137–166. Elsevier, Amsterdam.
Keller, H. U., and Cottier, H. (1982). *Prog. Clin. Biol. Res.* **85,** 415–423.
Keller, H. U., and Sorkin, E. (1965a). *Immunology* **9,** 241–247.
Keller, H. U., and Sorkin, E. (1965b). *Immunology* **9,** 441–447.
Keller, H. U., and Sorkin, E. (1966). *Immunology* **10,** 409–416.
Keller, H. U., and Zimmermann, A. (1985). *Cell Motil.* **5,** 447–461.
Keller, H. U., Borel, J. F., Wilkinson, P. C., Hess, M. W., and Cottier, H. (1972). *J. Immunol. Methods* **1,** 165–168.
Keller, H. U., Hess, M. W., and Cottier, H. (1977a). *Experientia* **33,** 1386–1387.
Keller, H. U., Wilkinson, P. C., Abercrombie, M., Becker, E. L., Hirsch, J. G., Miller, M., Ramsey, W. S., and Zigmond, S. H. (1977b). *Clin. Exp. Immunol.* **27,** 377–380.
Keller, H. U., Barandun, S., Kistler, P., and Ploem, J. S. (1979). *Exp. Cell Res.* **122,** 351–362.
Keller, H. U., Wilkinson, P. C., Abercrombie, M., Becker, E. L., Hirsch, J. G., Miller, M., Ramsey, W. S., Zigmond, S. H., Austen, K. F., Baum, J., Borel, J. F., Curtis, A. S. G., Dunn, G. A., Gallin, J. I., Goetzl, E. J., Harris, A. K., Sorkin, E., Trinkhaus, J. P., Vasiliev, J. M., Weiss, L., and Wissler, J. H. (1982). *Bull. W. H. O.* **58,** 505–509.
Keller, H. U., Zimmermann, A., and Cottier, H. (1983). *J. Cell Sci.* **64,** 89–106.
Kishimoto, T. K., Larsson, R. S., Corbi, A. L., Dustin, M. L., Staunton, D. E., and Springer, T. A. (1989). *Adv. Immunol.* **46,** 149–182.
Klebanoff, S. J., and Clark, R. A. (1978). "The Neutrophil: Functional and Clinical Disorders." North Holland, Amsterdam.

Kuroda, K. (1979). *In* "Cell Motility: Molecules and Organisation" (S. Hatano, H. Ishikawa, and H. Sato, eds.), pp. 347–362. Univ. Park Press, Baltimore.

Kuroda, K. (1990). *Int. Rev. Cytol.* **121,** 267–307.

Lackie, J. M. (1982). *In* "Cell Behaviour" (R. Bellairs, A. S. G. Curtis, and G. Dunn, eds.), Cambridge Univ. Press, Cambridge.

Lackie, J. M. (1986). "Cell Movement and Cell Behaviour." Allen & Unwin, London.

Lewis, W. H. (1934). *Bull. Johns Hopkins Hosp.* **55,** 273–279.

Lorch, J. (1973). *In* "The Biology of Amoeba" (K. W. Jeon, ed.), pp. 1–36. Academic Press, New York.

Low, M. G. (1989). *Biochim. Biophys. Acta* **988,** 427–454.

Maher, J., Martell, J. V., Brantley, B. A., Cox, E. B., Niedel, J. E., and Rosse, W. F. (1984). *Blood* **64,** 221–228.

Malawista, S. E., and de Boisfleury-Chevance, A. (1982). *J. Cell Biol.* **95,** 960–973.

Malawista, S. E., and de Boisfleury-Chevance, A. (1991). *J. Leukocyte Biol.* **50,** 313–315.

Marks, P. W., and Maxfield, F. R. (1990). *J. Cell Biol.* **110,** 43–52.

Mast, S. O. (1931). *Protoplasma* **14,** 321–330.

McCutcheon, M. (1923). *Am. J. Physiol.* **66,** 180–184.

McCutcheon, M. (1946). *Physiol. Rev.* **26,** 319–336.

Movat, H. Z. (1985). "The Inflammatory Reaction." Elsevier, Amsterdam.

Mudd, S., McCutcheon, M., and Lucke, B. (1934). *Physiol. Rev.* **14,** 210–275.

Müller-Eberhard, H. J. (1988). *In* "Inflammation: Basic Principles and Clinical Correlates" (J. I. Gallin, I. M. Goldstein, and R. Snyderman, eds.), pp. 21–54. Raven, New York.

Naccache, P. H. (1987). *Int. Rev. Cytol., Suppl.* No. 17, 457–492.

Nelson, R. D., Quie, P. G., and Simmons, R. L. (1975). *J. Immunol.* **115,** 1650–1656.

Nelson, R. D., Simmons, R. L., and Quie, P. G. (1976). *In "In Vitro* Methods in Cell Mediated and Tumour Immunity" (B. R. Bloom and J. R. David, eds.), pp. 663–672. Academic Press, New York.

Nelson, R. D., McCormack, R. T., and Fiegel, V. D. (1978). *In* "Leukocyte Chemotaxis: Methodology, Physiology and Clinical Implications" (J. I. Gallin and P. G. Quie, eds.), pp. 25–42. Raven, New York.

Newman, I., and Wilkinson, P. C. (1989). *Immunology* **66,** 318–329.

Odell, G. M., and Frisch, H. L. (1975). *J. Theor. Biol.* **50,** 59–86.

O'Flaherty, J. T., Kreutzer, D. L., and Ward, P. A. (1977). *J. Immunol.* **119,** 232–239.

O'Flaherty, J. T., Kreutzer, D. L., and Ward, P. A. (1978). *Am. J. Pathol.* **90,** 537–550.

Patarroyo, M. (1991). *Clin. Immunol. Immunopathol.* **60,** 333–348.

Pethica, B. A. (1961). *Exp. Cell Res., Suppl.* No. 8, 123–140.

Ponder, C. W. (1908). *Lancet* **ii,** 1746–1747.

Ramsey, W. S. (1972a). *Exp. Cell Res.* **72,** 489–501.

Ramsey, W. S. (1972b). *Exp. Cell Res.* **72,** 129–139.

Ramsey, W. S., Hertl, W., Nowlan, E. D., and Binkowski, N. J. (1984). *In Vitro* **20,** 802–808.

Rather, L. J. (1972). "Addison and the White Corpuscles: An Aspect of Nineteenth Century Biology." Univ. of California Press, Berkeley.

Rebuck, J. W., and Crowley, J. H. (1955). *Ann. N.Y. Acad. Sci.* **59,** 757–760.

Reid, G. G., and Newman, I. (1991). *Cell Biol. Int. Rep.* **15,** 711–720.

Rich, A., and Harris, A. K. (1981). *J. Cell Sci.* **50,** 1–7.

Rinaldini, L. M. J. (1958). *Int. Rev. Cytol.* **7,** 587–647.

Robineaux, R. (1964). *In* "Primitive Motile Systems in Cell Biology" (R. D. Allen and N. Kamiya, eds.), pp. 351–363. Academic Press, New York.

Roche e Silva, M. (1964). *Ann. N.Y. Acad. Sci.* **116,** 899–911.

Rolleston, H. (1934). *Proc. R. Soc. Med.* **27,** 1161–1178.

Roos, F. J., Zimmermann, A., and Keller, H. U. (1987). *J. Cell Sci.* **88,** 399–406.

Sabin, F. R. (1923). *Bull. Johns Hopkins Hosp.* **34,** 278–288.

Sadler, K. L., and Badwey, J. A. (1988). *Hematol. Oncol. Clin. North Am.* **2,** 185–200.

Schreiner, A., and Hopen, G. (1979). *Acta Pathol. Microbiol. Scand., Sect. C* **87C,** 333–340.

Schroeder, T. E. (1978). *Front. Biol.* **46,** 91–112.

Senda, N., Tamura, H., Shibata, N., Yoshitake, J., Kondo, K., and Tanaka, K. (1975). *Exp. Cell Res.* **91,** 393–407.

Sheetz, M. P., Vale, R., Schapp, B., Schroer, T., and Reese, T. (1986). *J. Cell Sci., Suppl.* No. 5, 181–188.

Sheterline, P. (1983). "Mechanisms of Cell Motility." Academic Press, New York.

Shields, J. M., and Haston, W. S. (1985). *J. Cell Sci.* **74,** 75–93.

Singer, S. J. (1990). *Annu. Rev. Cell Biol.* **6,** 247–296.

Singer, S. J., and Kupfer, A. (1986). *Annu. Rev. Cell Biol.* **2,** 337–365.

Singer, D. J., and Nicholson, G. L. (1972). *Science* **175,** 720–731.

Smith, C. W., Hollers, J. C., Patrick, R. A., and Haslett, C. (1979). *J. Clin. Invest.* **63,** 221–229.

Snyderman, R., Smith, C. D., and Verghese, M. W. (1986). *J. Leukocyte Biol.* **40,** 785–800.

Southwick, F. S., and Stossell, T. P. (1983). *Semin. Haematol.* **20,** 305–321.

Southwick, F. S., Dabiri, G. A., Paschetto, M., and Zigmond, S. H. (1989). *J. Cell Biol.* **109,** 1561–1569.

Stossell, T. P. (1982). *Philos. Trans. R. Soc. London, Ser. B* **299,** 275–289.

Stossell, T. P. (1988). *In* "Inflammation: Basic Principles and Clinical Correlates" (J. I. Gallin, I. M. Goldstein, and R. Snyderman, eds.), pp. 325–342. Raven, New York.

Stossell, T. P. (1989). *J. Biol. Chem.* **264,** 18261–18264.

Stossell, T. P. (1990). *Am. Sci.* **78,** 408–423.

Sullivan, J. A., and Mandell, G. L. (1983). *Cell Motil.* **3,** 31–46.

Svetina, S., and Zeks, B. (1990). *J. Theor. Biol.* **146,** 115–122.

Sykes, J. A., and Moore, E. B. (1959). *Proc. Soc. Exp. Biol. Med.* **100,** 125–127.

Taylor, A. C. (1961). *Exp. Cell Res., Suppl.* No. 8, 154–173.

Taylor, D. L., and Condeelis, J. S. (1979). *Int. Rev. Cytol.* **56,** 57–144.

Taylor, D. L., and Fechheimer, M. (1982). *Philos. Trans. R. Soc. London, Ser. B* **299,** 185–196.

Taylor, E. W. (1986). *J. Cell Sci., Suppl.* No. 4, 89–102.

Trueman, E. R., and Jones, H. D. (1977). *In* "Mechanics and Energetics of Animal Locomotion" (R. Alexander and G. Goldspink, eds.), pp. 204–221. Chapman & Hall, London.

Tsein, R. W., and Tsein, R. Y. (1990). *Annu. Rev. Cell Biol.* **6,** 715–760.

Valerius, N. H. (1983). *Acta Pathol. Microbiol. Scand., Sect. C* **91C,** 43–49.

Vasiliev, J. M. (1982). *Philos. Trans. R. Soc. London, Ser. B* **299,** 159–168.

Vasiliev, J. M. (1985). *Biochim. Biophys. Acta* **780,** 21–65.

Vasiliev, J. M. (1991). *J. Cell Sci.* **98,** 1–4.

Wallace, P. J., Weston, R. P., Packham, C. H., and Lichtman, M. A. (1984). *J. Cell Biol.* **99,** 1060–1065.

Ward, P. A. (1968). *Biochem. Pharmacol.* **17,** Suppl., 99–105.

Ward, P. A., Cochrane, C. G., and Müller-Eberhard, H. J. (1965). *J. Exp. Med.* **122,** 327–346.

Weatherbee, J. A. (1981). *Int. Rev. Cytol., Suppl.* No. 12, 113–176.

Weiss, L. (1960). *Int. Rev. Cytol.* **9,** 187–225.

Wessels, N. K., Spooner, B. S., and Luduena, M. A. (1973). *Ciba Found. Symp.* **14,** 53–82.

Westwick, J., and Poll, C. (1986). *Agents Actions* **19,** 80–86.

Wilhelm, D. L. (1971). *Annu. Rev. Med.* **22,** 63–84.

Wilkinson, P. C. (1976). *Exp. Cell Res.* **103,** 415.

Wilkinson, P. C. (1978). *Handb. Exp. Pharmacol.* **50/1,** 109–137.

Wilkinson, P. C. (1982). "Chemotaxis in Inflammation," 2nd Ed. Churchill-Livingstone, Edinburgh.

Wilkinson, P. C. (1988a). *J. Immunol. Methods* **110,** 143–144.

Wilkinson, P. C. (1988b). *J. Immunol. Methods* **110,** 149.

Wilkinson, P. C., and Lackie, J. M. (1979). *Curr. Top. Pathol.* **68,** 47–88.

Wilkinson, P. C., Haston, W. S., and Shields, J. M. (1982). *Clin. Exp. Immunol.* **50,** 461–473.

Wolpert, L. (1971). *Sci. Basis Med.* pp. 81–98.

Wolpert, L., Thompson, C. M., and O'Neill, C. H. (1964). *In* "Primitive Motile Systems in Cell Biology" (R. D. Allen and N. Kamiya, eds.), pp. 143–168. Academic Press, New York.

Wright, A. E., and Colebrook, L. (1921). "Technique of the Teat and Capillary Tube." Constable, London.

Wyman, M. P., Kernen, P., Bengtsson, T., Andersson, T., Bagglioni, M., and Deranleau, D. A. (1990). *J. Biol. Chem.* **265,** 619–622.

Zigmond, S. H. (1977). *J. Cell Biol.* **75,** 606–616.

Zigmond, S. H. (1978). *In* "Leukocyte Chemotaxis: Methodology, Physiology and Clinical Implications" (J. I. Gallin and P. G. Quie, eds.), pp. 87–96. Raven, New York.

Zigmond, S. H., and Hirsch, J. G. (1973). *J. Exp. Med.* **137,** 387–410.

In Search of Molecular Origins of Cellular Differentiation in *Volvox* and Its Relatives

Rüdiger Schmitt,* Stefan Fabry,* and David L. Kirk†
* Lehrstuhl für Genetik, Universität Regensburg, Regensburg, Germany
† Department of Biology, Washington University, St. Louis, Missouri 63130

I. Introduction: *Volvox* and Its Relatives as an Ontogenetic and Phylogenetic Model

> In the volvocine line, exemplified by the order Volvocales, a series of colonial forms has evolved from the *Chlamydomonas*-like type. . . . Although there are a variety of types, possibilities are limited; in the evolutionary sense, this line has apparently been a blind alley. (*Plant Diversity: An Evolutionary Approach,* Scagel *et al.,* 1969, p. 196)

> Interesting and illustrative as the progression from *Chlamydomonas* to *Volvox* may be, then, it appears to have been an evolutionary blind alley. (*Biology Today,* Kirk, 1975, p. 61)

> . . . increasing specialization in the members of [the volvocine line] is evident in . . . an increase in the cell number and in the size of the colonies . . . [and in] increasing specialization in cell morphology and function. . . . Nevertheless, this line clearly represents an evolutionary 'dead end,' in that it has not given rise to a more complex group of organisms. (*Biology of Plants,* Raven *et al.,* 1986, p. 276)

The only encounter that most nonspecialists will have with *Volvox* in a lifetime is in an introductory course in one of the life sciences, in which *Volvox* and its presumed relatives characteristically are presented as a "textbook example" of an evolutionary paradox.

The volvocine algae range in complexity from unicellular *Chlamydomonas,* through a series of colonial forms, each of which contains a characteristic number of equivalent *Chlamydomonas*-like cells, to *Volvox,* a multicellular organism in which there is a complete division of labor between nonmotile germ cells and terminally differentiated, biflagellate *Chlamydomonas*-like somatic cells. The unrivaled manner in which these green flagellates of the order Volvocales appear to illustrate an evolutionary progression in organismic size and developmental complexity is frequently cited as an indication that they might provide instructive insights into the pathway by which a multicellular organism with differentiated cell types has evolved from a unicellular ancestor (see, e.g., Scagel *et al.*, 1969; Kochert, 1973; Kirk *et al.*, 1975, 1991b; Starr, 1980; Kirk and Harper, 1986; Raven *et al.*, 1986; Kirk, 1988; Rausch *et al.*, 1989; Larson *et al.*, 1992). As the opening quotations illustrate, however, it has also become commonplace for textbooks to conclude a discussion of the Volvocales with a dismissal: discounting the group as an evolutionary cul de sac. If, indeed, it can be considered an evolutionary shortcoming that an extant lineage "has not given rise to a more complex group of organisms" (Raven *et al.*, 1986) it surely is a shortcoming that the volvocine algae share with some of the most conspicuous, successful, and intensively studied groups of modern organisms (including the mammals and the flowering plants), and it is by no means a reason to ignore any developmental and/or evolutionary insights that the group might have to offer.

Perhaps the primary reason that the volvocine algae are so widely dismissed as an evolutionary cul de sac is an assumption, which appears to have persisted over most of the past century, that any group with such a relatively simple set of body plans must be an extremely ancient group that has been subjected to the forces of natural selection for many eons, and has by now had adequate opportunity to test all of of its adaptive possibilities. However, this appears to be a misconception. Data indicate that the colonial and multicellular volvocine algae are relative newcomers to the evolutionary scene, perhaps rivaling the primates in their phylogenetic youthfulness. The apparent recency of the volvocine radiation is of operational significance because it provides reason to hope that the molecular genetic pathway(s) that led from unicellularity to multicellularity and cellular differentiation may still be discernible within the genomes of extant members of this group, having not yet been sufficiently blurred by the winds of genetic drift to have become untraceable.

It is toward this end—discerning the molecular genetic pathway by which the germ/soma dichotomy characteristic of *Volvox* arose during evolution of the volvocine algae from a *Chlamydomonas*-like unicellular ancestor—that much current research on *Volvox* is ultimately directed. Before discussing such studies, we will summarize the morphological and biochemical evidence that supports the volvocine lineage hypothesis (the hypothesis that the volvocine algae constitute something approximating a unidirectional evolutionary progression in size and developmental complexity) and more recent attempts to evaluate the validity of this hypothesis objectively with molecular phylogenetic methods.

Against that conceptual background, we will then review advances that have occurred in the following three areas of *Volvox* research since the field was last reviewed extensively (Kirk and Harper, 1986): (1) the genetic and molecular control of germ/soma differentiation, (2) the mechanism of action of the sexual pheromone, and (3) the comparative anatomy and evolution of selected structural genes of *Volvox* and *Chlamydomonas*.

As in the previous review in this series (Kirk and Harper, 1986), all statements about *Volvox* that are not otherwise qualified will refer to *Volvox carteri* forma *nagariensis*, which has become the standard subject of most *Volvox* research, owing in large measure to its genetic accessibility. A closely related organism, *V. carteri* f. *weismannia* was the first representative of the species to be used as an object of laboratory investigations (Kochert, 1968), but it has not been equally accessible to genetic analysis, and the only study of the past six years in which it figured importantly was a detailed study of flagellar regeneration (Coggin and Kochert, 1986). A third forma, *V. carteri* f. *kawasakiensis*, has now been described (Nozaki, 1988), but additional studies of it have yet to appear.

In one of the few series of comparative biochemical studies of *Volvox* species to have been reported during the period under review, Desnizky (1984, 1985a,b, 1986, 1987, 1990) has shown that the temporal pattern of embryonic cleavage divisions is different in *V. aureus* and *V. tertius* from that observed in *V. carteri* f. *nagariensis*, and that, consequently, cell division in the former two species exhibits significantly greater sensitivity to light deprivation and inhibitors of macromolecular biosynthesis than cell division in *V. carteri* does.

A. Organizational and Reproductive Patterns in the Volvocine Algae, and the Volvocine Lineage Hypothesis

In *C. reinhardtii*, which is the type species of the genus, and the most intensively studied of the more than 450 named species of *Chlamydomonas*, characteristic patterns of subcellular organization, cell wall structure, and reproduction are observed that recur (with certain important modifications) in the family Volvocaceae, which comprises all of the colonial and multicellular members of the so-called volvocine lineage. Because most of these features of *C. reinhardtii* have recently been reviewed in extraordinary depth and clarity (Harris, 1989), we will briefly outline only those features that are particularly relevant to discussions that will follow. These descriptions of *C. reinhardtii* will be followed by brief discussions of some of the similarities and differences that are seen in members of the family Volvocaceae generally and in *Volvox carteri* specifically. Additional comparative information about various aspects of volvocine biology is available from a number of sources (Starr, 1970, 1980; Kochert, 1973; Coleman, 1979; Bell, 1985; Nozaki, 1986; Koufopanou, 1990; Segaar, 1991). A wide range of more-or-less subtle variations in subcellular organization, cell wall structure, and division patterns are observed in other species of *Chlamydomonas,* and in other genera of unicellular green flagellates (Harris, 1989). However, because the equivalent features of members of the family Volvocaceae all appear to be variations on the theme established by *C. reinhardtii,* it is to that species that the description here is restricted.

1. Organization of Vegetative Cells

One of the most striking features of an interphase *C. reinhardtii* cell is the predictable regularity and polarity of its subcellular organization. This regularity, in turn, can be related to the way in which all major organelles are linked to the organizing center of the cell, the basal body complex (BBC), which defines the anterior end of each *C. reinhardtii* cell and from which the two anteriorly directed flagella emerge. The interphase BBC is composed of a pair of mature basal bodies (BBs), a pair of probasal bodies (PBBs), and a complicated, highly regular arrangement of fibers and "rootlets" of varying types that connect the BBs to one another, to the plasmalemma, to the nucleus, and to many other major organelles (Melkonian, 1984b). Of particular importance in establishing spatial relationships within both interphase and dividing cells are the members of a cross-shaped array of four microtubular rootlets (MTRs) that extend from highly specific attachment points on particular BBs and PBBs to traverse the anterior end of the cell just beneath the plasmalemma. MTRs containing four and two microtubules (MTs) alternate in this cruciate array, thus generating a 4-2-4-2 MTR pattern that is a highly characteristic feature of the order Volvocales (Melkonian, 1984a). Just posterior to the BBC lies a pair of contractile vacuoles, and posterior to them lies the nucleus, which is connected to the BBC by two fibers that are rich in the calcium-activated contractile protein, centrin (Wright *et al.*, 1985). The nucleus is surrounded by mitochondria, endoplasmic reticulum, Golgi complexes, peroxisomes, etc., in a rather predictable array. All of these membranous organelles lie in a cup-shaped depression of the chloroplast; the latter contains a single basal pyrenoid, fills the posterior end of the *C. reinhardtii* cell, and extends peripheral cusps nearly to the anterior end of the cell.

With one obvious exception, *C. reinhardtii* cells appear at first glance to exhibit bilateral symmetry, with the axis of symmetry defined by the symmetry of the BBC. Breaking this bilateral symmetry, however, is the eyespot, or stigma, an array of carotenoid-rich granules that is involved in phototaxis and that lies on one side of the cell, within a cusp of the chloroplast. It has been known for some time that the eyespot is invariably associated with one of the MTRs of the BBC (Melkonian, 1984b). However, it has been clearly demonstrated that this association is even more specific than had previously been recognized. During cell division BBs are distributed to daughter cells semiconservatively; that is to say, each cell receives one older BB (one that had been associated with an axoneme during interphase) plus one BB newly derived by elongation of a PBB. During division, old eyespots are broken down, and when new eyespots are formed in each daughter cell, each is formed in a specific association with the two-membered MTR that is attached to the younger of the two BBs of that cell (Holmes and Dutcher, 1989).

The adaptive significance of this invariant spatial relationship between the younger basal body and the eyespot is now becoming clear: it apparently underlies the ability of *Chlamydomonas* to swim toward moderate light and away from bright light. These directional swimming responses apparently result from the fact that in response to changing light intensity the flagella *cis* and *trans* to the eyespot change in opposite directions with respect to both beat frequency and flagellar waveform (Reufer and Nultsch, 1991). The photoreceptor mediating the phototactic responses of *Chlamydomonas* is a rhodopsin (Foster *et al.*, 1984; Beckmann and Hegemann, 1991), and when it is photostimulated it triggers an inward flux of calcium ions in both the eyespot and flagellar regions (Harz and Hegemann, 1991). However, the differential responses of the two flagella to light do not require that they be exposed to different levels of intracellular calcium; instead, they reflect an intrinsic physiological difference between the two flagella: the beat frequencies of the *cis* and *trans* flagella of fully permeabilized cells respond in opposite directions to a change in the calcium ion concentration (Kamiya and Witman, 1984). This asymmetric flagellar response to calcium ions can, in turn, be attributed to association of the two flagella with BBs of different developmental status. In each cellular generation the younger BB lies on the *cis* side of the cell (Holmes and Dutcher, 1989) and organizes the *cis*-flagellum, which beats more slowly than the *trans*-flagellum in one range of calcium concentrations and more rapidly than the *trans*-flagellum in others. But after one more round of cell division the BB matures, and from that time forward it will always lie on the *trans* side of the cell, and will organize a flagellum of opposite response characteristics to the flagellum it organized during its maiden voyage. But in order for this difference to be phototactically effective, it is essential that the two flagella of differing physiology have a regular spatial relationship to the eyespot, which they do (Holmes and Dutcher, 1989).

Holmes and Dutcher (1989) also demonstrated that the structures required for fusion of sexual *C. reinhardtii* cells are predictably and asymmetrically positioned with reference to the BBs of different developmental status. From these studies, reinforced by similar studies on other groups of green flagellates (see e.g., Melkonian *et al.*, 1987; Segaar, 1991), it is now clear that cells of *C. reinhardtii* and most other green flagellates are fundamentally asymmetric, and that this asymmetry is based on the different developmental status of the BBs, which have a maturation period that extends over two full division cycles. The potential importance of this fundamental cellular asymmetry for *V. carteri* development will be considered in a later section (II,F).

Organelles of the vegetative or somatic cells of all of the colonial and multicellular volvocaceans that have been examined ultrastructurally are organized in a fundamentally similar way as they are in vegetative cells of *C. reinhardtii* (Lang, 1963; Bisalputra and Stein, 1966; Pickett-Heaps, 1970, 1975; Deason and Darden, 1971; Birchem, 1978; Fulton, 1978; Viamontes, 1978; Dauwalder *et al.,* 1980; Hoops and Floyd, 1982; Hoops, 1984; Gruel and Floyd, 1985). However, an important exception to this general rule exists with respect to the interphase arrangement of the BBCs, and thus the flagella. In *C. reinhardtii* the two BBCs and flagellar axonemes of each vegetative cell are said to exhibit 180 degree rotational symmetry. That is to say, sister BBs face in opposite directions. As a consequence of this arrangement, the effective strokes of the two flagella are oriented in opposite directions within the same plane, generating a breast-stroke type of motility. But in the volvocaceans, the two basal bodies of each post-division cell are parallel in their orientation; thus, their flagella beat in the same direction in parallel planes (Hoops and Floyd, 1982; Hoops, 1984; Gruel and Floyd, 1985). The functional significance of this difference is readily apparent: breast-stroke motility moves a unicell like *C. reinhardtii* forward efficiently, but it would be totally ineffectual in moving a spherical multicellular organism, in which all cells are arranged so that their flagella project outward from the surface of the sphere. Such an organism can be moved forward efficiently, however, by flagella beating in parallel, provided that (as is the case) all cells are arranged so that their flagella direct their effective beats toward the posterior pole of the sphere. Naturally, these differences in orientation of the basal bodies between volvocacean vegetative cells and *C. reinhardtii* cells are accompanied by corresponding differences in the arrangements of various other BBC components, such as the MTRs and eyespots (Hoops and Floyd, 1982; Hoops, 1984; Gruel and Floyd, 1985).

Notwithstanding these functionally significant differences in BBC orientation in vegetative cells, the fact remains that in most regards the subcellular organization of all volvocacean vegetative or somatic cells is strikingly similar to that of *C. reinhardtii* vegetative cells. Indeed, the overall subcellular organization of *V. carteri* somatic cells resembles that in *C. reinhardtii* far more closely than the subcellular organization of many other species of *Chlamydomonas* does (Ettl, 1976).

2. Organization of Cell Walls and Extracellular Matrices

A noncellulosic cell wall, composed predominantly of hydroxyproline-rich glycoproteins (HRGPs) and complex polysaccharides, surrounds the interphase *C. reinhardtii* cell, interrupted only at the anterior end by two distinctive flagellar collars, through which the flagella emerge. This wall consists of two morphologically and chemically distinct regions: the outer wall, which is crystalline in its organization, and the inner wall, which is relatively amorphous (Roberts *et al.,* 1972; Hills *et al.,* 1973). Although quick-freeze/deep-etch electron-microscopic (EM) analysis (Goodenough and Heuser, 1985) and disassembly–reassembly studies (Adair *et al.*, 1987) reveal the crystalline layer to be composed of a highly regular latticework of self-assembling fibrous HRGPs, in conventional EM thin sections the crystalline layer frequently has a trilaminar appearance.

In colonial and multicellular volvocaceans, sister cells are held together by a relatively complex and extensive extracellular matrix (ECM). However most of the ECM of each organism is internal to a trilaminar layer, or boundry zone, that bears a strong resemblance to the crystalline layer of the *C. reinhardtii* wall (Kochert and Olsen, 1970; Burr and McCracken, 1973; Pickett-Heaps, 1975; McCracken and Barcellona, 1976; Birchem, 1978; Fulton, 1978; Dauwalder *et al.,* 1980; Kirk *et al.,* 1985). Indeed, detailed comparisons of the crystalline portion of the *C. reinhardtii* wall and the boundary zone of the volvocacean ECM provide some of the strongest biochemical evidence that these organisms are closely related. Early optical diffraction studies indicated that whereas there was sufficient variation in the crystal lattice structures among various species of *Chlamydomonas* to sort these species into five distinct groups, the group containing *C. reinhardtii* also contained all three volvocacean genera that were examined: *Pandorina, Eudorina,* and *Volvox* (Roberts, 1974). The concept that emerged from this study, namely, that the *C. reinhardtii* outer wall is more similar in structure to the boundary zone of the volvocacean ECM than it is to the outer walls of other species of *Chlamydomonas,* has been reinforced by more recent studies. Adair *et al.* (1987) demonstrated that reassembly of the HRGPs of *C. reinhardtii* in their original crystalline arrays could be nucleated by a *V. carteri* spheroid that had been stripped of its own crystalline layer by extraction with chaotropic agents. The relationship was reciprocal: stripped *C. reinhardtii* cells could also nucleate reassembly of *V. carteri* HRGPs in their characteristic crystalline array. In contrast, *C. reinhardtii* and *C. eugametos* are incapable of such reciprocal cross-nucleation of wall assembly (Adair *et al.*, 1987). In further confirmation of these relationships, Adair and Appel (1989) showed that a major HRGP of the crystalline layer is extremely similar in structure in

C. reinhardtii, Gonium pectorale (a small colonial volvocacean), and *V. carteri* but is without any recognizable homolog in *C. eugametos*.

In contrast to the marked similarities of their outer crystalline layers, the internal regions of the *C. reinhardtii* wall and the volvocacean ECM differ enormously. Particularly in the larger volvocaceans, the region internal to the trilaminar boundary zone contains a voluminous, complex, and species-specific set of fibrous and amorphous HRGP-rich ECM components that hold the volvocacean cells in fixed relationships to the surface of the sphere and one another (Kirk *et al.*, 1985; Ertl *et al.*, 1989) and inflate the sphere to generate an organismic volume that can reach or exceed 100 times the total cellular volume. It appears that extensive expansion, modification, and elaboration of the internal regions of the *C. reinhardtii* wall may have been one of the factors underlying diversification within the family Volvocaceae.

Analysis of these specialized internal components of the *V. carteri* ECM is underway in the laboratory of Manfred Sumper: An extreme example of an extracellular HRGP (62% hydroxyproline) has been characterized and shown to be preferentially synthesized by embryos toward the end of inversion, the final phase of embryogenesis (Schlipfenbacher *et al.*, 1986); the gene encoding ISG (as this inversion-specific sulfated glycoprotein is known) has recently been cloned and sequenced (M. Sumper, personal communication). [†]An unusual phosphodiester bond between arabinose units of the carbohydrate side chains has been shown to cross-link individual units of SSG 185 (sulfated surface glycoprotein of apparent molecular mass 185 kDa), the HRGP that forms the fibrous walls of the ECM compartments within which individual *V. carteri* somatic cells of the adult are suspended (Holst *et al.*, 1989). Moreover, the gene encoding SSG 185 has also been cloned and sequenced (Ertl *et al.*, 1989) (Section IV,C,2). Cloning of genes encoding other components of the *V. carteri* ECM is in progress (M. Sumper, personal communication).

It has been known for some time that young *Chlamydomonas* vegetative cells utilize autolysins (or V-lysins) to degrade and escape from the mother cell wall within which they have been formed by cell division, and that these autolysins are sufficiently species-specific in substrate range that they can be used to subdivide the genus *Chlamydomonas* into a limited number of presumably related species groups (Schlösser, 1976). G-lysins which are produced by gametes and exhibit similar species-specificity, but are capable of degrading the walls of both vegetative cells and gametes, have been described by a number of groups (Matsuda *et al.*, 1985; Jaenicke *et al.*, 1987; Buchanen and Snell,

[†] In this context, it should be noted that this laboratory has also described a novel glycosphingolipid that is synthesized selectively during late embryogenesis and appears to be involved in the inversion process (Wenzl and Sumper, 1986a).

1988). An analogous pair of lysins have also been described for *V. carteri* (Jaenicke and Waffenschmidt, 1979, 1981; Waffenschmidt *et al.*, 1990). Consistent with the fact that autolysins appear to act on the inner layers of the cell wall/ECM in which *C. reinhardtii* and *V. carteri* differ greatly, rather than the outer layers in which they appear so similar, no autolysin cross-reactivity is seen between these two species (Waffenschmidt *et al.*, 1990).

Thus, the outer regions of the ECM highlight the unity of the volvocine algae, whereas the inner layers highlight the diversity present within the group.

3. Patterns of Cell Division, Enlargement, and Differentiation during Asexual Development

Only under relatively adverse conditions does *C. reinhardtii* divide by the simple binary fission that one typically associates with unicellular organisms. Under optimum conditions, *C. reinhardtii* reproduces asexually by a division process that is analogous to the early embryonic cleavage divisions of invertebrate embryos: two, three, or (rarely) four cleavage divisions occur in rapid succession and, in the absence of intervening growth, generate a cluster of four or eight (or rarely 16) daughter cells that are temporarily held within the mother cell wall. Then, after the daughter cells have developed flagella and escaped from the mother wall to become free-swimming individuals, each cell enlarges 2^n-fold before once again resorbing its flagella, reorganizing its cytoplasm, rounding up, and executing a new round of cleavage divisions to form 2^n progeny cells.

In small colonial volvocaceans, such as *Gonium* and *Pandorina,* the division pattern during asexual development resembles that just described for *C. reinhardtii:* each of the (4–32) cells of one of these colonial flagellates enlarges 2^n-fold during the active, swimming phase and then rounds up to execute rapid cleavage divisions and produce 2^n daughter cells (Kochert, 1973; Starr, 1980). The principal way in which the asexual reproductive cycle of these colonial volvocaceans differs from that of *Chlamydomonas* is that in the volvocaceans all sister cells produced by cleavage of a single parental cell cohere to one another and swim as a unit after escaping from the parental ECM.

Under some conditions, *Eudorina elegans,* a slightly larger colonial volvocacean, reproduces in an identical manner as *Gonium* and *Pandorina:* all of the 16 or 32 cells in a colony enlarge and then cleave four or five times to form daughter colonies of 16 or 32 cells each. But under other conditions, the four anteriormost cells of a *E. elegans* colony often fail to enlarge or reproduce; they act as terminally differentiated cells, remain actively motile while the remaining cells enlarge and cleave, and then these anterior somatic cells eventually die (Kochert, 1973; Starr, 1980).

The segregation of a subpopulation of cells as terminally differentiated, biflagellate somatic cells that occurs conditionally in *E. elegans* is a constitutive feature of the *Pleodorina* asexual reproductive cycle (Kochert, 1973; Starr, 1980). In *P. californica,* for example, all 128 or 256 cells initially are the same size and have the same chlamydomonad morphology, but then approximately two thirds of the cells in the posterior of the spherical colony enlarge, modify their organization, and cleave, whereas the anterior one third of the cells remain small, terminally differentiated somatic cells that eventually die.

In most species of *Volvox* the asexual reproductive pattern resembles that of *Pleodorina,* except that much smaller numbers of cells (again, usually concentrated toward the posterior of the spheroid) enlarge, change morphology, and become asexual reproductive cells, or gonidia, that cleave to produce daughter spheroids in which all cells initially appear similar. However, in *V. carteri* and a few other (presumably closely related) species of *Volvox,* two categories of cells destined to become germ and soma are set apart by visibly asymmetric cleavage divisions, and a complete division of labor is established. In these species of *Volvox,* the large cells differentiate directly as nonmotile gonidia that rapidly enlarge further and reproduce, whereas all of the small cells become terminally differentiated somatic cells that execute vegetative functions, such as motility, phototaxis, and chemotaxis, and then undergo preprogrammed death (Kochert, 1968; Starr, 1969, 1970; Kirk and Harper, 1986).

To summarize, in *Chlamydomonas, Gonium,* and *Pandorina* all cells are structurally and functionally equivalent throughout the asexual reproductive phase, and each cell fulfills all of the vegetative and reproductive functions of the species sequentially. A partial division of vegetative and reproductive labors—in which all cells are initially similar in form and function, but then a subset of the cells enlarge and redifferentiate as reproductive cells—occurs in *Eudorina* conditionally and in *Pleodorina* and most species of *Volvox* constitutively. In contrast, in a few species of *Volvox,* including *V. carteri,* separate lineages that will become mortal somatic cells and immortal germ cells are set apart early in embryonic development, and the division of germ/soma labors is complete. In these species of *Volvox,* germ and soma are fully interdependent on one another, and neither cell type by itself could perpetuate the germ plasm in the wild. Therefore, *V. carteri* and its closest relatives are truly multicellular organisms, in distinction to the smaller volvocaceans, which are colonies of equivalent cells.

4. Sexual Reproduction and the Resting Zygote

Sexual reproduction in *Chlamydomonas* can be triggered by withdrawing nitrogen sources from the medium, which causes vegetative cells to transform into gametes capable of fusion with partners of opposite mating type. Following fusion, a diploid zygote, or zygospore, is formed that develops a thick, crenulated cell wall and becomes resistant to desiccation, heat, cold, and other environmental insults. Upon restoration of suitable conditions for growth, the *C. reinhardtii* zygote undergoes meiosis and hatches to release four haploid meiotic products that resume the asexual reproductive cycle. In *C. reinhardtii* and many other species of *Chlamydomonas,* gametes are of equal size and mating is said to be isogamous. In some *Chlamydomonas* species, however, motile gametes of different size are produced, and mating is said to be anisogamous. And in a few species of *Chlamydomonas,* small, motile cells of one mating type (''sperm'') swim to and fuse with large, nonmotile cells of the other mating type (''eggs''); in such cases mating is said to be oogamous.

In the volvocaceans similar sexual reproductive cycles occur, and they generate structurally and functionally similar resting zygotes. But several trends are observed in the sexual cycles in volvocaceans of increasing size. Whereas isogamous or anisogamous mating is observed in some of the smaller volvocaceans, in genera containing the larger forms, including *Volvox,* mating is always oogamous. Sex may be triggered by deteriorating nutritional conditions in some of the smaller volvocaceans, but in all species of *Volvox* examined, sexual development is triggered by species-specific pheromones, or sexual inducers. Upon germination, zygotes of the smaller volvocaceans may produce four viable meiotic products, but in the species of *Volvox* that have been studied, a single meiotic product and three inviable ''polar bodies'' are produced by meiosis (Starr, 1975).

5. The Volvocine Lineage Hypothesis

The volvocine algae (*Chlamydomonas* plus the members of the family Volvocaceae) all have a fundamentally similar and distinctive pattern of subcellular organization. However, it is possible to line them up conceptually in a series (such as *Chlamydomonas, Gonium, Pandorina, Eudorina, Pleodorina,* and *Volvox*), in which there is a progressive increase in each of the following parameters: (1) cell number, (2) organismic size, (3) ratio of ECM volume to cellular volume, (4) tendency for some cells to differentiate terminally as somatic cells, (5) ratio of somatic cell number to reproductive cell number, (6) emphasis on oogamous, as opposed to isogamous, sexual reproduction, and (7) asymmetric meiotic divisions, resulting in formation of a single viable product. The volvocine lineage hypothesis postulates that this conceptual series may actually reflect the evolutionary history of the group.

B. Volvocalean Relationships and the Volvocine Lineage Hypothesis

For most of the years that the volvocine lineage hypothesis has been perpetuated in one form or another, it has not been subject to any type of systematic testing. Vande Berg and Starr (1971) pointed out that "The concept of this evolutionary series is based entirely on subjective criteria, and as yet there are little or no cytological, genetic, immunological or physiological data which provide proof." In the intervening years, the concept that members of the so-called volvocine lineage constitute a closely related group and the concept that *C. reinhardtii* has closer affinities to the various volvocacean genera than it does to other members of its own genus have been reinforced by the cytological and biochemical studies cited in the preceding section. But such studies did not in any sense "provide proof" or even test the hypothesis that the volvocine lineage arose as a monophyletic progression in organismic size and developmental complexity. In recent years, however, molecular methods have been used in an attempt to address this question.

1. The Relationship of *V. Carteri* and *C. reinhardtii* to One Another and to More Distantly Related Taxa

The complete sequence of the nuclear rDNA encoding the small subunit (18S) rRNA of *V. carteri* was determined by Rausch *et al.* (1989) and compared with the equivalent sequences of *C. reinhardtii* (Gunderson *et al.*, 1987) and various other taxa to evaluate the phylogenetic relationships of these two algae to one another and to members of other major eukaryotic lineages (Rausch *et al.*, 1989). The aligned sequences of the 18S rRNAs of *V. carteri* and *C. reinhardtii* exhibited 99.2% sequence similarity, a value slightly greater than the 99.1% similarity observed between the equivalent molecules of two rather closely related cereal grasses, maize and rice. This was taken to indicate that perhaps "the evolution of *Volvox*, a multicellular organism with division of labor between germ and soma, from a *Chlamydomonas*-like unicellular ancestor may have occurred in as brief an interval as that during which the radiation of the cereal grasses has occurred [perhaps 50 million years (Myr) or less]." Although "molecular clock" methods (using estimates of the average rate of nucleotide substitution in 18S rRNA and of silent nucleotide interchanges in protein-coding genes) led Rausch *et al.* (1989) to a somewhat longer estimated period of divergence between *V. carteri* and *C. reinhardtii* (50–75 Myr), in no sense did they support the view that the volvocine radiation is an extremely ancient one.

A phylogenetic tree that was constructed from 18S rRNA sequences of 18 taxa, using distance–matrix methods (Rausch *et al.,* 1989), placed *V. carteri* and *C. reinhardtii* at the tip of a relatively long branch that diverged from the main trunk of eukaryotic evolution between the branches leading to the flowering plants and to the vertebrates but closer to the former (Fig. 1). Although this result certainly does not support the Nineteenth Century view that *Volvox* represents an evolutionary "missing link" between plants and animals, it is consistent with the combination of plant-like features (e.g., obligate photoautotrophy) and animal-like features (e.g., early and complete germ/soma differentiation) that *V. carteri* exhibits. In further support of this placement, various combinations of plant-like and animal-

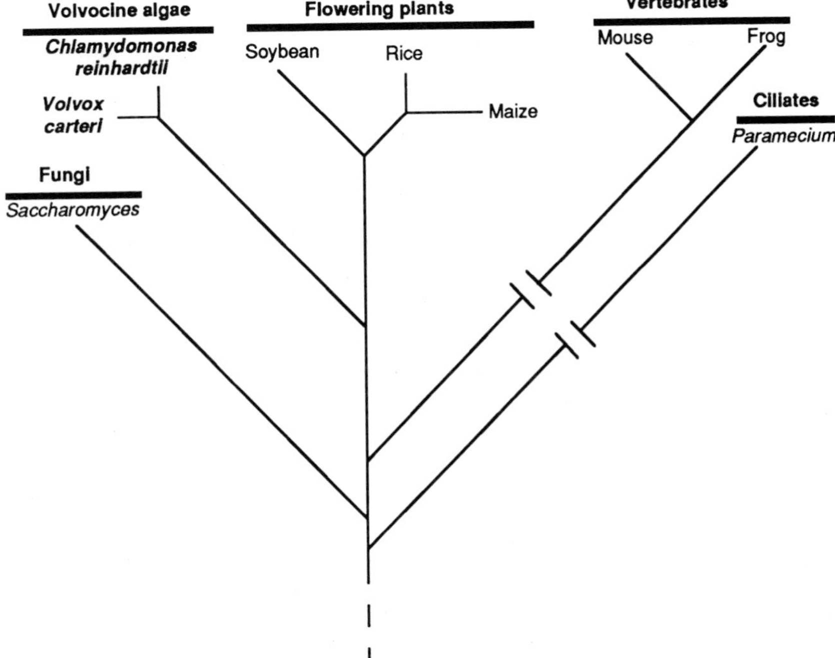

FIG. 1 A tree diagram indicating the phylogenetic relationship of the volvocine algae to other major taxa of eukaryotes, as inferred from a distance matrix analysis of small-subunit nuclear rDNA sequence data. Except where broken, the lengths of lines are proportional to the inferred number of changes (K_{nuc}) along each section of the dendrogram. (After Rausch *et al.,* 1989.)

like features have now been observed in the structures of certain protein-coding genes, as discussed in Section IV,B.

On the basis of cytochrome *c* sequence comparisons, Amati *et al.* (1988) estimated that lines leading to the modern Chlorophyta (green algae) and to the higher plants diverged about 700–750 Myr ago, about 500 Myr after plants and animals last shared a common ancestor. This scenario is consistent with the tree derived by Rausch *et al.* (1989) from rRNA sequences (Fig. 1), and reinforces the view that the chlorophytes last shared a common ancestor with the higher plants in deep antiquity, but that the radiation of the volvocine algae—at the tip of the chlorophyte branch—was a relatively recent event on the evolutionary time scale.

2. Relationships within the Order Volvocales and the Family Volvocaceae

In a specific attempt to evaluate the volvocine lineage hypothesis, *vis-á-vis* certain other hypotheses about the origins of the genus *Volvox,* Larson *et al.* (1992) examined aligned sequences in the variable regions of the large and small nuclear-encoded rRNAs of 15 taxa of green algae, including two species of *Chlamydomonas,* four species of *Volvox,* and representatives of several other volvocacean genera. Among the conclusions drawn by the authors of this study were the following:

1. The most parsimonious tree derived from the rRNA sequence data (Fig. 2) places *C. reinhardtii* with the members of the family Volvocaceae in what appears to be a closely related, monophyletic group. Earlier ideas that part (Fritsch, 1935) or all (Crow, 1918) of the genus *Volvox* might be more closely related to the Haematococaceae (another family of green flagellates) than to *Chlamydomonas* and the colonial Volvocaceae were not supported by these data.

2. The rRNA data supported conclusions drawn earlier that *C. reinhardtii* appears to have shared a common ancestor with members of the family Volvocaceae much more recently than it has shared a common ancestor with *C. eugametos*. Similar conclusions were reached when similar rRNA sequence comparisons were used first to examine relationships among numerous species presently grouped in the genus *Chlamydomonas* (Buchheim *et al.,* 1990) and then to examine the relationships between a number of unicellular and colonial green algae (Buchheim and Chapman, 1991).

3. Although the family Volvocaceae appers to be a coherent, closely related group, rRNA sequence data do not support the volvocine lineage hypothesis that the genus *Volvox* arose as the result of a simple, monophyletic progression from small and simple to larger and more complex organisms with sterile somatic cells. Indeed, the most parsimonious

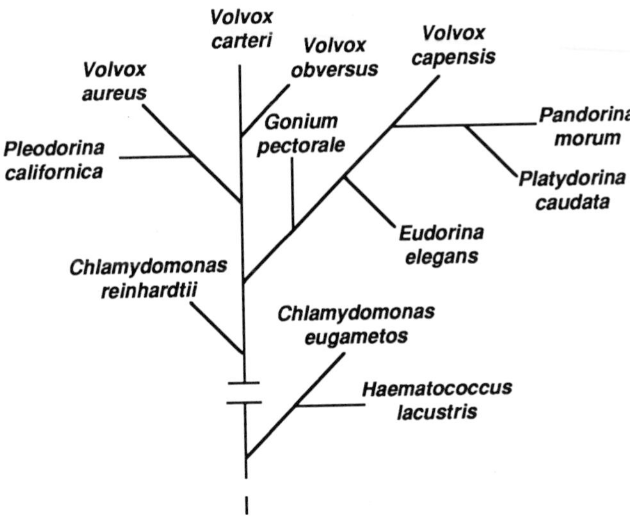

FIG. 2 A tree diagram indicating sister–taxon relationships among 12 volvocine algae, as deduced by maximum parsimony analysis of nt sequence data from the variable regions of both the large and small subunits of nuclear-encoded rRNAs. *V. capensis* appears to be part of a lineage that is distinct from that of the other species of *Volvox* that were analyzed. Also, *C. reinhardtii* and *C. eugametos* do not appear to be as closely related to one another as *C. reinhardtii* is to the colonial and multicellular volvocaceans. (After Larson *et al.*, 1992.)

rRNA-based tree placed members of the genus *Volvox* in two separate volvocacean lineages, each of which also contains colonial members of the family. It is noteworthy that the species of *Volvox* that was placed by rRNA sequence analysis on a separate branch of volvocacean evolution from all other species of *Volvox* that were studied is a representative of a section of the genus (the section Euvolvox) that Fritsch (1935) found so different from the rest of the genus that he proposed a separate evolutionary origin for it. He apparently was correct in this general conclusion although mistaken in his opinion that the Euvolvox were more closely related to the Haematococaceae than to the Volvocaceae.

The most novel suggestion derived from the study of Larson *et al.* (1992) was that the colonial life-style (in which all cells of an organism are similar in morphology and developmental behavior) and the multicellular life-style (with a division of labor between differentiated somatic and reproductive cells) "may represent different stable states, among which there may have been multiple transitions during the phylogeny of the group [and] . . . only a small number of genetic changes may be required to effect a transition, in either direction, between the colonial and multicellular . . . life history." Thus, although this study appears to have falsified the simple, traditional volvocine lineage hypothesis, at the same time it has reinforced the hope that the genetic origins of the *V. carteri* cellular differentiation program might be simple enough to be amenable to detailed molecular analysis.

II. Genetic and Experimental Analysis of Germ/Soma Differentiation in Asexual *V. carteri* Embryos

The aspect of *V. carteri* development that is most interesting from an evolutionary genetic standpoint is the aspect that most clearly differentiates it from other genera of volvocine algae, and from most other species of *Volvox,* namely, the early and complete segregation of separate germ and somatic cell lineages. The selective advantages and reproductive costs of dichotomous germ/soma differentiation in *V. carteri* and its relatives have been analyzed and discussed in considerable detail in recent years by life history biologists in search of its probable ultimate causes (Bell, 1985; Bell and Koufopanou, 1986; Koufopanou, 1990). These workers have proposed, on both theoretical and analytic grounds, that dichotomous differentiation has probably been selected for in certain environments because a sterile soma that remains motile throughout the life cycle makes it possible to obtain one major advantage of larger body size (i.e., decreased predation by small herbivores) without reaping one of its major potential disadvantages (i.e., sinking into the mud while reproduction is occurring). Presumably, in environments where the dominant predators are in an appropriate size range, the advantage to be gained from increased body size contributes more to fitness than is lost as a consequence of the increased reproductive costs of forming a sterile soma (i.e., the increased biomass and energy expenditure required per reproductive unit). Interesting as these thoughts on ultimate causes are, however, others have been more interested in elucidating the proximate causes of dichotomous germ/soma differentiation in terms of its genetic, cellular, and molecular bases.

A. Mutations That Disrupt the Germ/Soma Dichotomy

The major approach to identifying *V. carteri* genes that are involved in programming the germ/soma dichotomy has been the traditional one of identifying loci at which mutations cause a breakdown of that dichotomy. The first Mendelian mutation of this sort was the fertile somatic cell trait described by Starr (1970) in a clone that exhibited what has since become known as the somatic regenerator, or Reg, phenotype. In Reg mutants, somatic cells appear to differentiate normally at first, but then they dedifferentiate and redifferentiate as functional germ cells—gonidia or gametes, depending on whether or not the sexual pheromone is present (Starr, 1970; Huskey and Griffin, 1979). All (~60) strains with this phenotype that have been analyzed possess lesions that map to a locus known as *regA* (Huskey and Griffin, 1979; K. A. Stamer and D. L. Kirk, unpublished observations). However, second-site mutations that modify the Reg phenotype have been described; these map to at least one *ram* (*regA* modifier) locus that is unlinked to *regA* (Kirk, 1990; Kirk *et al.*, 1991b), and in which mutations are lethal in the absence of a *regA* mutation. It has been postulated that *regA* encodes a negative regulator of the genes required for reproductive development, whereas *ram* encodes a function required for the expression of such genes (Kirk, 1990; Kirk *et al.*, 1991b). It has been further postulated, on the basis of the extraordinary hypermutability of the *regA* gene that is observed just before and just after cleavage (Kirk *et al.*, 1987), that expression of the *regA* gene is regulated by a cyclic rearrangement that functions to prevent *regA* expression in developing gonidia while permitting its expression in developing somatic cells (Kirk *et al.*, 1987; Kirk, 1990).

In Reg mutants somatic cells behave like the cells of colonial volvocaceans: they first execute the vegetative functions of motility and phototaxis and later execute the reproductive functions of the species. Similar behavior is exhibited by the presumptive gonidia of Lag (late gonidia) mutants: in these mutants presumptive gonidia are set aside in a relatively normal fashion during cleavage, but they first differentiate as (large) somatic cells before redifferentiating as gonidia (Kirk, 1990; Kirk *et al.*, 1991b). This has been taken to indicate that the function of the wild-type *lag* loci (of which there are a number that appear to function in a common pathway) is to permit presumptive gonidia to bypass the vegetative portion of the life cycle and differentiate directly as gonidia. Because germ/some determination appears to occur relatively normally in Lag mutants (even though the subsequent differentiation of the germ cells is delayed), it has been concluded that the initial steps in germline determination must be under the control of some locus, or loci, other than the *lag* loci (Kirk *et al.*, 1991b). Mutations of such loci have not been detected yet, however, possibly because they are invariably lethal.

Mutations at the *regA* locus have no discernible effects on the development of gonidia, and *lag* mutations have no discernible effects on the development of somatic cells. This has been interpreted as indicating that these two types of loci are expressed in a mutually exclusive fashion, with the *regA* locus being expressed in presumptive somatic cells but not gonidia and the *lag* loci being expressed in gonidia but not somatic cells (Kirk, 1988, 1990; Kirk *et al.*, 1991b).

The first visible step in *V. carteri* germ/soma differentiation is a set of asymmetric divisions that set apart large cells destined to become gonidia from smaller sister cells that are destined to produce only somatic cells

(Starr, 1970). A number of *mul* (multiple gonidia) loci have been defined at which mutations modify asymmetric division patterns and, hence, the number and distribution of gonidia (for review see Kirk *et al.*, 1991a). In all cases, however, the numbers and locations of gonidia in the adult can be traced back to the temporal and spatial patterns of asymmetric cleavage divisions in the embryo. In contrast to the *mul* mutations that modify the timing and location of asymmetric divisions, *gls* (gonidialess) mutations can result in a failure to execute any asymmetric divisions. Symmetric divisions occur without any discernible abnormality in Gls embryos. Therefore, it is postulated that *gls* encodes a function that is required to shift the division apparatus away from the middle of the cell at the times and places specified by the *mul* genes (Kirk, 1990; Kirk *et al.*, 1991a). Putative second-site suppressors of *gls* have been identified and inferred to encode additional functions involved in the asymmetric division process (Kirk, 1990; Kirk *et al.*, 1991a).

Because a tight *gls* mutation abolishes the formation of true gonidia, such a mutation can only be selected and maintained on a *regA* background, in which reproductive potential is conferred on the somatic cells. Gls/Reg double mutants exhibit a complete breakdown of the germ/soma division of labor; they resemble colonial volvocaceans in the respect that all cells execute the vegetative and reproductive functions of the species in a cyclic manner.

B. A Model for the Genetic Control of Germ/Soma Differentiation

A model based on the observations described in the preceding section, and which postulates how all of the loci that have been mentioned function in the *V. carteri* life cycle, is presented in Figure 3. The key features of this model are as follows: (1) The basal program for volvocelean cellular differentiation is sequential, with all cells differentiating first as vegetative cells and then redifferentiating as reproductive cells (Fig. 3A). (2) This basal program is converted to a program for dichotomous germ/soma differentiation by the action of a small number of genes: (a) the *gls* locus acts (at the times and places dictated by the *mul* loci) to generate large and small cells via asymmetric division; (b) the *lag* loci are then expressed in large cells to shunt them directly into the reproductive pathway, bypassing the vegetative phase of development; (c) meanwhile, the *regA* locus is activated in the small cells suppressing reproductive functions and causing the small cells to differentiate terminally as somatic cells. In *gls*/*regA* double mutants the *gls* and *regA* functions are not expressed because of mutations, and the *lag* functions are not expressed because of an ab-

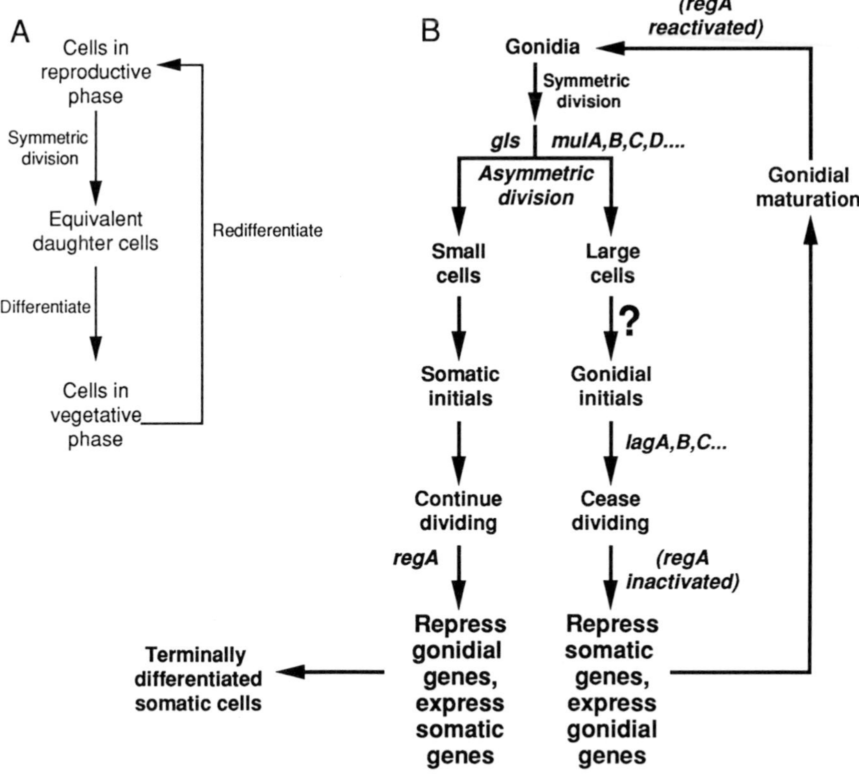

FIG. 3 Diagrammatic representations of the asexual life cycles of volvocine algae. (A) In unicellular and colonial volvocine algae, all cells execute vegetative and reproductive functions sequentially. (B) In *V. carteri,* vegetative and reproductive functions are divided between somatic cells and gonidia, respectively. Symbols beside arrows indicate where and when during development certain control genes are postulated to act to convert the sequential pathway of cellular differentiation shown in A to the germ/soma dichotomy of *Volvox*. After a number of symmetric divisions have occurred, the *gls* gene acts (at times and places specified by the various *mul* genes) to cause a set of asymmetric cleavage divisions that generate large/small sister cell pairs. In the small cells, the *regA* gene is expressed; this acts to suppress expression of genes required for gonidial differentiation; therefore, these cells become terminally differentiated as somatic cells. In the large cells, meanwhile, the *lag* genes are expressed and the *regA* gene is inactivated; this leads to repression of the genes required for differentiation of somatic cell features and expression of the genes required for gonidial differentiation. The question mark next to one arrow in the gonidial pathway indicates that it has been inferred (Kirk, 1990) that one or more genes must act in the large cells before the *lag* genes do to commit these cells to the gonidial pathway. However, such a gene has not yet been defined in terms of a mutant phenotype. [Modified from Kirk (1987)].

sence of cells of sufficient size to activate them; the result is a return to the basal pathway of cyclic development seen in the more primitive members of the family Volvocaceae.

C. The Role of Asymmetric Division and Cell Size in Germ/Soma Specification

In both asexual and sexual embryos, and in both wild-type and mutant embryos, there is a one-to-one correspondence between asymmetric division and the formation of germ cells: large cells generated by asymmetric division produce one germ cell each, whereas their smaller sister cells divide symmetrically to produce a population of somatic cells. This one-to-one relationship is maintained in Gls/Reg mutants (in which the absence of large cells produced by asymmetric division leads to an absence of true germ cells) and in a number of other cleavage mutants not described above. For example, in a mutant with a temperature-sensitive defect of cleavage plane orientation *(cleA*$^{ts\text{-}1}$*)*, as the culture temperature is raised cleavage planes become increasingly randomized, large cells are produced in unpredictable numbers and positions, and all cells above a certain size go on to become gonidia (Kirk *et al.*, 1992).

Conflicting interpretations of the way in which asymmetric division leads to germ/soma specification in *V. carteri* have appeared in the literature (for review see Kirk and Harper, 1986). Some have concluded that visibly asymmetric division is accompanied by distribution of cytoplasmic determinants that are the cause of gonidial specification (Kochert and Yates, 1970; Kochert, 1975). Others have concluded that it is the size difference generated by asymmetric division, or some property derived from the difference in size, that is the cause of germ/soma specification (Pall, 1975).

Recent studies have strongly supported the latter hypothesis: that it is the size of the large cells that are produced by asymmetric division, and not any special quality of the cytoplasm that they possess, that is causally important in their specification as presumptive germ cells (Kirk *et al.*, 1992). This conclusion was initially drawn by Pall (1975), when he observed that *pcd* (premature cessation of division) mutant embryos that stopped cleaving while cells were still relatively large had both a relative and an absolute superabundance of germ cells and a corresponding deficiency of somatic cells. Among the most important recent observations supporting such a conclusion are results obtained when heat shock was used to interrupt the cleavage programs of wild-type and Gls/Reg mutant embryos (Kirk *et al.*, 1992). When wild-type embryos were heat-shocked after asymmetric division (and hence at a time when any cytoplasmic determinants involved in germ/soma specification should have already been segregated), and were thereby caused to cease cleaving at the 128-cell stage, presumptive somatic cells were left 16-to 32-fold larger than they would have been if they had completed the usual 11 or 12 rounds of division. Most such cells differentiated as gonidia and cleaved to give normal progeny. However, when the heat shock was applied at progressively later stages (so that somatic cell initials were permitted to complete additional divisions) progressively fewer cells differentiated as gonidia. Gls/Reg embryos normally cleave symmetrically 8 or 9 times to give 256 or 512 equal cells, all of which first differentiate as somatic cells and then redifferentiate as gonidia. However, when Gls/Reg embryos were heat-shocked to interrupt cleavage at the 32-cell stage, thus leaving the cells 8- to 16-fold larger than they would normally be at the end of cleavage, all cells bypassed the somatic phase and differentiated directly as gonidia. Taken together, such studies have led to the conclusion that *V. carteri* cells that remain above a threshold size at the end of cleavage—wherever or however they may have been produced—differentiate as germ cells, whereas cells below this threshold size develop as somatic cells (Kirk *et al.,* 1992).

This conclusion differs from the one drawn recently with respect to a related species of *Volvox*. Following his experimental studies of the gonidial specification process in *V. obversus,* Ransick (1988, 1991, 1992) concluded that in this species specialized cytoplasm from the anterior region of the embryo plays a crucial role in gonidial specification. In *V. obversus* the large, presumptive gonidial cells generated by asymmetric division are always the eight anteriormost cells of the fully cleaved embryo. When Ransick isolated individual blastomeres at the 8- or 16-cell stage, only the anterior cells went on to cleave asymmetrically and produce gonidia. When he subsequently used an ingenious form of micromanipulation to generate very large cells in the posterior of the embryo, these cells never developed into gonidia (Ransick, 1988, 1991). In contrast, when other forms of micromanipulation were used to distribute anterior cytoplasm to more than the usual number of blastomeres, supernumerary gondia could be produced.

However, although Ransick's studies clearly indicated that anterior cytoplasm is *necessary,* they also demonstrated that it is not *sufficient* for gonidial specification in *V. obversus*. In these studies cells developed as gonidia only if they *both* inherited anterior cytoplasm *and* remained significantly larger than somatic initials at the end of cleavage. Cells that inherited anterior cytoplasm but were (as a consequence of one of two different types of experimental intervention) below ~7 μm in diameter at the end of cleavage developed as somatic cells (Ransick, 1988, 1991). It is noteworthy that this size threshold is similar to the threshold established for gonidial specification in *V. carteri* (Kirk *et al.,* 1992).

Ransick has now shown that operations similar to ones initially performed with *V. obversus* produce very different results when performed with *V. carteri* embryos: Posterior *V. carteri* blastomeres isolated at the 16-cell stage regularly divide asymmetrically and produce one or more gonidia, which they would never do *in situ*. Moreover, when micromanipulation was used to generate large posterior cells in otherwise intact *V. carteri* embryos, these large cells produced gonidia (Kirk *et al.*, 1992).

The apparent difference between *V. carteri* and *V. obversus* with respect to gonidial specification is given added significance by the fact that another closely related species of *Volvox, V. powersii,* clearly generates its gonidia by some other mechanism. In *V. powersii* there are no visibly asymmetric divisions, at the end of cleavage all cells are of equivalent size, all cells initially develop the morphological features of somatic cells, and only after the juvenile individual has hatched out of the parental spheroid and has become free-swimming do certain cells toward the posterior end of the spheroid enlarge and differentiate into gonidia (Vande Berg and Starr, 1971). [Volvocacean embryos turn inside out at the end of embryogenesis in a process called "inversion." In this process the anterior end of the embryo becomes the posterior end of the adult.] Gonidial specification in *V. powersii* obviously involves some mechanism (presumably cytoplasmic localization) that does not depend on initial differences in cell size. The same situation occurs in *Pleodorina californica,* the colonial volvocacean most closely related to these three species of *Volvox* (Larson *et al.*, 1992). In *P. californica* all cells are initially the same size, but later in development all of the cells in the posterior end of the colony enlarge and become gonidia, whereas all cells in the anterior end remain terminally differentiated somatic cells.

The foregoing observations lead to the postulate that: (1) a cell size-independent, cytoplasmic localization mechanism (as presumably occurs in *P. californica* and *V. powersii*) is the primitive method of germ cell specification in the Volvocaceae; (2) a mechanism involving combined actions of cytoplasmic localization and cell size (as in *V. obversus*) is a more derived method of germ cell specification; and (3) a mechanism involving the effects of cell size alone (as in *V. carteri*) is the most derived method of germ cell specification in this group. This hypothesis predicts that *V. powersii* will be found to be more closely related to *P. californica* than *V. obversus* and *V. carteri* are. Tests of this prediction are underway (A. Larson *et al.*, unpublished observations).

The selective advantage of a germ cell specification mechanism that is independent of localized cytoplasmic determinants is far from clear. However, it is noteworthy that the volvocaceans are not the only group in which evolution in this direction appears to have occurred: in most of the metazoa (at least up to and including insects and amphibians) germ cell determination is dependent on localized germ cell determinants in the zygote, but there is no evidence that any form of cytoplasmic localization is involved in germ cell specification in mammals (Gilbert, 1991).

D. Differential Gene Expression in Germ and Soma

As a complement to the mutational approach that attempts to identify the control genes that ultimately regulate germ/soma differentiation and the experimental approach that attempts to identify physiological factors regulating expression of those control genes, the Kirk laboratory has recently initiated molecular studies designed to identify downstream genes that are expressed in a cell type-specific manner, presumably under the influence of the control genes.

By differential screening of a cDNA library derived from RNAs isolated from a mixture of *V. carteri* developmental stages and cell types, 31 distinct families of genes that are expressed in a cell type- and stage-specific manner were identified (Tam and Kirk, 1991a). Twelve of these gene families are expressed exclusively in somatic cells, whereas 19 are expressed preferentially in gonidia and/or embryos. Northern-blot and *in situ* hybridization analysis indicated that for each category of cell type-specific genes two stage-specific patterns of transcript accumulation could be identified (Tam and Kirk, 1991a; Tam *et al.*, 1991). These patterns are summarized graphically in Figures 4A and B.

Five of the cloned cDNAs identify early somatic genes, from which transcripts begin to accumulate in presumptive somatic cells of the embryo shortly after the asymmetric clevage divisions has been completed and several hours before these cells begin to develop the morphological features of somatic cells (Tam *et al.*, 1991). The virtual absence of these early somatic transcripts from presumptive gonidia during the embryonic period is remarkable because it is known that throughout the embryonic period all cells of the embryo are linked to one another by numerous cytoplasmic bridges that are more than 200 nm in diameter (Green and Kirk, 1981). Apparently, some mechanism must exist to prevent free movement of transcripts between neighboring embryonic cells through these numerous and rather massive-looking channels. Although previously unsuspected, such restricted intercellular communication is significant, as it obviously permits the process of differential gene expression to begin in the two presumptive cell types long before the cytoplasmic bridges are to be broken in the postembryonic period. Transcripts of all early somatic genes remain abundant in somatic cells and virtually undetectable in gonidia throughout most of the asexual life cycle. Only as the somatic cells approach senescence and programmed death do these transcripts begin to

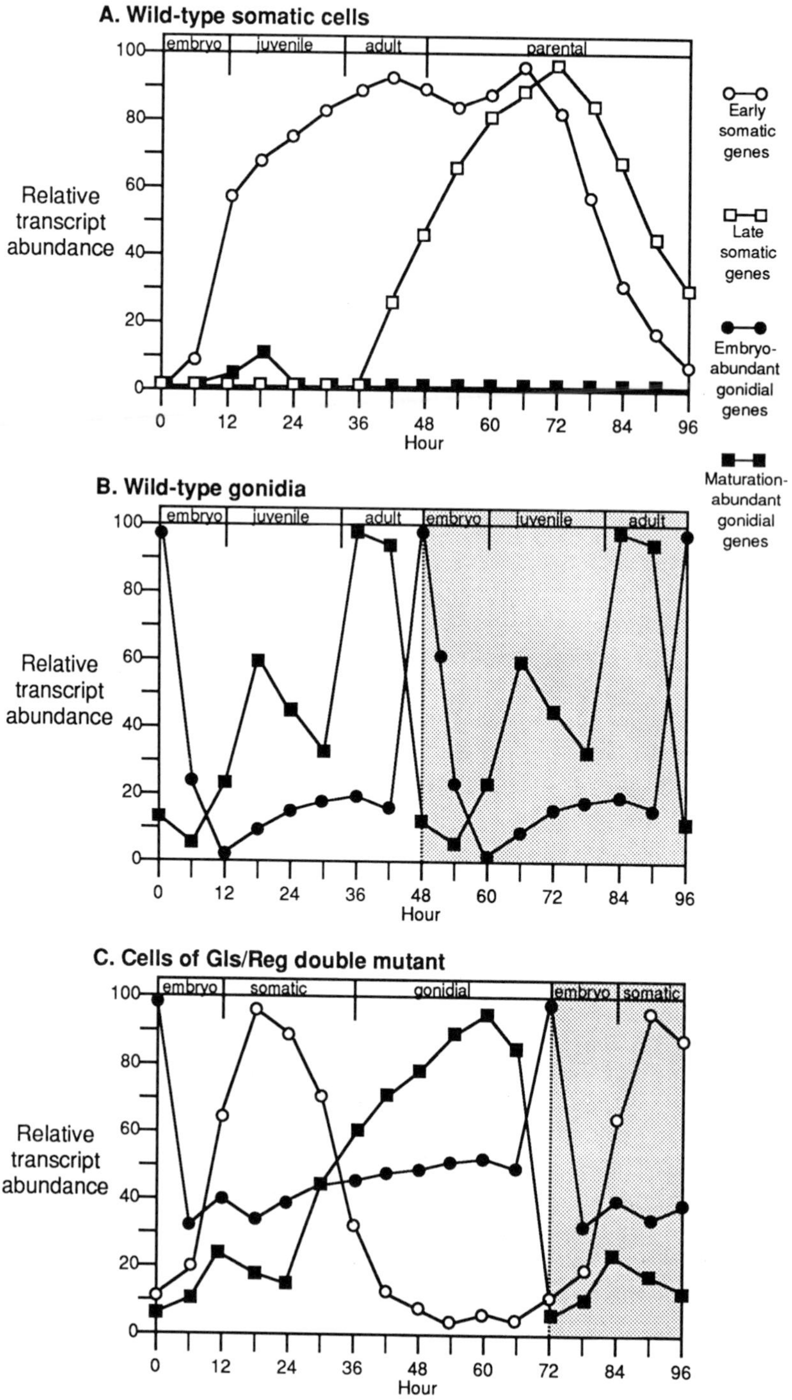

A. Wild-type somatic cells

embryo juvenile adult parental

Relative transcript abundance

0 20 40 60 80 100

0 12 24 36 48 60 72 84 96
Hour

Early somatic genes

Late somatic genes

Embryo-abundant gonidial genes

Maturation-abundant gonidial genes

B. Wild-type gonidia

embryo juvenile adult embryo juvenile adult

Relative transcript abundance

0 20 40 60 80 100

0 12 24 36 48 60 72 84 96
Hour

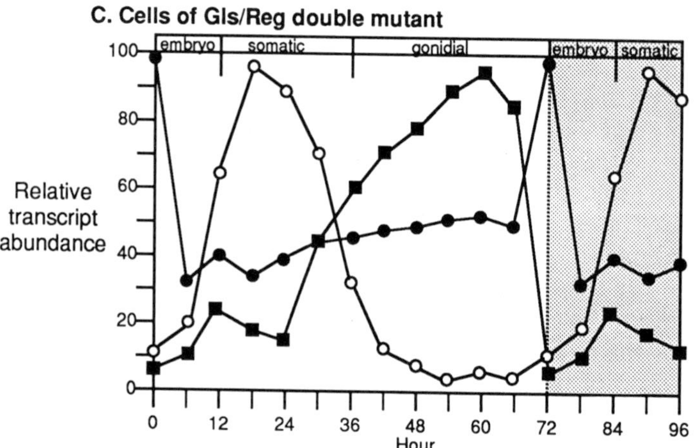

decline markedly in their abundance (Tam *et al.*, 1991). Presumably, these gene products are required for development and/or maintenance of the somatic cell phenotype. Genes encoding tubulins (which are required through most of the life cycle for flagellar elongation) and SSG 185 (the glycoprotein that forms the walls of the continuously growing chambers in which individual somatic cells are located) are expressed in the same pattern as the early somatic genes defined by cloned cDNAs (Tam and Kirk, 1991a).

In contrast to the early somatic genes, the seven late somatic genes studied do not begin to be expressed until more than a day later, after the cells have completed their morphological differentiation and are mature (Tam *et al.*, 1991). The functions of these gene products are unknown, but obviously they must not be required in abundance for morphological differentiation; whether these genes might encode proteins that are required for apoptosis (i.e., programmed death) of the somatic cells (Pommerville and Kochert, 1981, 1982) remains to be determined.

Only one of the 19 gonidium-specific genes exhibits the embryo-abundant pattern of expression; its transcript is most abundant in early cleavage, falls to low levels after the asymmetric cleavage divisions, and then returns to moderate levels in differentiating and maturing gonidia. Transcripts of this gene are never detected in somatic cells. The suggestion

---

FIG. 4 Accumulation patterns for transcripts of cell type-specific genes in three types of *V. carteri* cells. (A) In wild-type somatic cells, transcripts of early somatic genes appear during mid-embryogenesis and are maintained at relatively high levels throughout most of the life history of the somatic cells. Transcripts of late somatic genes, in contrast, are accumulated only after differentiation of somatic cells has been completed. Transcripts of maturation-abundant gonidial genes are transiently accumulated in young somatic cells, but they quickly disappear from these cells. Transcripts of embryo-abundant gonidial genes are never detected in the somatic cells. (B) Whereas somatic cells survive for more than 4 days, gonidia cleave to form new embryos when they are only 2-days old. Thus, the pattern of gonidial transcript accumulation shown for hours 49–96 (stippled) is a repeat of the pattern shown for hours 0–48. In gonidia none of the somatic cell transcripts is ever detected at significant levels. Transcripts of embryo-abundant gonidial genes are most abundant prior to the asymmetric division; thereafter, they decline rapidly, but they are present at moderate levels throughout the gonidial maturation period. Transcripts of the maturation-abundant gonidial genes exhibit one abundance peak during gonidial differentiation and a second higher peak during gonidial maturation. (C) Gls/Reg cells have a 3-day life history; during the first half of this period they exhibit the features of somatic cells, but then they redifferentiate as gonidia, and by 72 hours they cleave to repeat the cycle. Therefore, the pattern shown for hours 72–96 (stippled) is a repeat of the pattern shown for hours 0–24. In contrast to wild-type somatic cells, Gls/Reg cells express both categories of gonidial genes, as well as the early somatic genes, while they are developing and expressing the features of somatic cells. Between 24 and 48 hours, the levels of maturation-abundant gonidial transcripts increase as the levels of early somatic transcripts decrease. Transcripts of the late somatic genes are not detected at significant levels at any time. (Data from Tam and Kirk, 1991a,b; Tam *et al.*, 1991.)

(Tam and Kirk, 1991a) that the product of this gene might be involved somehow in the programming of asymmetric division, although intriguing, is untested.

Transcripts of the remaining 18 gonidial genes all exhibit a maturation-abundant pattern of accumulation: they appear after the end of embryogenesis, exhibit one abundance peak during early gonidial differentiation, a second, higher abundance peak during gonidial maturation, and then fall precipitously just before the gonidia initiate the next round of embryogenesis (Tam and Kirk, 1991a).

It is noteworthy that in the first few hours of cytodifferentiation, as the maturation-abundant gonidial transcripts slowly begin to accumulate in presumptive gonidia, they also begin to accumulate in presumptive somatic cells. But a short time later, as these transcripts accumulate in an accelerating fashion in the gonidial cells, they all disappear from the somatic cells (Tam and Kirk, 1991a). The stage at which these gonidial transcripts disappear from somatic cells is approximately the stage at which the *regA* gene is believed to be expressed (Huskey and Griffin, 1979); their disappearance at this time is consistent with the hypothesis that the function of the *regA* product is to suppress in somatic cells the expression of genes required for reproduction.

E. Expression of Cell Type-Specific Genes in the Absence of Asymmetric Division

The cell type-specific cDNA probes identified in the studies just described were subsequently used by Tam and Kirk (1991b) to examine transcription patterns in a Gls/Reg mutant. As noted, Gls/Reg embryos fail to divide asymmetrically; thus, they produce cells of only a single size and type: regenerating somatic cells that first develop the morphological features of somatic cells and then dedifferentiate and redifferentiate as gonidia. Although Gls/Reg cells and wild-type somatic cells are initially indistinguishable in terms of development of visible morphological features, their transcript abundance patterns are strikingly different (Fig. 4C). Wild-type somatic cells contain only traces of gonidial gene transcripts—and these only transiently during the very earliest stage of cytodifferentiation. In contrast, during the period when they are developing the morphological features of somatic cells, Gls/Reg cells accumulate fairly substantial amounts of gonidial gene transcripts, along with abundant levels of early somatic gene transcripts. Then, about a day after the onset of cytodifferentiation, Gls/Reg cells exhibit three additional, dramatic differences from wild-type somatic cells in their transcript abundance patterns: (1) levels of early somatic gene transcripts (which would remain elevated in wild-type) fall rapidly to vanishingly low levels; (2) late somatic gene transcripts (which would accumulate to high levels at this time in wild-type somatic cells) fail to appear, and (3) gonidial transcripts accumulate to substantially elevated levels. Shortly after this transition, the differentiated features of somatic cells (such as flagella and eyespots) begin to be resorbed, and the morphological features of gonidia begin to appear.

The results just described have been interpreted (Tam and Kirk, 1991b) as being consistent with, and therefore reinforcing, the model for control of *Volvox* cytodifferentiation that was presented herein (Fig. 3). The failure of Gls/Reg cells to suuppress expression of gonidial-specific genes while they are differentiating as somatic cells was attributed to the mutant character of the *regA* function in this strain. However, the initial failure to suppress the expression of the early somatic genes in any of these cells was attributed to the fact that (because of the mutant *gls* function and the consequent absence of asymmetric division) no cells were present at the end of cleavage that were large enough to activate expression of the *lag* loci. However, the fact that expression of both the early and the late somatic genes was suppressed a day later requires an explanation that is not explicitly provided by the model in Figure 3. It was postulated (Tam and Kirk, 1991b) that there are two ways in which somatic gene expression can be suppressed in *V. carteri*: (1) by the action of one or more of the gonidial gene products that are accumulated during the first day of Gls/Reg cytodifferentiation and (2) by the action of the *lag* gene products. It was further postulated that the former mechanism might be the more ancient one (and the molecular basis for the sequential pattern of vegetative and reproductive development that is observed in colonial volvocaceans), whereas the latter (*lag*-dependent) mechanism has been added to the genetic repertoire more recently and has the effect of preempting the ancient mechanism to permit an early and complete division of labor between germ and soma in *V. carteri*.

The preceding interpretations make numerous testable predictions about the patterns of gene expression to be expected in various colonial relatives of *Volvox*. For example, *Eudorina* cells and the posterior cells of *Pleodorina* (both of which follow a pathway of visible differentiation similar to that of Gls/Reg cells) should exhibit gene expression patterns similar to that of Gls/Reg cells, whereas the anterior cells of *Pleodorina* (which are terminally differentiated) should more closely resemble the wild-type somatic cells of *V. carteri*. Genomic sequences capable of cross-hybridizing to some of the cell type-specific cDNAs of *V. carteri* have been detected in *Eudorina* and *Pleodorina* (D. L. Kirk *et al.*, unpublished observations); however, tests of these predictions remain to be made.

F. Cytological Control of Division Symmetry in *V. carteri*

All of the results summarized so far in Section II indicate that asymmetric division plays an absolutely central role in the program of dichotomous germ/soma differentiation of the *V. carteri* embryo: it not only generates the spatial pattern in which germ and somatic cells will be distributed in the adult, it is essential for initiating the program of differential gene expression required to establish these two distinct cell types. Thus, an understanding of the way in which the genes regulating division symmetry exert their effects at the cytological level will be essential for a full understanding of the way in which germ/soma differentiation is preprogrammed in the *Volvox* genome. In turn, however, a prerequisite for understanding how asymmetric division is accomplished is a rather detailed understanding of the structure of the division apparatus of *V. carteri* and the mechanism by which the division plane is established in the much more numerous symmetric divisions of the embryo.

Studies over many years, at a variety of levels of microscopic resolution, and with a variety of related green algae, including several species of *Volvox* (Zimmerman, 1921; Kater, 1929; Metzner, 1945; Johnson and Porter, 1968; Deason and Darden, 1971; Pickett-Heaps and Marchant, 1972; Cavalier-Smith, 1974; Coss, 1974; Triemer and Brown, 1974; Pickett-Heaps, 1975; Stewart and Mattox, 1975; Marchant, 1977; Floyd, 1978; Birchem and Kochert, 1979; Green *et al.*, 1981; Hoops and Floyd, 1982; Huang *et al.*, 1982; Hoops, 1984; Mesquita and Fátima Santos, 1984; Adams *et al.*, 1985; Wright *et al.*, 1985; Aitchison and Brown, 1986; Harper and John, 1986; LeDizet and Piperno, 1986; Domozych, 1987; Doonan and Grief, 1987; Gaffal, 1988; Salisbury *et al.*, 1988; Holmes and Dutcher, 1989; Seegar and Gerritsen, 1989; Seegar *et al.*, 1989; Seegar, 1990), combine to indicate that the cell division apparatus has many conserved features throughout the green algae and that some of these are significantly different from the better-known features of the cell division apparatus in either higher plants or animals. Details of these characteristic features of the volvocalean division apparatus have been reviewed recently at some length elsewhere (Kirk *et al.*, 1991a), and in more abbreviated form in Section I,A,3. Here it will suffice to reiterate briefly only those structural features that are thought to be key to the control of division symmetry.

Perhaps the most distinctive cytological feature of the volvocaleans is the clarity with which the BBA (the BBs plus their associated MTRs, striated fibers, etc.) can be seen to function as the organizing center of the cell in both interphase and cell division (Section I,A,1). Prior to cell division, two of the four MTRs of the BBA define the plane in which the mitotic spindle will form, while the remaining two rootlets define the plane in which the metaphase plate will form and in which the cell will divide.

During division the BBs (having lost their flagella) act as centrioles while still remaining attached to the plasmalemma. They do this by sliding in the plane of the membrane while attached to the spindle poles by fibers that include the calcium-activated contractile protein, centrin. By the end of anaphase, the BBs begin to move back toward the incipient cleavage furrows; the sister nuclei follow. By telophase two parallel sets of cleavage MTs are organized by the sister BBAs; these cleavage MTs extend deep into the cell in the internuclear region and define the plane in which the furrow will ingress. From these structural considerations alone, it is clear that a more detailed understanding of the organization and functioning of the BBA is essential for an understanding of the control of division symmetry in the Volvocales.

Preliminary immunocytological studies have confirmed that the BBAs and the remainder of the division apparatus of *V. carteri* possess most of the features just outlined, as well as certain subtle variations that have not yet been reported for other species (Kirk *et al.*, 1991a). Extensions of these immunocytological studies have revealed at least one interesting aspect of the *V. carteri* division apparatus: a G-protein (*Ypt*V1; Section IV,B,7), appears to be localized within the cell division apparatus in close association with centrin, between the basal bodies and the spindle poles (R. Keeling *et al.*, unpublished observations). This observation is particularly interesting in light of the role of small G-proteins in establishing the site at which the asymmetric division of budding yeast will occur (Chant and Herskowitz, 1991; Chant *et al.*, 1991; Johnson and Pringle, 1990; Powers *et al.*, 1991).

Although the primary emphasis of our present immunocytological studies of *V. carteri* cleavage is on the structural and dynamic features of the cytokinetic apparatus involved in the early, symmetric divisions of the *Volvox* embryo, one interesting feature of the later asymmetric divisions has recently been deduced. The younger of the two BBs of each asymmetrically dividing cell goes to the smaller daughter cell (the presumptive somatic initial), whereas the older BB goes to the larger cell (the gonidial initial) (R. M. Keeling and D. L. Kirk, unpublished observations). As mentioned, studies of cell division and flagellar physiology have indicated that unicellular volvocaleans have a fundamental asymmetry that is based on the difference in developmental age of the two basal bodies within each cell (Section I,A,3). Recent analysis indicates that the asymmetric division that underlies germ/soma differentiation in *V. carteri* may well be based on this fundamental volvocalean asymmetry.

Once a more complete picture is developed of the organizational and dynamic features of the division apparatus involved in the symmetric cleavage divisions of the *Volvox* embryo and the modifications that characterize the later asymmetric divisions, the ultimate challenge will be to analyze the genetic basis for these features through the analysis of appropriate mutants with modified division patterns.

III. Experimental Analysis of Sexual Induction in *V. carteri*

Attempts to elucidate the mechanism by which *V. carteri* is diverted from the asexual to the sexual pathway of development have constituted one of the most intensive areas of *Volvox* research over more than two decades (Kochert, 1968; Starr, 1969; for review of early studies see Kirk and Harper, 1986).

The *V. carteri* f. *nagariensis* sex-inducing pheromone is one of the most powerful regulatory molecules in the living world: it exerts its full effects at a concentration below 10^{-16} *M* (Gilles *et al.*, 1984). Gonidia within spheroids that have been exposed to effective concentrations of inducer for sufficient time modify their cleavage patterns (which would otherwise result in production of more asexual progeny) and produce sexual progeny containing eggs or sperm (Starr, 1969; Starr and Jaenicke, 1974; for review see Jaenicke, 1991).

As discussed in more detail (Section IV,C,1), the genes encoding the polypeptide chain of the sexual pheromone have now been cloned and sequenced (Tschochner *et al.*, 1987; H.-W. Mages *et al.*, 1988). Although earlier cyanogen bromide cleavage studies had led to the conclusion that at least two variants of the pheromone exist that differ by at least one amino acid interchange (Gilles *et al.*, 1987), all genes in this family that have been sequenced to date encode the same deduced amino acid sequence, although they fall into at least three categories in terms of insertions and/or deletions in noncoding regions (M. Sumper, personal communication).

The secreted pheromone has an apparent molecular mass of ~30 kDa, is about 40% carbohydrate, and occurs as a mixture of differentially glycosylated isoforms that contain both O-linked xylogalactan and arabinogalactan residues, and N-glycosidic, chitobiose-linked xylomannan side chains (Günther *et al.*, 1987; Balshüsemann and Jaenicke, 1990a). Selective removal of the N-glycosidic chains inactivates the molecule (Jaenicke, 1991).

The sexual pheromone is produced most abundantly by sexual males, late in the process of sperm differentiation (Balshüsemann and Jaenicke, 1990b). As it releases its sperm packets, a single sexual male releases enough pheromone to induce about 5×10^8 other males and females in a cubic meter of water to enter the sexual pathway. The first sexual male to appear in a pond thus can convert the entire population to sexual activity, and such first males have been shown to appear in laboratory populations as the result of a mutation to constitutive sexuality (Weisshaar *et al.*, 1984). Alternatively, elevated temperature (in the range that might be encountered in the shallow, temporary ponds where *V. carteri* is found in the wild) can trigger a sexual orgy by causing somatic cells of both asexual males and females to produce and secrete the sexual pheromone (Kirk and Kirk, 1986).

Despite numerous and sustained efforts over the past decade to elucidate the mechanism by which the sexual inducer works to divert gonidia from the asexual into the sexual pathway, the process remains highly enigmatic.

A. Four Paradoxes of Inducer Action

1. The Exquisite Sensitivity versus the Sluggish Kinetics of Induction

In view of the extremely small quantities of pheromone that are required, it is astonishing how long an exposure to the pheromone is required before *V. carteri* gonidia become committed to initiating sexual development. Fifty percent induction of the gonidia in a population of spheroids requires exposure to the pheromone for 5 hours prior to the cleavage cycle in which the effects of the inducer will be visibly expressed, and full induction requires exposure for at least 8–10 hours (Gilles *et al.*, 1984; Jaenicke and Gilles, 1985). This response time is not reduced by pheromone concentrations that are orders of magnitude greater than the minimum required for complete induction. Moreover, the induced state that is eventually elicited by the pheromone decays with kinetics similar to the induction kinetics when the pheromone is removed from the external medium prior to the initiation of cleavage; induction becomes fully irreversible only when cleavage to form a sexual embryo has begun (Gilles *et al.*, 1984). These observations suggest that for the sexual pattern of embryogenesis to be elicited, either some slowly accumulated product of inducer action must reach a critical threshold level or some lengthy chain of inducer-triggered events must be completed. It is clear that continuous presence of the pheromone is required for both the establishment and the maintenance of the induced state throughout this long period.

2. The Presumed Target for the Pheromone versus Its Apparent Site of Action

Clearly, the ultimate target of sexual induction is the maturing gonidium that will eventually divide to form either an asexual or a sexual offspring.

Although it has been asserted that certain modified biochemical activities can be detected in gonidia from spheroids that have been exposed to the pheromone (Gilles *et al.*, 1984, 1987), specific date substantiating this claim remain to be published. All of the biochemical changes that have been reported to occur in response to the pheromone take place outside the gonidium, in either the somatic cells or the extracellular matrix (ECM). This is made all the more puzzling in light of observations that gonidia can actually be induced by the pheromone in the absence of either somatic cells or ECM.

Changes in the ^{35}S-sulfate–labeling patterns of four ECM sulfated glycoproteins (SGs) have been observed within the first hour of exposure to the pheromone (Wenzl and Sumper, 1982, 1986b, 1987):

1. Labeling of a novel component, SG 70, is detected within 10 minutes of pheromone exposure, peaks at 50 minutes, and declines after 150 minutes.
2. Incorporation of label into FSG (female surface glycoprotein) begins at 30 minutes and continues through the beginning of embryogenesis.
3. Between 30 and 90 minutes incorporation of label into SSG 140 (a component of the asexual ECM) declines.
4. Meanwhile, incorporation of label into SSG 110 (a presumed replacement for SSG 140) rises.

All of these changes in ^{35}S incorporation exhibit about the same sensitivity to limiting pheromone concentrations as sexual induction itself and occur identically in intact spheroids and isolated somatic cells; however, no ^{35}S labeling of any ECM components can be detected in isolated gonidia (Wenzl and Sumper, 1987). Similar changes in ^{35}S labeling patterns are observed whether the pheromone is added at the time of maximum inducibility or at later stages when the gonidia have become noninducible; this is cited as evidence that these SSGs do not have any direct or immediate effect on the state of determination of the gonidia (Wenzl and Sumper, 1986b).

Somewhat similar changes in ^{32}P labeling of ECM phosphoproteins (pps) are also observed following addition of pheromone (Gilles *et al.*, 1983; Jaenicke and Gilles, 1985). Within the first hour after pheromone addition, incorporation of ^{32}P into one pp (pp 290) drops transiently. By hour 3 incorporation into two new pps (pp 240 and pp 120) is detected. Once again, however, these biochemical responses have been attributed to somatic cells because no labeled pps can be detected when highly concentrated extracts of ^{32}P-incubated gonidia are analyzed (Wenzl and Sumper, 1987).

3. Elevation of cAMP Levels by Pheromone versus Inhibitory Effects of cAMP

Kochert (1981) stated that 3′-5′ cyclic AMP (cAMP) and phosphodiesterase (PDE) activity could be detected in isolated gonidia of *V. carteri* f.

weismannia and postulated (strictly by analogy with animal systems) that the sexual pheromone acts by binding to the gonidial membrane and activating an intracellular cAMP cascade. Since that time, several reports have appeared implicating cAMP and/or a cAMP-triggered cascade in the chain of events linking pheromone exposure to sexual morphogenesis in *V. carteri* f. *nagariensis* (Gilles *et al.,* 1984, 1987; Jaenicke and Gilles, 1985; Moka, 1985, 1988; Colling *et al.,* 1988; Jaenicke, 1991).

Gilles *et al.* (1984) reported that isobutyl methyl xanthine (IBMX, an inhibitor of PDE) and cAMP inhibited sexual induction by the pheromone and that exogenous PDE acts as a sexual inducer. Based on the assumption that PDE is too large to penetrate cells and therefore must be working extracellularly, it was postulated that a key step in pheromone action is a pheromone-triggered decline in the levels of extracellular cAMP, a decline that could be prevented by IBMX or generated by exogenous phosphodiesterase. Subsequent analytical data, however, have not supported this hypothesis. It turns out that what is observed is not a decrease but two increases in both intracellular and extracellular cAMP levels in induced spheroids (Gilles *et al.,* 1985; Jaenicke, 1991). The first cAMP increase occurs shortly after adding inducer and the second occurs while the responding gonidia prepare for cleavage (Moka, 1988). These data, combined with the earlier observation that the pheromone elicits changes in protein phosphorylation (Gilles *et al.*, 1983), have led to the hypothesis that cAMP plays a positive role in the induction process via its effects on a cAMP-regulated protein-kinase chain (Jaenicke, 1991). If the hypothesis is correct, however, we are left without any obvious explanation for the observation that IBMX and cAMP inhibit sexual induction by the pheromone (Gilles *et al.,* 1984). It should be noted in this context that induction of sexuality by phosphodiesterase reported by Gilles *et al.* (1984) has been attributed to contamination of the enzyme solutions with the sexual pheromone (Wenzl and Sumper, 1987).

4. Stimulatory versus Inhibitory Roles of the ECM

Many lines of evidence from different laboratories indicate that the ECM plays a key role in the sexual induction process. But very different views have been expressed on what that role is.

It was demonstrated that at limiting pheromone concentration the ECM binds much more inducer than either somatic cells or gonidia do and that it thereby concentrates the inducer to a level 100-fold higher than that in the surrounding medium (Gilles *et al.,* 1984). In this same study it was reported that isolated gonidia could not be induced by the sexual pheromone, unless ECM was added back to the culture medium. Subsequently, however, it was found that isolated gonidia could be induced, but that this required 100-fold higher concentrations of pheromone than were required to induce the gonidia of intact spheroids (Wenzl and Sumper, 1986b, 1987; Gilles *et al.*, 1987). Two different interpretations of these results have been offered.

Based in large part on the earlier studies of Gilles *et al.* (1984), Wenzl and Sumper (1986b, 1987) proposed that the role of the matrix is as "an amplification system, accumulating the pheromone and raising the actual pheromone concentration up to 100-fold at the surface of the gonidium." They further proposed that SG 70 (the novel HRGP that appears within 10 minutes of pheromone addition; Wenzl and Sumper, 1986b) might constitute a major component of this "pheromone transport system." This hypothesis implies that the first target of the pheromone is the somatic cells, which respond by producing a novel sulfated glycoprotein that acts to concentrate the pheromone to levels that are adequate to elicit a later, secondary response by the gonidia.

In contrast, Gilles *et al.* (1987) postulated that although the ECM does concentrate the inducer, the really essential role of the matrix is as an "intraspheroidal signal transducing . . . system," presumably involving cAMP and the protein kinase cascade. In support of this idea that a second messenger generated in the ECM must be accumulated to some critical concentration for sexual induction to occur, Gilles *et al.* (1987) reported that the percent induction observed in isolated gonidia fell as the volume of medium in which they were exposed to the pheromone was increased. They postulated that the reason isolated gonidia can respond to the pheromone at all is due to the fact that they retain a layer of ECM (the gonidial vesicle, or CZ1 layer; Kirk *et al.*, 1985) even after they have been released from the spheroid by protease digestion.

More recently, however, any unified view of the role of the ECM in sexual induction has become greatly complicated by reports that, instead of (or in addition to) its presumed stimulatory effect that had been deduced from the results discussed, the ECM may play an inhibitory role in the sexual induction process. Specifically, it was reported that gonidia produce sexual progeny in the absence of *any* exogenous pheromone if (but only if) they are isolated during the first hour of illumination on the second day of the 2-day life cycle (Starr and Jaenicke, 1989). This result is obtained whether the release of the gonidia at this critical time point is accomplished by pronase digestion of the spheroid (Starr and Jaenicke, 1989) or by mechanical disruption of the spheroid (R. C. Starr, personal communication). Two other categories of presumably unrelated treatments were found to cause sexual development in the absence of exogenous pheromone when applied to intact spheroids at this same critical time point in development; these were sublethal ultraviolet (UV) irradiation and treatment with various aldehyde/amine mixtures (Starr and Jaenicke, 1988, 1989). Because sexual development in all of these cases is prevented if antipheromone antibody is added to the culture medium (Starr and Jaenicke, 1989; R. C. Starr, personal communication), all of these disparate treatments are presumed to act *via* autoinduction, rather than by some pheromone-independent pathway. The authors propose, as an explanation of these results, that gonidia normally secrete the sexual pheromone in small amounts but that the ECM contains a neutralizer of the pheromone that prevents self-induction, and that can be inactivated by UV irradiation or treatment with aldehydes and amines (Starr and Jaenicke, 1989). Consistent with the idea that the ECM may contain an inhibitor instead of, or in addition to, an enhancer of pheromone action, is the observation that gonidia of a noninducible mutant strain do not respond to any concentration of pheromone as long as they are retained within the parental spheroid but that they become inducible when they are removed from the spheroid prior to pheromone exposure (R. C. Starr, personal communication).

The hypothesis that the wild-type ECM contains an inhibitor, or neutralizer, of the pheromone does not provide an obvious explanation for the fact that gonidia isolated from the spheroid at any time other than in the first hour of the light cycle only develop sexually if exposed to much higher levels of the pheromone than are required when they are left in the intact spheroid. Nor does it account for the fact that several treatments that elicit sexuality in this 1-hour period are without effect during the reaminder of the 48-hour life cycle. However, published data do not rule out the possibility that sublethal UV irradiation, aldehyde/amine mistures, and gonidial isolation all generate stresses that simulate the effect of heat shock (which has been shown to elicit production and secretion of the sexual pheromone at nearly any point in the life cycle; Kirk and Kirk, 1986) and that cells are simply more sensitive to such stresses during the period while they are adapting to a dark-to-light transition than they are at other times.

B. Competing Theories of Inducer Action

Far from resolving the mechanism by which the pheromone acts to induce sexual development, studies have left us with many apparently conflicting observations and with several competing theories, none of which seems to explain all of these observations. The one concept on which consensus appears to have been reached is that the somatic cells and/or the ECM play some important role(s) in the induction process. But consensus remains to be reached about whether this role is to generate a cAMP/protein kinase cascade that either inhibits sexual development (Gilles *et al.*, 1985) or elicits sexual development (Jaenicke, 1991), or whether it is merely to concentrate the inducer and present it in an effective concentration to the responding gonidia (Wenzl and Sumper, 1986b, 1987), or whether it is to suppress a continuous, weak autoinductive stimulus (Starr and Jaenicke, 1989).

It is now abundantly clear that either directly or indirectly the pheromone does elicit a number of changes in both the somatic cells and the ECM. But it remains true, as stated by Gilles *et al.* (1987), that "The sexual induction system of *Volvox carteri* is far from being understood on a biochemical level . . . but many modifications of the extracellular matrix have been found." It is to be hoped that it will soon become clear which of these may be causally important in the sexual induction of the gonidium (the ultimate target of the pheromone) and which may be merely epiphenomena.

IV. Molecular Analysis of Gene Structure and Expression in *V. carteri*

A major goal of our laboratories has been to develop tools that will facilitate isolation and detailed analysis of the genes that play key roles in controlling *Volvox* development (Section II). To that end, we have cloned the gene that encodes nitrate reductase (Gruber *et al.*, 1992; Section IV,B,6) and have succeeded in using it both as a trap for transposons that should be useful for gene tagging, and as a selectable marker in a transformation system (Schmitt *et al.*, and Kirk *et al.*, unpublished observations). These methods hold great promise for future analysis of genes whose developmental roles can presently be defined only in terms of mutant phenotypes. Meanwhile, however, significant advances have been made in the molecular analysis of the structure and/or expression of a number of other genes of *V. carteri*.

With its 1.2×10^8 bp distributed among 14 chromosomes (Kirk and Harper, 1986), the haploid *V. carteri* genome is about the same size as the haploid *Drosophila* genome and probably has about the same degree of complexity. Restriction fragment length polymorphisms (RFLPs) are being used to expand the genetic map of the species (Harper *et al.*, 1987; Adams *et al.*, 1990), which had previously been based almost exclusively on loci defined by morphological mutants (Huskey *et al.*, 1979a).

The *V. carteri* genes that have been cloned and sequenced so far (Table I) fall into three major categories: rRNA genes, genes encoding proteins of wide-spread distribution and known function, and genes encoding *Volvox*-specific proteins.

A. Ribosomal Genes

The nuclear rRNA genes of *V. carteri* are organized similarly to those of other eukaryotes: as tandemly repeated copies, each of which contains segments that encode the 18S, 5.8S, and 28S rRNAs, in that order, and that are separated from one another by short (0.3kb) transcribed spacers (Rausch *et al.*, 1989). None of the copies analyzed contains an intron, and each 6.3 kb transcription unit is separated from its neighbor by a 3.3 kb nontranscribed spacer, yielding a total length of 9.6 kb per rDNA repeat unit (Rausch, 1988).

Titration experiments indicated that there are about 1200 rDNA copies per haploid *V. carteri* genome; this value is 2–12 times greater than that of most animals (100–600) but an order of magnitude smaller than that of some higher plants (Gerlach and Bedbrook, 1979). rDNA, therefore, occupies $\sim 1.2 \times 10^7$ bp, or ~10%, of the *Volvox* genome. This relatively high proportion of the genome devoted to rDNA presumably reflects the need for efficient translation machinery in the juvenile phase of the asexual life cycle (Yates and Kochert, 1976; Hagen and Kochert, 1980).

Phylogenetic inferences drawn from comparative studies of rRNA sequences are discussed in Section I,B.

B. Genes Encoding Ubiquitous Proteins

Studies of *V. carteri* versions of genes encoding ubiquitous eukaryotic proteins of known function have been of interest for several reasons. First, by virtue of the unique position of the green algae as a branch emerging from the main trunk of eukaryotic evolution between branches leading to the higher plants and the animals (Rausch *et al.*, 1989), analysis of volvocalean genes may be expected to provide some insights into possible origins of differences in gene structure that have been observed between plants and animals. Second, through comparisons of homologous genes of *C. reinhardtii* and *V. carteri*, one may hope to gain some insights into molecular–evolutionary trends that have accompanied diversification of the volvocaceans. Third, any recurring, idiosyncratic features observed in the structural genes of *V. carteri* may forecast features to be expected in *Volvox* genes of developmental significance that remain to be cloned and characterized.

1. Tubulin Genes

Tubulins play a pivotal role in the volvocalean cell as essential constituents of the cytoskeleton, the mitotic spindle, and the flagellar axoneme (for review see Schmitt and Kirk, 1992) (Section I,A,2). Both *V. carteri* (W. Mages *et al.*, 1988; Harper and Mages, 1988; Mages, 1990) and *C. reinhardtii* (Youngblom *et al.*, 1984; Silflow *et al.*, 1985) possess two α-and two β-tubulin genes, all of which have been sequenced. In both organisms the β-tubulin loci are genetically linked, but the α-tubulin genes are not.

TABLE I

Sequenced Nuclear Genes of *V. carteri*

Gene family	Gene product	Copy number/ number sequenced	Coding region (bp)	Polypeptide (aa)	Number of introns	Transcript abundance	Signals	Reference
rrnA	18 S rRNA	~1250/1	1788	—	0	High		Rausch *et al.* (1989)
rrnB	5.8 S rRNA	~1250/1	158	—	0	High		Rausch (1988)
rrnC	28 S rRNA	~1250/1[a]	3500	—	0	High		Larson *et al.* (1992)
tubA	α-Tubulin	2/2	1353	451	3	High, whole life cycle	Conserved promoter; UGUAA poly(A) signal	W. Mages *et al.* (1988); Mages (1990); Schmitt and Kirk (1992)
tubB	β-Tubulin	2/2	1329	443	3	High, whole life cycle	Conserved promoter; UGUAG poly(A) signal	Harper and Mages (1988); Mages (1990); Schmitt and Kirk (1992)
hstH2A	Histone H2A	15/2	387	128[b]	0	High, embryogenesis	20-bp promoter elements; 3′ palindrome; no poly(A) tail	Müller *et al.* (1990)
hstH2B	Histone H2B	15/2	471 465	156[b] 154[b]	0 0	High, embryogenesis	20-bp promoter elements; 3′ palindrome; no poly(A) tail	Müller *et al.* (1990)
hstH3	Histone H3	15/2	405	134[b]	1[c]	High, embryogenesis	14- and 20-bp promoter elements; 3′ palindrome; no poly(A) tail	Müller and Schmitt (1988)
hstH4	Histone H4	15/2	312	102[b]	0	High, embryogenesis	14- and 20-bp promoter elements;	Müller and Schmitt (1988)

							3′ palindrome; no poly(A) tail	
hstH1	Histone H1	≥4/2	783 723	260[b] 240[b]	3 3	High, embryogenesis	20-bp promoter elements; 3′ palindrome; no poly(A) tail	Lindauer *et al.* (1992)
ubqA	Polyubiquitin	≥1/1	1143	381[d]	5	High, whole life cycle	TATA box; two UGUAA poly(A) signals	Schiedlmeier and Schmitt (1992)
actA	Actin	1/1	1131	377	9[e]	Moderate, whole life cycle		Cresnar *et al.* (1990)
nitA	Nitrate reductase	1/1	2592	864	10	Induced by NO_3^- repressed by NH_4^+	UGUAA poly(A) signal	Gruber *et al.* (1992)
yptV1	YptV1 (G-protein)	1/1	609	203	8	Low	Three UGU(C/A)(A/G) poly(A) signals	Fabry *et al.* (1992b)
yptV2	YptV2	1/1	648	217	5	Low		Fabry *et al.* (1992a)
yptV3	YptV3	1/1	609	203	6	Low	UGUAG poly(A) signal	Fabry *et al.* (1992a)
yptV4	YptV4	1/1	639	213	7	Low		Fabry *et al.* (1992a)
yptV5	YptV5	1/1	615	205	5	Low	UGUAA poly(A) signal	Fabry *et al.* (1992a)
ssg185	SSG 185	1/1	1455	485	7[e]	High in somatic cells	UGUAA poly(A) signal	Ertl (1989); Ertl *et al.* (1989)
sexI	ECM Sexual inducer	~6/1	624	208	4	Low, inducible	UGUAA poly(A) signal	Tschochner *et al.* (1987); H.-W. Mages *et al.* (1988)

[a] Partial sequences amounting to about 2000 nucleotides.
[b] Numbering according to Wells (1986); initiator methionine not counted.
[c] Intron position shifted by 1 nt between the two H3 genes.
[d] Processed to four ubiquitin polypeptides 76-aa long and one 77-aa long.
[e] First intron located within 5′-untranslated region.

The related quadriflagellate alga *Polytomella agilis* (Conner *et al.*, 1989) contains three β-tubulin genes, all of which have been sequenced; the α-tubulin genes of this alga have not been studied yet.

The 1356-bp coding regions of all four *V. carteri* and *C. reinhardtii* α-tubulin genes encode polypeptides of 451 amino acid residues each. It has now been established for both *V. carteri* (Mages, 1990) and *C. reinhardtii* (C. Silflow, personal communication) that the αl-tubulin and α2-tubulin genes encode identical polypeptides, but between the two species there are two amino acid exchanges: Val_{252} and Ala_{331} of *V. carteri* are exchanged for Ile_{252} and Ser_{331} in *C. reinhardtii.* As in most other species, a carboxy-terminal Tyr (that can be reversibly removed to expose a subterminal Glu residue) is encoded by the volvocalean α-tubulin genes.

In contrast to the highly conserved aa sequences encoded by the volvocalean α-tubulin genes (>99.5%), there are 10–11% (mostly silent) nt interchanges between the two algae. Even more interesting are differences in the extent of intraspecfic nt exchanges in the two species, which amount to 2% between the two *C. reinhardtii* α-tubulin genes but 12% between the two *V. carteri* α-tubulin genes. This curious difference will be discussed in a later section (V,B,3), as it has been found to be a recurring feature in the comparative analysis of *Volvox* and *Chlamydomonas* genes.

Exon–intron arrangements of the two α-tubulin genes are identical within both species but exhibit one difference between species: the *V. carteri* genes contain three introns, following Glu_{15}, within Gln_{90}, and within Asp_{211}, whereas the *C. reinhardtii* genes contain only the first two introns, at the identical positions.

The β1- and β2-tubulin genes of *V. carteri* and *C. reinhardtii* closely resemble each other in their exon–intron organization and in the polypeptides they encode (Youngblom *et al.*, 1984; Harper and Mages, 1988). All four 1332-bp coding regions specify 443-residue polypeptides that are identical within species but differ in the C-terminal residue between species: an Asn (AAC) in the *V. carteri* genes is exchanged for Ala (GCC) in *C. reinhardtii.* As with the α-tubulins, this high conservation (99.8%) of deduced aa sequence is in contrast to a 9.5–11.5% difference in nt sequence between homologous β-tubulin coding regions of the two algae. And, again in parallel to the α-tubulin genes, the nucleotide sequences of the *Volvox* β1- and β2-tubulin genes differ by 11%, whereas the corresponding *Chlamydomonas* genes differ by only 1.4%.

The three β-tubulin genes of the related quadriflagellate *P. agilis* also encode 443 aa residues. The β1- and β3-tubulin genes are much more similar to one another than either is to the β2-tubulin gene, leading to the suggestion that β1 and β3 result from a fairly recent duplication (Conner *et al.*, 1989). β1 and β3 differ by 28 (2%) silent nt exchanges, but from β2 by 64 (5%) exchanges, one of which results in a Leu_{37} to Ile_{37} exchange in β2-tubulin. All three *P. agilis* β-tubulins share the C-terminal Asn of *V. carteri* and differ from *V. carteri* and *C. reinhardtii* β-tubulins by five and six residues, respectively.

All four β-tubulin genes of *Volvox* and *Chlamydomonas* contain three introns at identical positions: following Gln_8, within Gly_{57}, and within Gly_{132}. The three *P. agilis* β-tubulin genes share two of these intron positions (at Gly_{57} and Gly_{132}), but a third intron following Gln_{15} in all three *P. agilis* genes, plus a fourth following His_6 in the β1 and β3 genes, are unique to this species. Particularly interesting is the observation (Youngblom *et al.*, 1984) that the third introns of the *C. reinhardtii* β-tubulin genes not only exhibit 89% sequence similarity to one another but also 81% similarity to a short segment of the second intron of the α2-tubulin gene. No such similarities are detected among the various introns of the *V. carteri* or *P. agilis* tubulin genes.

Codon usage in all of these volvocalean tubulin genes is strongly biased in favor of C and against A in the third codon position, with the preferred order being C>G>U>>A, and with many codons not being used at all (W. Mages *et al.*, 1988; Schmitt and Kirk, 1992). These biases, which are most extreme in *C. reinhardtii,* are seen in varying degrees in volvocalean genes encoding other proteins and, thus, are commented on at some length later (Section V,B).

Sequence elements that regulate tubulin gene expression have been located by S1 mapping and primer extension, functional tests in *Xenopus* oocytes, and by comparison of the 5′- and 3′- untranslated regions of all volvocalean tubulin genes (Schmitt and Kirk, 1992). A GC-rich 9- to 12-bp sequence is located between the TATA box and the site of transcription initiation. This highly conserved element is absolutely required for transcription (Bandziulis and Rosenbaum, 1988). Coordinate transcription of all tubulin genes is presumably directed by a 16-bp motif, with a conserved 10-bp core, that is present in three or more copies upstream of each TATA box. A third element, 400–800 bp upstream of transcription initiation serves as an enhancer (Bandziulis and Rosenbaum, 1988). A 28-bp sequence motif with partial dyad symmetry common to all α- (but not β-) tubulin genes was identified between the transcription and translation starts (Mages, 1990). In a computer-generated secondary-structure model, this element is exposed in a particular stem–loop configuration and may, therefore, play a role in regulating translation (W. Mages, personal communication). Preceding the ATG start codon, all tubulin genes share an AACC consensus sequence thought to indicate the correct translation start. Finally, a specific polyadenylation signal, UGUAA/G, is found in the 3′-untranslated region of tubulin genes, approximately 20 bp upstream of the polyadenylation site. This deviation from the corresponding animal consensus (AAUAAA; Proudfoot, 1982) appears to be a general feature of the volvocalean mRNA processing machinery.

The high degree of primary structure conservation (>98%) among the analogous tubulins supports the close phylogenetic relationship among the volvocalean algae *V. carteri, C. reinhardtii,* and *P. agilis*. Although they are only ~90% similar in aa sequence to higher plant β-tubulins, the volvocalean β- tubulins do share certain plant-specific residues not found in animal or fungal equivalents, such as Cys-Met-Val-Leu (positions 199–202) and Trp_{294}.

2. Nucleosomal Histone Genes

Histones H2A, H2B, H3 and H4 are highly conserved basic proteins that form the nucleosomal core of the eukaryotic chromatin structure. Unlike most other structural genes, nucleosomal histone genes are generally expressed in concert with DNA synthesis during the S-phase. The high requirement for histone protein (in strict stoichiometry with DNA) is met by increased gene dosage as reflected by a variety of tandem gene clusters coding for the histones (Hentschel and Birnstiel, 1981; Gigot, 1988).

The first algal histone genes have recently been characterized by the Schmitt laboratory (Müller and Schmitt, 1988; Müller *et al.,* 1990) and found to exhibit a combination of features not previously described. The *V. carteri* and *C. reinhardtii* genomes contain ~15 copies of each type of nucleosomal histone gene; these are all arranged in clusters of H2A–H2B and H3–H4 pairs, in which each gene pair shows outwardly divergent transcription from a short (<300bp) intercistronic region that contains the promoter signals (Fig. 5). In contrast to the histone mRNAs of higher plants (Chaboute *et al.,* 1988), histone mRNAs of *V. carteri* (and presumably *C. reinhardtii*) are not polyadenylated (Müller *et al.,* 1990). Instead, transcription is terminated in each case by a 3′ palindrome, as in vertebrates and sea urchins (Hentschel and Birnstiel, 1981).

In *V. carteri* two nonallelic copies of each of the four nucleosomal histone genes have now been sequenced (Table I), and two H3–H4 clusters and one H2A–H2B cluster have also been sequenced in *C. reinhardtii* (Müller *et al.,* 1992). As in other eukaryotes, aa sequence conservation is highest in histones H4 and lowest in the H2B proteins. The two deduced *Volvox* H4 and H3 proteins are identical, there is one aa difference (out of 128 aa residues) between the two H2A proteins, and 16 exchanges (in 156 aa) between the two H2B proteins. Differences between the homologous *Volvox* and *Chlamydomonas* histone sequences are slightly higher: 1/104 for H4, 0/136 for H3, 4–5/128 for H2A, and 29–32/156 for H2B. Codon

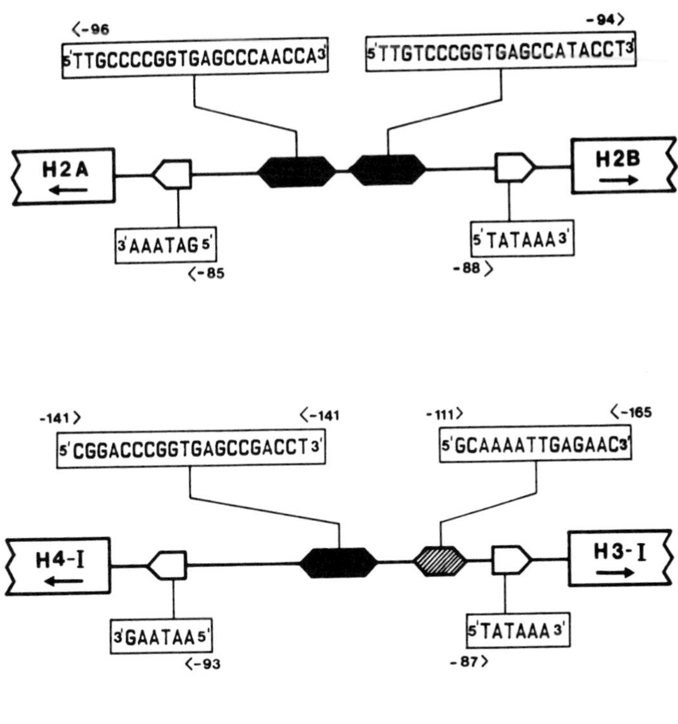

AAAATCCGGTGTTTTTCAACACCACCA

FIG. 5 Organization of nucleosomal histone–gene pairs, and location of presumptive promoter elements in the intercistronic region. Below: 3′ palindromic transcription terminator (consensus) succeeding each histone–gene. Diverging arrows indicate partial dyad symmetry. (Modified from Müller *et al.*, 1990.)

usage is strongly biased in the histone genes of these volvocaleans as it is in their tubulin genes (see Section V,B).

All *Volvox* and many *Chlamydomonas* H3 genes have an intron (Müller and Schmitt, 1988; Müller *et al.*, 1992). This is a feature that is shared by fungal and protist H3 genes and by the vertebrate replacement H3.3 genes (Wu *et al.*, 1986) but not by the classical, replication-type H3 genes of either animals (Wells *et al.*, 1986) or plants (Chaboute *et al.*, 1988). Nevertheless, gene expression studies, codon usage patterns, certain diagnostic aa positions, and the absence of polyadenylation all argue strongly for a structural and functional relationship of the *Volvox* H3 genes to the highly expressed, replication-dependent class of animal histone genes (Müller and Schmitt, 1988).

This unusual combination of structural features in the algal H3 genes reflects an interesting phylogenetic situation. It is widely assumed that higher plants evolved from a *Chlamydomonas*-like algal ancestor, and thus one might have expected to find preserved in the Volvocales an intermediate in the evolution of the plant H3 genes, which are intron-free and produce polyadenylated mRNAs. But such a relationship is not readily apparent. The aa sequences of the volvocalean histones do, indeed, exhibit greater similarity to plant histones than to those of other lower eukaryotes or to those of animals. But by virtue of their possession of a 3′ palindrome and their lack of polyadenylation, volvocalean H3 genes resemble animal H3 genes. However, in their split-gene organization they resemble fungal and protist H3 genes and the vertebrate H3.3-like genes (all of which, however, are polyadenylated and lack 3′ palindromes). One might postulate that on the evolutionary route to higher plants the introns and 3′ palindromes of algal H3 genes were both lost, and polyadenylation was readapted from nonhistone genes. Alternatively, two types of H3 genes may have been present in a common ancestor of algae, plants, and animals—as they are in modern vertebrates (Wells and Kedes, 1985). In that case, early divergence of volvocalean algae and higher plants would have been accompanied by the loss of different forms of H3 genes in the two lineages, resulting in intron-free, palindrome-free histone genes in higher plants and intron-containing, 3′ palindrome-terminated histone genes in the Volvocales (Müller and Schmitt, 1988).

Stage-specific transcription of the nucleosomal histone genes of *Volvox* is indicated by Northern-blot analysis of synchronized cultures: histone transcripts are abundant during embryogenesis, from the onset of cleavage through inversion, but they are not detectable thereafter, suggesting that the turnover of message is rapid and, thus, that histone gene expression in *Volvox* is regulated both by control of transcription initiation and by temporal control of mRNA stability. It is known that the stability of nonpolyadenylated mRNA molecules processed at the 3′ palindrome is controlled through factors that act on this symmetrical termination sequence (Birnstiel *et al.*, 1985). The appearance of histone transcripts at the beginning of the embryonic cleavage period suggests that DNA synthesis also takes place at this stage and that cell division and replication are coordinate events. Similarly, histone mRNA in *Chlamydomonas* is entirely restricted to the period of exponential growth, when the cells are dividing (Müller *et al.*, 1992).

The search for conserved sequence motifs that might act as signal sequences in the control of histone transcription in *V. carteri* is facilitated by the compact arrangement of the histone loci. Each of the four loci analyzed contains two divergently transcribed coding regions (H2A and H2B or H3 and H4), and the promoters for both genes of a pair are located within the short (<300bp) untranscribed region that separates them. A comparative search revealed a conserved 20-bp element that is always present between the divergent TATA boxes (Müller *et al.*, 1990). This element, which presumably functions as an enhancer, is present in tandem repeats in the H2A–H2B loci, and as a single copy in the H3–H4 loci (Fig. 5). The latter loci also contain an AT-rich, 14-bp element that resembles a yeast upstream activating sequence (UAS). It is assumed that specific nuclear proteins binding to these elements regulate the stoichiometric and stage-specific transcription of the histone genes. *Chlamydomonas* exhibits the same pairwise arrangement of divergently polarized histone genes, and the intercistronic regions contain conserved sequence elements. However, the latter are different from those in *Volvox*, suggesting that rather different nuclear proteins may be involved in the control of histone gene transcription in *C. reinhardtii* (Müller *et al.*, 1992).

3. Histone H1 Genes

The lysine-rich histone, H1, has been implicated in the organization of nucleosomes into higher orders of chromatin structure. Of all the histones, H1 exhibits the highest degree of heterogeneity, and indications are that a family of H1 isotypes exist that differ in their ability to condense chromatin *in vitro* and *in vivo* (Cole, 1984); moreover, it has been suggested that different isotypes of H1 may bind specifically to different chromosomal regions (Mohr *et al.*, 1989). In contrast to the extensive literature that exists on animal H1 proteins, the literature on higher plant H1 proteins is sparse (Ivanchenko *et al.*, 1987), but limited protein and cDNA sequence data have been reported on maize (Razafimahatratra *et al.*, 1991; Hurley and Stout, 1980), pea (Gantt and Key, 1987), and wheat (von Holt and Brandt, 1986).

The first algal H1 genes have now been isolated and characterized (Lindauer *et al.*, 1992). There are at least four different H1 genes in the *V. carteri* genome, two of which have been sequenced. Each H1 gene occurs as a singlet, and none are present in clones containing nucleosomal histone genes, suggesting that they may occupy separate map positions. H1 clones have recently been obtained from a *C. reinhardtii* library also, but they have not yet been characterized.

The derived H1 peptide sequences of *V. carteri* contain 260 (H1–I) or 240 (H1–II) residues, and they exhibit the typical three–domain pattern of H1 proteins (Doenecke, 1988), with a highly conserved central globular core flanked by variable N- and C-terminal domains. There are extensive differences between the variable regions of these two deduced H1 proteins in both length (N: 54 versus 20 residues; C: 129 versus 143 residues) and composition. As expected, the C-terminal domains of both contain lysine-rich repetitive elements (H1–I: seven KKATP repeats; H1–II: eight PKKAA[K]A repeats), but neither possesses the conserved octapeptide (TKKPAKKP) that is found near the middle of the C-terminal domain in most higher eukaryotes (Eick *et al.*, 1989). The central hydrophobic domain (implicated in binding DNA at the entry and exit sites of nucleosomes) is well conserved in length and composition between the two H1 proteins (only 16/77 aa exchanges), whereas differences from higher plant H1 (33–35/77 aa exchanges) and from the animal H1 consensus sequence (49–52/78 aa exchanges) are dramatically higher. Like the nucleosomal histone and tubulin genes, histone H1 genes exhibit strongly biased codon usage (see Section V,B).

The *Volvox* H1 genes possess a distinctive structural feature: each coding region is interrupted by three introns between 93 and 195 bp in length. We are aware of only two other organisms in which an intron (a single one in each case) has been found in an H1 gene, namely, *Tetrahymena* (Wu *et al.*, 1986), and *Caenorhabditis* (Sanicola *et al.*, 1990). The first and third introns occupy different positions in the two sequenced *Volvox* genes, whereas the location of the second intron (near the transition between globular and C-terminal domains) is conserved. Like the nucleosomal histone genes, each *Volvox* H1 gene possesses a 3′ palindromic termination signal, and cDNA sequence analysis indicates that the transcripts are not polyadenylated.

The transcription of both sequenced H1 genes initiates about 30 nt downstreat of a TATA box. Further upstream, a 20-bp element is seen that closely resembles the conserved enhancer motif associated with the *Volvox* nucleosomal gene pairs. H1-specific upstream elements resembling those found in animal H1 genes (Eick *et al.*, 1989) have not been detected, suggesting that different modes of transcriptional control may exist for algal H1 genes. The two H1 genes are efficiently transcribed during embryogenesis together with the nucleosomal histone genes, but it has yet to be determined whether H1 transcription is entirely confined to this part of the *Volvox* life cycle.

4. Ubiquitin Genes

Ubiquitin, a 76-aa polypeptide, is the most conserved eukaryotic protein known (Finley and Varshavsky, 1985; Hershko and Ciechanover, 1986), but it is not found in prokaryotes. It exists in cells either free or covalently attached to a variety of cytoplasmic, nuclear, and integral membrane proteins. Biochemical and genetic evidence indicates that conjugation of ubiquitin to intracellular proteins is essential for their selective degradation. Ubiquitin–conjugation activity is also required for transition across the S/G2 boundary of the cell cycle, and it may be involved in DNA repair (Jentsch *et al.*, 1987). Furthermore, the presence in chromatin of ubiquitinated histones H2A and H2B has led to the suggestion that ubiquitin could well modulate protein functions (West and Bonner, 1980). A long ubiquitin extension has been identified on the ribosomes of various eukaryotes (Redman and Rechensteiner, 1989). Finally, ubiquitin can be detected immunochemically on cell surfaces bound to various receptor proteins (Siegelman *et al.*, 1986). In view of the importance of this protein, ubiquitin gene structure and expression have also been studied in *Volvox*.

Two classes of ubiquitin genes are found in many organisms: monoubiquitin genes encoding a single ubiquitin polypeptide with a C-terminal extension and polyubiquitin genes consisting of tandem repeats of the coding unit, the protein products of which are subsequently processed by specific proteolysis to yield multiple copies of the 76-aa long active polypeptide. *V. carteri* contains both types of ubiquitin genes: a polyubiquitin gene that will be described here and three or more monoubiquitin genes not closely linked to the polyubiquitin locus (Schiedlmeier *et al.*, 1992). The polyubiquitin gene consists of five tandem repeats of the 228-bp coding unit in a head-to-tail, spacerless array. Five introns split the coding region into six exons, with each intron interrupting a ubiquitin unit within the codon for Gly_{35}, and with the last repeat ending with an extra Leu codon (Fig. 6). Selective forces maintaining ubiquitin primary structure are obviously robust because the *Chlamydomonas* monoubiquitin (Callis *et al.*, 1989) and *Volvox* polyubiquitin polypeptide sequences are identical to one another and differ by only one aa residue from higher plant ubiquitins and by two from animal ubiquitins.

Whereas the deduced aa sequences of the *Volvox* ubiquitin repeats are identical, there are 15 silent nt exchanges that are grouped in a remarkable way: 13 exchanges distinguish coding unit I from the other four units

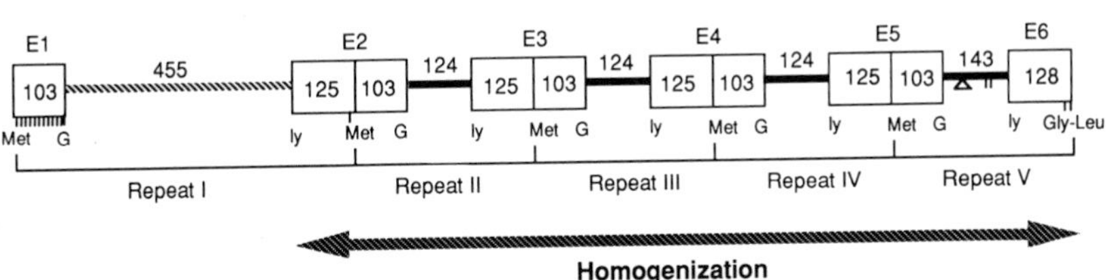

FIG. 6 Organization of the polyubiquitin gene. Boxes represent exons (E1–E6); bold lines represent introns. Their respective sizes are drawn to scale and indicated in bp (numbers). The five individual ubitiquitin repeats (I–V) are delineated below. The start codons (Met) and intron positions (Gly) for each of these repeats, plus the end of the coding region (Gly-Leu), are shown. The locations of 15 silent nucleotide exchanges in the coding region are marked by short vertical lines; a 19-bp insertion in the fifth intron is symbolized by a triangle. The nonidentical intron I is distinguished from the identical introns II–V by shading.

(which are all identical except for two exchanges at the end of unit V, just upstream of the 3′ terminal CUG that codes for the extra C-terminal Leu residue). Moreover, 12 of these 13 exchanges in unit I are concentrated in the first exon. Equally remarkable is the degree of similarity of four of the five introns. The 124-bp introns II–IV are identical and the 143-bp intron V differs from II–IV by only a 19-bp insertion and two nt exchanges. The 455-bp intron I is distinctively different in sequence, however (Fig. 6).

This organization of the *Volvox* polyubiquitin gene suggests that it was generated by duplication of the coding region of a monoubiquitin gene after it had acquired an intron. This conclusion is supported by the observation that its intron position (at Gly_{35}) is exactly the same as that of the intron found in mammalian and higher-plant monoubiquitin genes (Baker and Board, 1991). It seems likely that a single duplication occurred first and persisted long enough for introns I and II to diverge in sequence, before the second repeat was amplified three more times. Assuming that the latter event was not an *extremely* recent one, some form of gene homogenization must be acting to keep these repeats (four of five introns included) so similar in sequence. Sharp and Li (1987), who analyzed polyubiquitin sequenes from various other organisms as a paradigm of concerted evolution (Dover, 1982), concluded that the divergence observed between successive repeat units is usually great enough to indicate that homogenizing events occur only infrequently. Such a conclusion could not be supported by the *Volvox* gene, where the 4 1/2 repeats trailing intron I (including introns II–V) are virtually identical (Fig. 6), a circumstance that must demand frequent homogenization. Interestingly, the 19-bp insertion in intron V is the only portion of introns II–V that bears any significant sequence similarity (64%) with a portion of intron I, suggesting a possible exchange of sequence information between these two most distantly located regions during homogenization. The structure of the *V. carteri* polyubiquitin gene leads us to suggest that concerted evolution may be operating in this species by an efficient survey mechanism, possibly involving unequal crossing over. As a test of this interpretation, it will be of particular interest to perform a comparative study of the polyubiquitin gene of a genetically distant strain of *V. carteri* (Adams *et al.*, 1990) and one or more of the monoubiquitin genes of *V. carteri*.

Two transcription start sites were mapped in the *Volvox* polyubiquitin gene, 24 and 28 bp downstream of a TATAA motif; no other promoter signals have been identified yet in the upstream region. The 3′ untranslated sequence contains two potential UGUAA polyadenylation signals at a distance of 99 and 156 bp from the translational stop.

It has been reported that stress by light or heat shock alters the pattern of ubiquitin conjugation to cellular proteins in *C. reinhardtii* (Wettern *et al.*, 1990); in *Volvox* a labeled polypeptide tentatively identified as free ubiquitin was seen following heat shock (M. M. Kirk and D. L. Kirk, unpublished observations). However, no change in the abundance of *Volvox* polyubiquitin or monoubiquitin transcripts has yet been observed following heat shock (Schiedlmeier and Schmitt, 1992), indicating that in this species heat stress may effect ubiquitin metabolism at some level other than transcription.

5. Actin Gene

Actin, originally recognized as a major component of the contractile apparatus of muscle cells, is now known to be one of the most ancient and ubiquitous eukaryotic proteins and a key component of the cytoskeleton of all eukaryotic cells. However, volvocalean actins may be involved in certain functions that are specific to members of this group: it has been shown that actin assembly is essential for formation of the *Chlamydomonas* mating structure required for gamete fusion (Detmers *et al.*, 1985), and studies in *Volvox* have indicted that actin-based microfilaments have important functions in several species-specific aspects of embryogenesis (Viamontes *et al.*, 1979; Green and Kirk, 1981; Green, 1982).

Volvox contains a single actin gene (Cresnar *et al.*, 1990). In this respect it differs from higher multicellular organisms (which generally contain actin multigene families) and resembles certain unicellular eukaryotes, such as yeast (Gallwitz and Sures, 1980) and *Tetrahymena* (Hirono *et al.*, 1987), both of which also contain a single actin gene. The 1131-bp coding region of the *Volvox* gene is interrupted by eight introns, more than in any other known actin gene. Moreover, an additional intron has been identified in the 5′-untranslated region (Cresnar *et al.*, 1990). [An intron is also present in the 5′-untranslated region of the human β-actin gene, where it is required for efficient promoter activity (Frederickson *et al.*, 1989).] Like the actin genes of insects and vertebrates, the *Volvox* actin gene contains, in addition to a TATA box at -31, a CCAAT motif at -75 nt that might play a distinctive role in the regulation of gene expression (Cresnar *et al.*, 1990).

The deduced 377-aa polypeptide sequence of *Volvox* actin is highly similar (79–94%) to actins from other eukaryotic phyla. Interestingly, similarities to higher plant actins (about 83%) are distinctly lower than those to most animal and fungal actins (about 90%), a fact supported by the presence of Ala_9 and Ser_{16} in two diagnostic positions typical of animal and fungal actins, instead of the Pro_9 and Thr_{16} found in higher-plant actins. This closer resemblance to animals and lower eukaryotes parallels observations on the organization of *Volvox* histone genes (Section IV,B,2), further illustrating the interesting phylogenetic position of *Volvox* between higher plants and animals.

6. Nitrate Reductase Gene

Nitrate reductase (NR), the enzyme that catalyzes the first step in the nitrate assimilatory pathway, has been characterized in higher plants, in fungi, and in green algae, such as *Chlorella* and *Chlamydomonas* (for review see Wray and Kinghorn, 1989). The active enzyme is a dimer, each subunit comprising about 900 aa residues that fold into three major domains, each containing a redox prosthetic group (Caboche and Rouzé, 1990). Both the synthesis and the activity of NR are highly regulated by factors such as light and the available nitrogen sources. The cloned *Chlamydomonas* NR gene has been successfully used as a selectable marker for transformation (Fernandez *et al.*, 1989; Kindle, 1990). Its potential utility as a homologous transformation marker and the possibility of using it as a transposon trap for studying the movement of endogenous transposable elements were reasons for cloning and characterizing the NR structural gene of *Volvox* (Gruber *et al.*, 1992).

The *V. carteri* genome contains a single copy of the NR structural gene, and restriction fragment length polymorphisms in the structural gene have now been used (Gruber *et al.*, 1992) to map this gene to linkage group IX, at the locus previously defined as *nitA* (Huskey *et al.*, 1979a,b), or *chl* (Adams *et al.*, 1990).

The 5.9-kb coding region of *nitA* contains 10 introns separating 11 exons that encode an 864-aa polypeptide. Codon usage is much less extensively biased in NR than in the genes encoding abundant structural proteins of the Volvocales, such as tubulins and histones. Although the deduced *Volvox* NR is about 50 aa residues shorter than the higher plant NR monomers, functional domains for the FAD-, heme-, and molybdenum cofactor-binding sites are fully conserved; variations in the aa sequence are restricted to the N-terminal end and to the three interdomain regions (Gruber *et al.*, 1992). Sequence similarities of these three functional domains of NR to cytochrome b5-reductase, cytochrome b5, and sulfite oxidase, respectively, support the contention that the active domains of such enzymes were once fused, along with two intervening hinge regions, to generate the prototypic eukaryotic NR (Crawford *et al.*, 1988). These observations are consistent with the theory that many proteins of novel function were primitively produced by shuffling and recombining exons encoding active centers of existing proteins (Gilbert *et al.*, 1986). This hypothesis will be discussed further in Section V,A,3.

Transcription of *Volvox* NR is both induced by nitrate and repressed by reduced nitrogen compounds, such as ammonia or urea (Gruber *et al.*, 1992). However, although transcription start sites (439 and 452 bp upstream of the initiation codon) and a polyadenylation signal can be identified in the untranslated regions of the gene (Gruber *et al.*, 1992), *cis* regulatory elements involved in the induction and repression of NR transcription remain to be defined.

7. G Protein Genes

Many genes have been characterized, in a wide variety of organisms, that encode members of the so-called GTPase, or G-protein, superfamily. G-proteins generally act as molecular switches that go from the inactive to the active form when bound GDP is replaced by GTP; they then return to the inactive form when the bound GTP is hydrolyzed to GDP (Bourne *et al.*, 1990, 1991). Among the several subdivisions of this superfamily, an exclusively eukaryotic group of small G-proteins, also called the Ras family, exhibits by far the largest number of different proteins and is the group that has been most widely discussed in terms of its possible roles in the control of cellular activities and presumed mechanisms of action (Barbacid, 1987; Chardin, 1988). The high degree of sequence conservation (30% to >90% similarity of aa sequence) among the various members of the family has facilitated the isolation of many *ras*-like genes from mammals, insects, fungi (Chardin, 1988), and, more recently, higher plants (Matsui *et al.*, 1989; Diefenthal, 1989; Palme *et al.*, 1992). According to Valencia *et al.* (1991), an early gene duplication led to the ancestors of at least four different subgroups (the *ras, rho, ypt/rab* and *tc4* groups), which then diverged to encode a diversity of proteins.

Site-directed mutagenesis and gene disruption experiments have indicated that nearly every member of the *ras* gene family examined to date is indispensable to cell function. However, it has so far been possible to assign specific roles to Ras-like gene products in only a few cases, such as the following: The mammalian Ras protein appears to be important in the control of cell growth and differentiation (Barbacid, 1987). The RAS protein of *S. cerevisiae* is responsible for adenylate cyclase activation (Tamanoi, 1988). And products of various *ypt* (yeast protein two) genes (which are homologs of the mammalian *rab* genes; Gallwitz *et al.*, 1983), are implicated in various stages of intracellular vesicle traffic, from the endoplasmic reticulum, through the Golgi, to the plasmalemma (Bourne, 1988; Rexach and Schekman, 1991). In view of the importance of this latter pathway for the synthesis and secretion of the large quantities of ECM glycoproteins that characterize volvocalean spheroids, it was of particular interest to isolate and characterize the *ypt* gene family of *V. carteri* and its relatives.

Two groups, in Regensburg and Cologne, have recently identified and characterized five *ypt* genes of *V. carteri* that have been named *yptV1* to *yptV5* (Fabry *et al.*, 1992a,b) and one in *C. reinhardtii* that is called *ypt C1* (Dietmaier *et al.*, 1992).

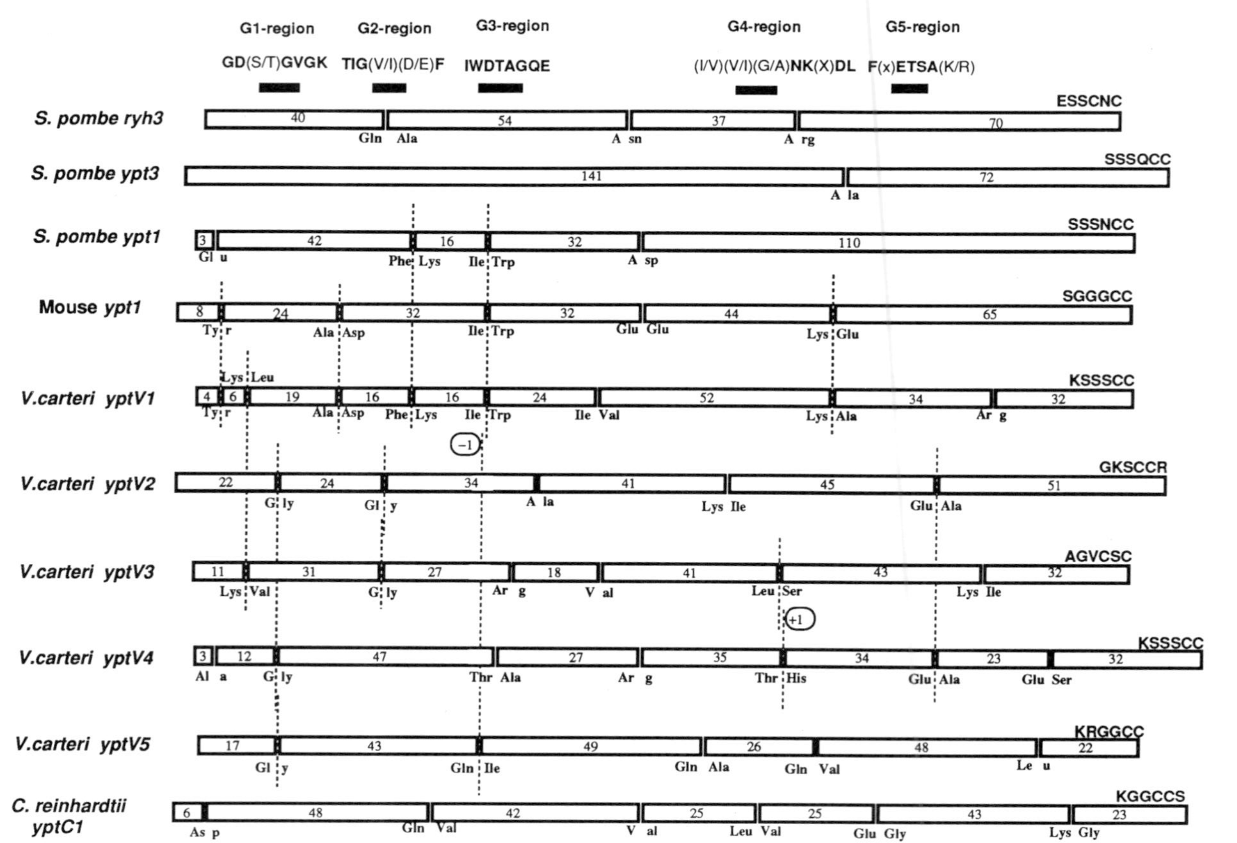

G1-region
G2-region
G3-region
G4-region
G5-region
GD(S/T)GVGK
TIG(V/I)(D/E)F
IWDTAGQE
(I/V)(V/I)(G/A)NK(X)DL
F(x)ETSA(K/R)
S. pombe ryh3
40
54
37
70
Gln Ala
A sn
A rg
ESSCNC
S. pombe ypt3
141
72
A la
SSSQCC
S. pombe ypt1
3
42
16
32
110
Gl u
Phe Lys
Ile Trp
A sp
SSSNCC
Mouse ypt1
8
24
32
32
44
65
Ty r
Ala Asp
Ile Trp
Glu Glu
Lys Glu
SGGGCC
V.carteri yptV1
4
6
19
16
16
24
52
34
32
Lys Leu
Ty r
Ala Asp
Phe Lys
Ile Trp
Ile Val
Lys Ala
Ar g
KSSSCC
−1
V.carteri yptV2
22
24
34
41
45
51
Gl y
Gl y
A la
Lys Ile
Glu Ala
GKSCCR
V.carteri yptV3
11
31
27
18
41
43
32
Lys Val
G ly
Ar g
V al
Leu Ser
Lys Ile
AGVCSC
+1
V.carteri yptV4
3
12
47
27
35
34
23
32
Al a
G ly
Thr Ala
Ar g
Thr His
Glu Ala
Glu Ser
KSSSCC
V.carteri yptV5
17
43
49
26
48
22
Gl y
Gln Ile
Gln Ala
Gln Val
Le u
KRGGCC
C. reinhardtii yptC1
6
48
42
25
25
43
23
As p
Gln Val
V al
Leu Val
Glu Gly
Lys Gly
KGGCCS

The *Volvox* genes *yptV1*, *yptV3*, *yptV4*, and *yptV5* have been sequenced completely on both the genomic and cDNA levels, *yptV2* only from a genomic clone. The coding regions specify polypeptides 203–217 aa residues long (Table I), a size consistent with that of small G-proteins (20–27 kDa). cDNA sequencing of four *Volvox ypt* genes indicated that polyadenylation of mRNAs takes place at different distances from the translation stop site: 260 bp downstream in *yptV5*, 800 bp downstream in *yptV3*, and, in the case of *yptV1*, even at three different positions within one gene, 210, 420, and 550 bp downstream of the stop codon. Putative polyadenylation signals have been found to precede the polyA-tails by 12–16 nt. Transcription initiation signals and transcription patterns during the life cycle remain to be determined.

The deduced proteins encoded by these five genes contain typical sequence motifs that unequivocally classify the genes as members of the *ypt/rab* subgroup of the *ras* family (Fig. 7). Features common to all G-proteins, such as the GTP-binding and GTPase domains (consisting of five islands of 4–8 contiguous residues distributed over the N-terminal two thirds of the sequence) are seen. So are the structural motifs that are diagnostic for the Ypt/Rab-subgroup, such as the sequence IT(S/T)xYYRGA around position 70, and the two C-terminal Cys residues implicated in membrane binding (Valencia *et al.*, 1991). Moreover, the N-terminal portion (except for the very first 5–10 residues) bears the most conserved sequence element in the entire Ras family. In contrast, clear differences among individual members of the group are found in the hypervariable C-terminal 30–50 residues; these define various Ypt isotypes.

In *C. reinhardtii* one *ypt* gene *(yptC1)* has recently been sequenced (Dietmaier *et al.*, 1992); three additional *ypt* clones have been isolated but not yet analyzed. The sequenced gene encodes a typical Ypt protein but a novel isotype not yet found in *Volvox*. This is the first *Chlamydomonas* gene encoding a G-protein, but it follows closely the recent observation that a GTPase (which can be immunoprecipitated with an anti-Ras antiserum) is present in *C. reinhardtii* eyespots (Korolkov *et al.*, 1990). Future

FIG. 7 Exon–intron organization of aligned volvocalean and nonvolvocalean *ypt* genes. Boxes represent exons (with the number of codons indicated); interruptions indicate introns (not drawn to scale). Intron borders are defined by the encoded amino acids in 3-letter code; C-terminal amino acid sequences with typical cysteine residues are shown in 1-letter code. Conserved domain regions are symbolized by the solid bars near the top of the figure, and the corresponding amino acid motifs are given in 1-letter code. Domain assignments (G1–G5) are those previously defined (Bourne *et al.*, 1991).

experiments may show whether the eyespot GTPase is a Ypt protein or whether some other G-protein is involved.

Southern-blot analysis indicates that each of the five *Volvox ypt* genes exists as a single copy (Kirk *et al.*, unpublished). It is not clear, however, whether the five *ypt* genes characterized to date represent the complete set of *ypt/rab*-like genes in *Volvox* or whether this family contains additional members. In *Schizosaccharomyces pombe* only four *ypt* genes were identified (Miyake and Yamamoto, 1990; Haubruck *et al.*, 1990; Hengst *et al.*, 1990), but in mammals there are at least 11 different *rab* genes known (Chavrier *et al.*, 1990). Because *Chlamydomonas yptC1* clearly is not a close homolog of any of the *Volvox ypt* genes characterized so far, it is conceivable that at least this isotype may be detected as a sixth *Volvox ypt* gene. Although it would be particularly interesting if these closely related algae turned out to have completely different populations of *ypt* genes (in view of their organizational differences), this possibility seems rather unlikely.

The concept that the different *ypt* genes encode functionally distinct protein isotypes is supported by comparative studies: virtually every Ypt protein of a single species is more similar to one of the Ypt proteins of some other species than it is to any of the other Ypt of its own species. For example, the highest aa sequence similarity to be found among the five *Volvox* Ypt's is only 57% (for YptV1 and YptV2), and YptV5 exhibits no more than 35% similarity to any of the other four *Volvox* Ypt. In contrast, for each *V. carteri* Ypt (except YptV3) a homolog with greater aa sequence similarity can be found in some other species. For example, YptV5 exhibits about 70% similarity to the Rab 7 protein from dog and to the Ypt protein called BRL-ras from rat liver (Bucci *et al.*, 1988). The similarity of YptC1 of *C. reinhardtii* and canine Rab 11 is even greater, 84%. A similar relationship is reflected in intron positions; the largest number of common intron positions is shared not by any *intra*specific *ypt* gene pairs but by those *inter*specific gene pairs that also exhibit the highest levels of sequence similarity. For example, when the five *Volvox ypt* genes are compared with one another and with all other *ypt* genes for which intron data are available, the *Volvox yptV1* and the mouse *ypt-1* genes are found to have both the largest number of shared intron positions (four) and the highest level of aa sequence similarity (81%) (Section IV,A,5). This pattern of intraspecific divergence versus interspecific conservation of sequences and intron–exon structure is the pattern observed in every protein family (e.g., the tubulins and globins) in which gene duplication has been followed by divergence to produce two or more functionally distinct polypeptide isotypes, which are then conserved in sequence across species lines.

The functions of the various Ypt isotypes in *Volvox* remain largely unexplored. However, when *Volvox* cells were fractionated biochemically and tested for their reactivity with an antiserum raised against Ypt1V1, only the membrane fraction was positive. However, as indicated in Section II,F, in dividing embryos, embryo YptV1 is localized between the poles of the mitotic spindle and the basal bodies, in close association with the centrin-rich nuclear basal body connectors (R. Keeling *et al.*, unpublished observations). This observation is particularly interesting in light of the demonstrated role of small G-proteins in establishing the site of asymmetric division in yeast (Chant and Herskowitz, 1991; Chant *et al.*, 1991; Johnson and Pringle, 1990; Powers *et al.*, 1991) and the demonstrated importance of asymmetric division in controlling cell specification in *Volvox* (Section II,C).

C. Genes Encoding *Volvox*-Specific Proteins of Known Function

Two categories of genes that encode proteins important in *Volvox* development, but that are without known homologs in other organisms, have now been cloned and characterized. These encode the sexual pheromone (Section III) and a major component of the extracellular matrix (Section I,A,2).

1. The Sex-Inducer Gene

In a series of experiments, M. Sumper and his colleagues succeeded in purifying the sex-inducing pheromone of *V. carteri* (Section III), determining a partial amino acid sequence for it and using this sequence information to clone the corresponding gene (Tschochner *et al.*, 1987). Because male cultures tend to produce enough pheromone to go sexual (and thereby self-destruct) while still at very low density, the critical experimental step that led to cloning of the gene was finding a way to prevent the culture from undergoing auto-induction, in order to build the population to a sufficiently large size that milligram quantities of inducer would be produced when sexual development was finally permitted to occur. This goal was achieved by treating the culture once each generation (except for the last one) with pronase. Pronase does not degrade the inducer, but it does degrade a matrix glycoprotein that is produced as one of the earliest responses to inducer (Section III,A,2), and that is believed to play an essential role in the sexual induction process (Wenzl and Sumper, 1986b).

Using probes designed on the basis of the partial aa sequence obtained from the purified inducer, a genomic clone encoding the pheromone was first isolated from a genomic library, and a partial nucleotide sequence was derived (Tschochner *et al.*, 1987). However, the complex exon–intron and tandem repeat structure of the (~40 kb) *sexI* genetic locus complicated the assignment of coding regions. The nucleotide sequence of the coding region and the deduced aa sequence of the inducer protein were, therefore, ultimately derived from a cDNA (H.-W. Mages *et al.*, 1988). This information was then used to complete the analysis of the genomic clones and to attempt to reconstruct the organization of the entire *sexI* locus.

The 627-bp coding region of each *sexI* gene is interrupted by four introns, varying from 99 to 1526 bp in length (H.-W. Mages *et al.*, 1988). As expected for a secreted glycoprotein, the deduced 208-aa sequence of the sexual inducer includes a typical 11-aa, amino-terminal signal sequence and six Asn-X-Ser(Thr) potential *N*-glycosylation sites (H.-W. Mages *et al.*, 1988). Of all *Volvox* genes analyzed so far, *sexI* exhibits the least codon usage bias. The *sexI* mRNA is polyadenylated 48 nt downstream of a UGUGA motif, but little information is available on upstream promoter signals.

Many genomic clones recovered from genomic libraries contain two or more similar copies of the *sexI* gene, separated by nontranscribed spacers; the fact that most such clones have been recovered repeatedly, and from different libraries, indicates that they represent faithful gene duplications that are present in the genome (M. Sumper, personal communication). However, parental RFLPs that have been detected on Southern blots with the cloned *sexI* cDNA as a probe segregated in >200 progeny as alleles at a single Mendelian locus, indicating that all copies of the *sexI* gene must reside at a single locus; this locus is not linked to mating type (C. Velloff and D. L. Kirk, unpublished observations). Taken together, the sequences of the recovered clones and restriction mapping of genomic DNA have led to the following tentative model (M. Sumper and D. L. Kirk, unpublished observations). The *sexI* locus consists of six similar (~7kb) repeats, each of which contains a (~4 kb) transcribed coding region preceded by an untranscribed (~3 kb) leader sequence. The coding regions are distinguishable from one another only by the absence or presence (or, when present, the length) of an $(ATT)_n$ insertion in intron IV. The 5′-leaders are distinguishable only by virtue of the fact that the second and sixth copies contain a 290-bp insertion that contains 49-bp terminal repeats flanked by 12-bp direct repeats, suggestive of a transposable element (M. Sumper, personal communication). However, no transposition of this element has yet been observed, suggesting that it may be defective and require trans-acting functions for its transposition.

The nearly complete sequence identity of the six *sexI* repeats, including introns and nontranscribed leader regions, suggests either that tandem replication of the locus has occurred recently or that the locus is subjected to efficient homogenization, or both. The possibility that duplication of the *sexI* locus may have been a recent event is suggested by the observation that a strain of *V. carteri* isolated from a different geographic region than the common laboratory strains were, and which is marginally interfertile with them (Adams *et al.*, 1990), appears to carry only a single copy of the *sexI* gene (C. Velloff and D. L. Kirk, unpublished observations). However, the existence of an efficient homogenization process in *Volvox* is suggested by the structure of other genes, such as the polyubiquitin gene (Section IV,B,5), and may be occurring at the *sexI* locus also.

2. Gene Encoding the ECM Glycoprotein SSG 185

M. Sumper and colleagues have also succeeded in cloning the gene encoding SSG 185 (sulfated surface glycoprotein of apparent molecular mass 185 kD). SSG 185 is the monomeric form of the HRGP that polymerizes to form the ECM component called CZ3, a honeycomb-like set of compartments within which the individual somatic cells of a *V. carteri* spheroid are suspended (Section I,A,2). Monoclonal antibodies were raised against a core glycopeptide of SSG 185 and shown to prevent expansion of *Volvox* spheroids *in vivo,* presumably by inhibiting the insertion and polymerization of SSG 185 into the CZ3 region (Ertl *et al.*, 1989). By screening cDNA expression libraries with this same monoclonal antibody and then using the recovered partial cDNAs to screen genomic libraries of *V. carteri,* these authors cloned and analyzed the single gene responsible for SSG 185 synthesis.

This unique, *Volvox*-specific glycoprotein has an unusual composition: the most conspicuous feature of the 485-aa polypeptide chain is a central domain, ~80 residues long, that is composed almost entirely of hydroxyproline (Hyp) residues. Most of these Hyps appear to be glycosylated with 1,2-linked di- and tri-arabinosides, consistent with the observation that SSG 185 is largely protease-resistant. The Hyp-rich portion of the peptide chain is thought to form a polyproline II helix, the most extended helix conformation known for proteins; this view is supported by electron-microscopy of the core fragment revealing an extended, rod-like structure ~29 nm long (Ertl *et al.*, 1989). The amino- and carboxy-terminal regions exhibit rather conventional amino acid composition, except for unusually high Cys content (13/165) in the C-terminal domain. SSG 185 differs from extensins, the HRGPs of plant cell walls, by virtue of its less basic nature and its long run of hydroxyprolines, in lieu of the Ser(Hyp)$_4$ repeats of the plant protein (Cassab and Varner, 1988).

The other striking feature of SSG 185 is an extensively sulfated polysaccharide that is covalently attached near the center of the Hyp domain. The polysaccharide consists of an oligomannose backbone of at least 40 monosaccharide units, each with a di-arabinoside side chain. A fully sulfated mannose di-arabinoside unit contains five sulfate ester groups, one on the mannose backbone and two on each of the arabinose residues. The extent of sulfation of this polysaccharide is under an interesting developmental control. Even though SSG 185 is a product of adult somatic cells, it exhibits enhanced sulfation of the mannose residues in close synchrony with the cleavage behavior of the embryos (Wenzl *et al.*, 1984).

Its extreme density of negative charges makes SSG 185 a strong cation exchanger, capable of binding to positively charged proteins. It has therefore been proposed (Ertl *et al.*, 1989) that SSG 185 may provide the nucleation centers required for *in vitro* reassembly of positively charged glycoproteins derived from the crystalline layers of *Chlamydomonas* and *Volvox* (Adair *et al.*, 1987; Goodenough and Heuser, 1988) (Section I,A,2). Finally, it should be noted that polymerization of SSG 185 appears to involve formation of phosphodiester bridges connecting two arabinose residues at their C5 atoms (Holst *et al.*, 1989), which is a novel type of cross-link among glycoproteins.

The organization and genetic control of the locus encoding this vital and highly unusual polypeptide are of great interest. Southern-blot hybridization indicates the presence of a single *ssg 185* gene in *V. carteri,* more than 6 kb in length. The 1455-bp region coding for the SSG 185 is interrupted by six introns; a seventh is located in the 5′-untranslated region (Ertl, 1989). Upstream promoter signals remain to be defined; however, the 3′-untranslated region contains a UGUAA signal preceding the polyA-tail by 13 nt.

V. Evolutionary Implications of the Structural Features of Volvocalean Genes

Certain recurring structural features of the volvocalean protein-coding genes that have been sequenced to date, plus the unique phylogenetic position of the volvocalean algae *vis-á-vis* the major taxa of eukaryotes that have hitherto received much greater attention (i.e., fungi, animals and higher plants; Section I,B,1), combine to provide useful new insights into two aspects of eukaryotic gene evolution that have been the subject of considerable speculation and debate. The first of these relates to the origins and significance of introns, the second to the significance of nonrandom codon usage.

A. Intron–Exon Structure: Introns Early or Introns Late?

One of the most striking general features of the *V. carteri* and *C. reinhardtii* genes that have been sequenced to date is the fact that they tend to have more introns‡ than the corresponding genes of most other organisms for which data are available (Table I). This fact alone indicates that it may be worth considering what the exon–intron structure of volvocalean genes may imply about the origin and significance of introns in eukaryotic protein-coding genes.

In simplified form the two major hypotheses that have been advanced to account for the exon–intron structures of modern eukaryotic protein-coding genes are as follows:

1. The intron-early hypothesis (Darnell, 1978; Doolittle, 1978; Gilbert *et al.*, 1986): The primordial gene from which any particular category of modern eukaryotic genes was derived possessed introns at all of the locations where they are found in various eukaryotes today; differences in intron number and/or position among various modern representatives are the result of differential intron loss during the process of evolutionary streamlining. An important corollary of this hypothesis is that introns were and are preferentially located at the boundaries of regions encoding protein domains, and by virtue of this position, they played an important role in protein evolution by facilitating exon shuffling, by which the functional domains of various extant proteins were recombined to produce proteins with novel properties.
2. The intron-late hypothesis (see, e.g., Cavalier-Smith, 1978, 1991; Rogers, 1989): Introns have been added to eukaryotic genes at many different times during the course of evolution, via insertion of various kinds of mobile elements. Shared intron positions for cognate genes in two organisms merely imply that an intron had been inserted at this position by the time that that these two organisms last shared a common ancestor, and there is no evidence that primordial genes encoding any intracellular proteins (as opposed to more recently derived genes of animals that encode extracellular proteins) were assembled by exon shuffling (Cavalier-Smith, 1991).

Observations that self-splicing introns are present in tRNA genes of certain eubacteria (Kuhsel *et al.*, 1990; Xu *et al.*, 1990) were widely interpreted as providing support for the intron-early hypothesis. However, others discount this because they visualize different evolutionary histories for the self-splicing introns of tRNAs and the spliceosome-processed introns of protein-coding genes (Cavalier-Smith, 1991).

Details of the exon–intron structures of volvocalean protein-coding genes that have been sequenced to date appear to speak to various aspects of this controversy.

‡ Wiebauer, *et al.* (1988) have noted that the polypyrimidine tracts that precede the 3' splice sites of animal introns (and appear to influence splice site selection) are replaced in the introns of plant genes by AU-rich regions; in this regard, the sequenced *Volvox* and *Chlamydomonas* genes resemble the genes of animals rather than those of plants.

1. Actin Introns

The exceptional diversity of exon–intron structures represented by the actin genes from different eukaryotes (Wildemann, 1988) stands in sharp contrast to the extensive conservation of the actin-encoding sequences. Whereas actin genes from a few unicellular eukaryotes (such as *Oxytricha, Dictyostelium,* and *Schizosaccharomyces*) are uninterrupted, all other previously sequenced actin genes had been found to contain from one *(Saccharomyces)* to seven (human aortic muscle type) introns in highly taxon-specific positions. This intron diversity was highlighted by the discovery of eight introns within the coding region of the *Volvox* actin gene, plus a ninth in the 5′-untranslated region (Cresnar *et al.,* 1990). Of the eight coding region introns, only three are in shared locations: one *Volvox* intron position is shared with higher plants and nematodes, one with fruit flies, and one with sea urchins and vertebrates, but the other five are all in novel locations (Cresnar *et al.,* 1990). In a similar pattern, the actin genes of *Aspergillus* and *Thermomyces* share one intron position with fruit flies but also define four taxon-specific intron positions (Fidel *et al.,* 1988; Wildemann, 1988). And although higher plant actin genes share one intron position with vertebrates and one with nematodes and *Volvox,* they all have one intron in a plant-specific position. Indeed, it requires the actin gene sequences of only 10 species to define 30 distinct positions at which introns are present in modern actin genes (Cresnar *et al.,* 1990). Under the unlikely asumption that the relatively few actin genes that happen to have been sequenced to date define all of the positions in which introns were present in the primordial actin gene postulated by the intron-early hypothesis, that gene must have had at least 30 exons, of an average length of only about 12 codons.

2. α-Tubulin Introns

The α-tubulin genes of *V. carteri* each contain three introns, two of which are also present in the *C. reinhardtii* homologs. The three introns found in rat α-tubulin genes (Cleveland and Sullivan, 1985), a different three found in *Zea mays* (Montolin *et al.,* 1990), and six of seven introns seen in *Physarum* (Monterio and Cos, 1987) are all located at positions different from the introns present in *Volvox*. Thus, the α-tubulin genes of only four species define 15 different positions at which introns would have to have been present in a primordial α-tubulin gene to conform to the predictions of the intron-early hypothesis.

3. Nitrate Reductase Introns

Among NR genes, the 10 introns interrupting the *Volvox* NR coding sequence are surpassed only by the 14 present in the *Chlamydomonas* homolog (P. Lefebvre, personal communication). Only seven of these introns are at equivalent positions in the two algal NR genes, indicating that a minimum of 10 introns have been added to and/or deleted from the NR gene in the relatively brief period since *Chlamydomonas* and *Volvox* last shared a common ancestor. Moreover, none of the five introns seen in the *A. nidulans* gene (Kinghorn and Campbell, 1989) matches the position of any of the algal introns. Thus, in just two species of algae and a fungus, 22 different intron positions have been identified within the otherwise largely conserved NR gene.

Similarities of the three functional domains of NR with sulfite oxidase, cytochrome b5, and cytochrome-b5 reductase suggest that the active domains of such enzymes were once fused to generate NR. Previous workers have used these regions of sequence similarity plus two hinge regions of proteolytic sensitivity to subdivide NR primary structure into five domains (Crawford *et al.*, 1988). Even though these domains are rather well defined in the aligned polypeptide sequences, it is difficult to match any intron positions in existing NR genes with any of the domain boundaries. Therefore, if exon shuffling (Gilbert *et al.*, 1986) were operative in creating NR, the extant intron positions give no indication of this. However, the multitude of diverse intron positions in extant NR genes (one half varying even among closely related algae) seem to be at least as easily accommodated by the theory that most modern introns were later additions to an already functioning gene.

4. Histone H3 Introns

An interesting feature of the first two *Volvox* H3 genes sequenced is that although they are otherwise similar, they possess an intron, the position of which is shifted by one base pair (Müller and Schmitt, 1988). The introns of these two genes are different in length, but the shorter one bears a strong sequence similarity to the 3′-end of the longer one. In view of this relationship, it was proposed that the introns of these two genes may have initially been extremely similar until intron sliding occurred in one of them by two successive, compensating frameshift mutations: (1) a deletion that removed the 5′-end of the intron plus one nucleotide from the coding region at the 5′-splice junction and (2) a compensating nucleotide insertion immediately downstream of the 3′-splice junction (Müller and Schmitt, 1988). However, subsequent study has called this proposal into question. Three additional *Volvox* H3 genes have now been partially sequenced, and each has been shown to have introns at one or the other of these two positions, implying that both types of H3 gene underwent duplication after the two intron positions became fixed (Müller *et al.*, 1992). However, there is so much less sequence similarity among the introns of these recently sequenced genes that, had they been sequenced first, they would not have lent themselves readily to an explanation for intron sliding like that mentioned earlier. Therefore, an alternative possibility cannot be dismissed: that independent intron insertions into two adjacent positions occurred following the initial duplication of the H3 locus. It is difficult to imagine a primordial H3 gene with two introns flanking an exon that contained only one base pair (as a rigorous form of the intron-early hypothesis would require).

PCR amplification of H3 introns with primers complementary to the flanking exon sequences indicates that each of the 15 odd *Volvox* H3 genes possesses an intron, but that some *C. reinhardtii* H3 genes have an intron and others do not. (The two sequenced *Chlamydomonas* H3 genes do not; Müller *et al.,* 1992). Southern-blot analysis suggests that the introns in *Chlamydomonas* genes are in approximately the same locations as in the *Volvox* H3 genes, but it remains to be determined whether there is only one intron position or two in the *Chlamydomonas* H3 genes.

5. Introns of *ypt* Genes

In view of the relatively short coding region (about 600bp), all volvocalean *ypt* genes are interrupted by a sizable number of introns (5–8; Fig.7). There are no obvious similarities between intron sequences interrupting the coding region of different *ypt* genes at identical positions. With the exception of two, all *ypt* introns contain the canonical GU . . . AG boundaries. Those exceptions (intron III of *yptV5* and intron II of *yptC1*) have a GC instead of a GU at the splice–donor site. This variant was previously observed in the β1-tubulin gene of *V. carteri* (Harper and Mages, 1988), at which time such a splice junction was without precedent. However, the discovery of two additional examples strongly suggests that such an intron boundary sequence must be acceptable to the *Volvox* splicing machinery.

Information on intron positions in other *ypt* genes is scarce because most published sequences were derived from cDNAs. However, as shown in Figure 7, the few *ypt* genes for which intron information is available contain 1–8 introns. Intron lengths vary from 70 bp (intron IV of *yptV1*) to more than 13 kb (intron I of *ypt1* from mouse; Wichmann *et al.,* 1989). All of the *Volvox ypt* genes share at least one intron position with at least one other *Volvox ypt* gene, suggesting common derivation from a duplicated ancestral gene. Nevertheless, these five genes define 24 different intron positions, only one of which is shared by more than two genes. The *Chlamydomonas yptC1* gene defines six additional unshared intron positions, underscoring the uniqueness of this gene. The mouse and *S. pombe ypt* genes define six additional unshared intron positions. In none of these genes are exon–intron boundaries congruent with the borders of conserved domains. Instead, introns appear to be located preferentially within the most highly conserved sequences (Fig. 7), thus rendering evolution of *ypt* genes by exon shuffling rather improbable. Moreover, if the 36 intron locations defined by these 10 *ypt* genes were all occupied in the promordial *ypt* gene, the exons of that gene would have averaged about five codons in length, and two exons would have comprised a single base pair (Fig. 7).

To summarize, introns are unusually abundant in the volvocalean genes that have been sequenced to date. And nearly all of the details of exon–intron structure revealed by these studies appear to fit much more comfortably within the framework of the intron-late rather than the intron-early, hypothesis.

B. Codon Usage Patterns: Concerted Evolution or Convergent Evolution?

As pointed out earlier (Section IV,B,1), codon usage in many volvocalean genes sequenced so far is strongly biased in favor of C and against A in the third codon position, with the preferred order being C>G>U>>A, and with many codons not being used at all (W. Mages *et al.*, 1988; Schmitt and Kirk, 1992).

1. Codon Bias in Tubulin and Histone Genes

Codon bias is particularly striking in the *C. reinhardtii* tubulin genes, in which 25–28 codons are not used in any given gene and 20 are not used in any of the four tubulin genes (Fig. 8). The *C. reinhardtii* histone genes exhibit a bias that is only slightly less uniform. Ignoring the codons for cysteine and tryptophan in cases in which these amino acids are not present, one finds that individual histone genes exclude between 20 (H3–I) and 29 (H2A) codons and that 10 codons are not used in any of the six histone genes sequenced to date.

Codon bias in *V. carteri,* although qualitatively similar, is both less extreme and less regular (Fig. 8). For example, although the α1-tubulin gene exhibits a moderately strong bias (17 codons not used), the α2-tubulin gene exhibits much more relaxed codon usage (only seven codons not used). Although the α1- and α2-tubulin mRNAs are detectable at nearly the same level in *Volvox* (Mages, 1990), it is possible that the α2-tubulin mRNA is translated at lower efficiency than α1-tubulin mRNA, because of the presence in the sequence of rarely used codons (AUA, GUA, AAA) that presumably require rare tRNA species (Ikemura, 1985). However, because these two genes encode identical polypeptides, this possibility is

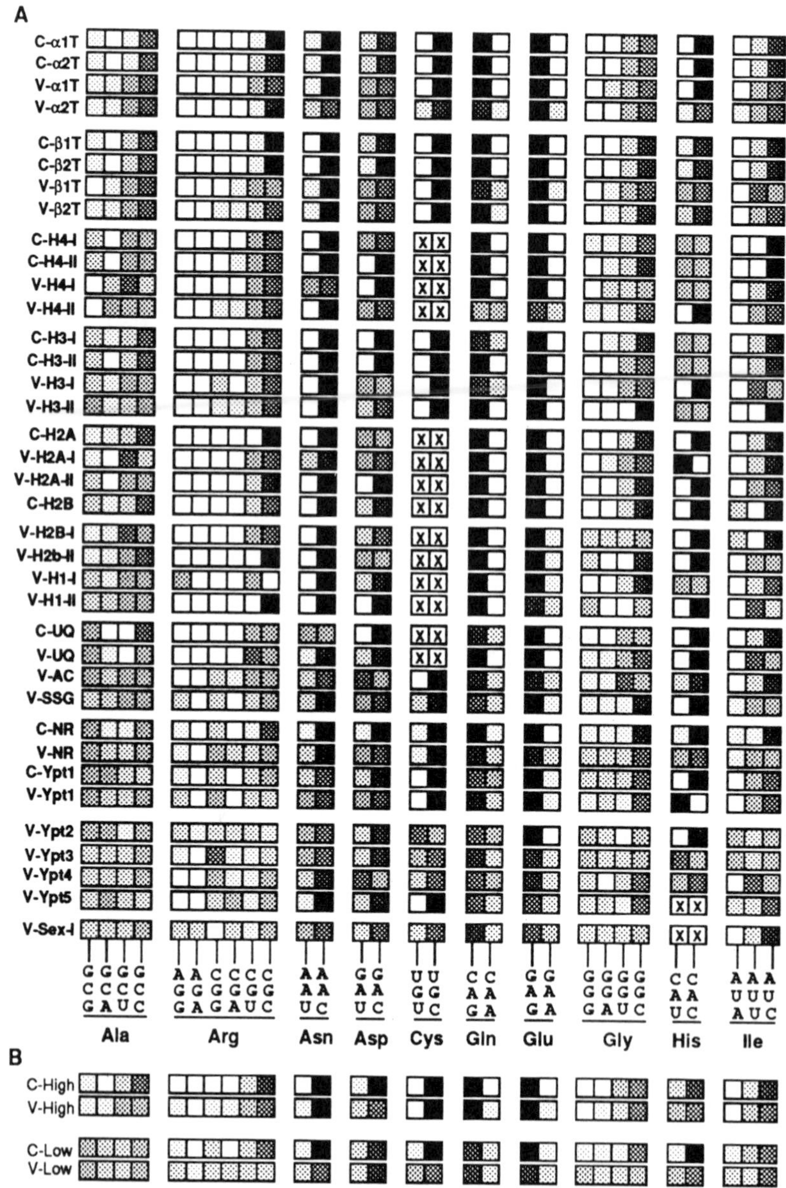

FIG. 8 Codon usage patterns for *Volvox* and *Chlamydomonas* genes. Codons are grouped by encoded amino acids, and the nt sequences of all codons are given. Stop codons and codons for which synonyms are not available are not shown. (A) Codon usage in each of the *V. carteri* genes that have been sequenced to date, and all of the equivalent *C. reinhardtii*

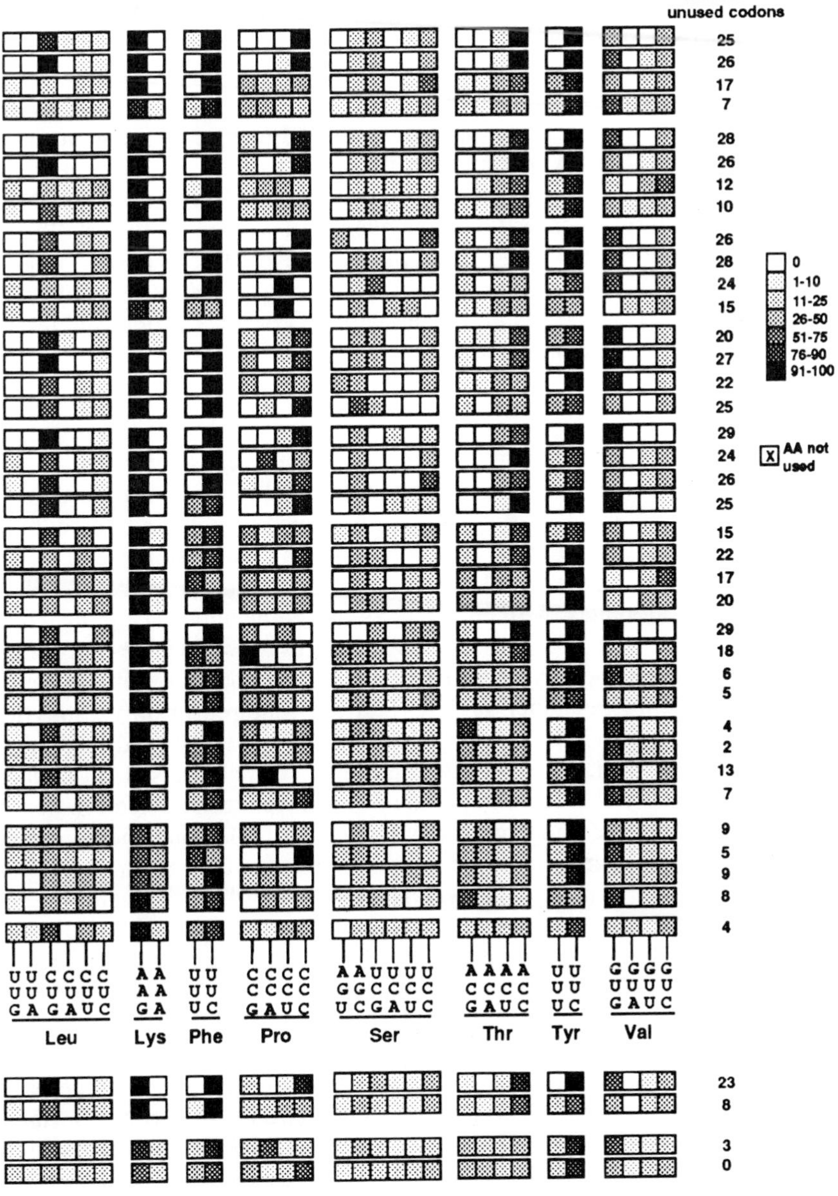

genes for which sequence data are available. Gene abbreviations: C, *C. reinhardtii;* V, *V. carteri;* α1T, α2T, β1T, β2T, α1-tubulin, etc.; H, histone; UQ, ubiquitin; AC, actin; SSG, SSG 185. (B) Codon usage averaged for selected *Chlamydomonas* and *Volvox* genes from the above group that show the highest and lowest bias in their codon usage patterns.

not readily tested. Although the individual histones genes of *Volvox* fail to utilize between 15 (H4–II) and 26 (H2A–II) codons, only three codons go completely unused in the eight nucleosomal histone genes of *Volvox* that have been sequenced to date.

Despite some quantitative variations, all of these genes show a very strong C>(G or U)>>A preference with respect to the third codon position. Biased codon usage patterns have been found to be matched to tRNA availability patterns in cases that have been well studied (Ikemura, 1985). Based on this finding and on details of the wobble hypothesis (Crick, 1966), the codon bias of volvocalean genes can be interpreted as reflecting an abundance of tRNA species in the cell with a G or a C (as opposed to an I or U) in the first anticodon position.

2. Codon Bias in Other Genes

The two volvocalean ubiquitin genes that have been sequenced so far reiterate the codon bias patterns observed in the tubulin and histone genes: the *C. reinhardtii* and *V. carteri* ubiquitin genes leave 29 and 18 codons unused, respectively. Of these unused codons, a large majority end in A.

In contrast, the NR genes of both *C. reinhardtii* and *V. carteri* exhibit low levels of codon bias, with only 2–4 codons going unused (Fig. 8). This highlights a difference that is seen in many species between genes encoding proteins that are generally required in abundance (such as tubulins and histones) and enzymes (such as NR) that are required in only catalytic quantities (Ikemura, 1985). Genes in the former category tend to be the most strongly biased, and the latter the least strongly biased, in codon usage. Presumably this tendency is due to differences in selection pressure, with genes encoding proteins required in abundance being more strongly selected to maintain codon usage patterns that are matched to the abundance patterns of isoaccepting tRNAs and that can therefore be translated with high efficiency (Ikemura, 1985).

In this context, it is not surprising to find that the sex-inducer gene of *Volvox* (the product of which is extraordinarily potent, and hence not required in high quantities, Section III) also uses all but four codons and that *ypt* genes (which are believed to encode molecular switches rather than major structural proteins) have relatively relaxed codon usage patterns. But it is somewhat surprising to find that two presumably important structural proteins of *Volvox,* namely, actin and SSG 185, a major structural component of the ECM, are encoded by messengers that include all but 5 or 6 codons (Fig. 8).

3. Different Levels of Sequence Variation in *Volvox* and *Chlamydomonas*

Perhaps the most unexpected observation to come from the study of volvocalean gene sequences to date is that, in all cases studied so far,

TABLE II

Percentage of All Codons Exhibiting Silent Third-Position Nucleotide Exchanges

	α-1 and α-2 tubulins	β-1 and β-2 tubulins	H3-I and H3-II histones	H4-I and H4-II histones
V. carteri versus *V. carteri*	13	11	17	27
V. carteri versus *C. reinhardtii*	17–24	18–23	15–18	19–29
C. reinhardtii versus *C. reinhardtii*	3	2	6	8

duplicate genes of equivalent function are much more similar to one another in nt sequence in *C. reinhardtii* than they are in *V. carteri*. For example, the two α-tubulin genes of *C. reinhardtii* differ in nt sequence from one another by only 2%, whereas the α-tubulin genes of *V. carteri* differ in nt sequence by 12%. Most of these sequence differences between cognate genes represent silent, third-codon–position nt exchanges. When the frequencies of intraspecific and interspecific silent nt interchanges are tabulated for all four cases in which two cognate genes have been sequenced in both species (Table II), it can be clearly seen that in each case the two *Volvox* genes exhibit substantially more silent interchanges than the equivalent genes of *C. reinhardtii* do. In some cases the equivalent genes of *Volvox* are nearly as different from one another as either of them is from either of the corresponding genes of *Chlamydomonas*.

What does this curious relationship imply about the evolutionary history of *Volvox* and *Chlamydomonas* genes? If, for the sake of discussion, one grants provisionally the assumption that the rate of silent substitutions is approximately equal in all lineages (Ochman and Wilson, 1987), the marked differences that these two algae exhibit in the number of silent interchanges within each sibling–gene pair imply that the gene pairs of *C. reinhardtii* have been diverging in sequence from one another for a much shorter time than they have been diverging from the equivalent genes of *V. carteri*, or than the members of each *V. carteri* gene pair have been diverging in sequence from one another. This possibility was considered in a recent discussion of only the tubulin gene sequences (Schmitt and Kirk, 1992) and was judged highly improbable because to explain the known facts it requires the following three assumptions, each of which seems less likely than the one before: (1) The α- and β-tubulin genes were present as single copy genes in the last common ancestor of *C. reinhardtii* and *V. carteri* and underwent duplication independently, but at different times, in these two lineages. (2) Duplication of the α- and the β-tubulin genes occurred at nearly the same time within each lineage, and (3) The duplication processes were extremely similar in the two lineages but different for the two types of genes, thereby accounting for the fact that the two β-tubulin genes remain linked but the α-tubulin genes are unlinked in both species (Section IV,B,1). The improbability of this sort of scenario becomes substantially greater when one considers the parallel assumptions that would be required to explain the current status of the volvocalean nucleosomal histone genes, of which there are presently 10–15 copies of each type in each species.

Altogether, any explanation of the interspecific differences in sequence similarities of sibling genes that is based on the assumption of equal rates of silent substitution in the two lineages appears completely unsatisfactory. Therefore, alternatives must be considered that assume that the duplicated genes now present in *V. carteri* and *C. reinhardtii* were inherited as such from a common ancestor and that similarity differences in the two lineages are the result of differences in the rates at which silent substitutions accumulated in the two lineages.

One way to account for current differences in sequence similarities in the two species would be to postulate that a process of gene homogenization (Dover, 1982) acts more efficiently in *C. reinhardtii* than in *V. carteri* to keep the sequences of sibling genes similar in sequence. A highly efficient gene homogenization process in *V. carteri* was invoked earlier as a possible explanation of the extreme similarities of exon and intron sequences in the tandem repeats that are present within both the polyubiquitin gene (Section IV,B,4) and the *sexI* locus (Section IV,C,1). However, it can be imagined that *intra*locus and *inter*locus homogenization mechanisms might be different and that the latter might be much more efficient in *C. reinhardtii* than in *V. carteri.* One observation that appears to support such a possibility is the fact that the third introns of the two *C. reinhardtii* β-tubulin genes exhibit 89% sequence similarity, whereas no sequence similarities can be detected in the equivalent introns of *V. carteri.* However, the other two introns of the two *C. reinhardtii* β-tubulin genes exhibit no significant sequences similarities, and it is not obvious why a homogenization process should act to maintain a high degree of sequence similarity in one intron but not in the other two.

An alternative way to account for the marked differences in sequence divergence in sibling genes of the two species would be to postulate that much more stringent selection is exerted in *C. reinhardtii* than in *V. carteri* over the use of various synonymous codons.

These two hypotheses make different predictions about the differences that should be observed currently between *C. reinhardtii* and *V. carteri* in codon usage patterns. If more stringent selection regarding the use of alternative third-position nucleotides is the force keeping sibling sequences more similar in *C. reinhardtii,* then the differences between codon usage patterns in *Chlamydomonas* and *Volvox* genes should not only be decided nonrandom, they should be similar in nature for all gene pairs. However, if some homogenization process is simply acting more efficiently in *Chlamydomonas* to keep the two members of each gene pair similar to one another—and without more stringent selection being exerted over the use of various synonymous codons—random interspecific differences in codon usage should be observed.

The observed differences in codon usage patterns between the two species are decidedly nonrandom (Fig. 8). As reviewed, in most cases codon usage is significantly more strongly biased in *C. reinhardtii* genes than in the equivalent *V. carteri* genes, and the nonrandom nature of these codon biases in the *C. reinhardtii* tubulin and histone genes can be clearly highlighted when the nature of all silent exchanges between eight cognate tubulin and histone genes of *C. reinhardtii* and *V. carteri* are tabulated (Table III). In all eight comparisons, the most abundant type of interspecific difference (42–84% of the total exchanges) involves replacement of some other nucleotide in the *V. carteri* gene by a C in the *Chlamydomonas* equivalent. Averaged over all eight genes, two thirds of all interspecific sequence differences are of this type. For simplicity, all interspecific third-position interchanges are arbitrarily displayed in Table III as changes required to go from the *Volvox* sequence to the *Chlamydomonas* equivalent. In actuality, the directionality of the interchanges cannot be determined from the existing data set. Most likely, changes in both directions have occurred, but this would not weaken the conclusion that differences between the two species in third position bases are decidedly nonrandom.

These data strongly suggest that sequence similarities among sibling genes in *C. reinhardtii* are primarily the result of strong directional and/or stabilizing selection in favor of codons containing C in the third position. Whether selection in the *Chlamydomonas* lineage has been primarily stabilizing (preserving biases that were present in ancestral genes) or directional

TABLE III
Third-Position Nucleotide Exchanges, *Volvox* to *Chlamydomonas*

	α1-T[a]	α2-T	β1-T	β2-T	H3-I	H3-II	H4-I	H4-II
N→C	56	79	61	57	13	8	16	16
N→G	16	27	42	22	1	5	2	10
N→U	3	0	0	0	9	6	1	3
N→A	1	0	0	0	2	0	0	1
%N→C	74	75	59	72	52	42	84	55

[a] Abbreviations: α1-T, etc., α1-tubulin, etc.; H3-I, etc., histone H3-I, etc.

(resulting in convergence of gene sequences that were once more dissimilar) cannot be decided in the absence of additional gene sequences from other related species of green algae.

VI. Overview and Conclusions

In the 6 years since *Volvox* biology was last reviewed in this series (Kirk and Harper, 1986), some progress has been made in elucidating the genetic program for germ/soma differentiation of *V. carteri* and in analyzing the way in which this program appears to control the differential expression of cell type-specific genes. During this period *V. carteri* has also shown itself to be highly amenable to detailed molecular genetic analysis, and the sequences (and certain highly unusual features) of a sizeable number of structural genes of both *V. carteri* and its closest unicellular relative, *C. reinhardtii,* have been described. Moreover, the phylogenetic relationship of these two species to one another and to other volvocalean algae has begun to be clarified by studies at the molecular level. It may be hoped that before another 6 years have passed, these three lines of inquiry will come together, and deeper insights will be available regarding the molecular mechanisms by which germ/soma differentiation is presently effected in *V. carteri,* and the manner in which the genetic programs for multicellularity and cellular differentiation evolved in this interesting model system that nature has so graciously provided us.

Acknowledgments

We wish to thank the colleagues who have shared with us some of their unpublished data, and given us permission to discuss it in this review. Results from our own laboratories that are reported here were made possible by grants from the Deutsche Forschungsgemeinschaft to Rüdiger Schmitt (SFB 43) and to Stefan Fabry (Fa232/1-1), from the National Institutes of Health (GM 27215) and the National Science Foundation (DCB 8615691) to David L. Kirk, and by a travel grant from the North Atlantic Treaty Organization (870065) to Rüdiger Schmitt and David L. Kirk.

References

Adair, W. S., and Appel, H. (1989). *Planta* **179,** 381–386.

Adair, W. S., Steinmetz, S. A., Matson, D. M., Goodenough, U. W., and Heuser, J. E. (1987). *J. Cell Biol.* **105,** 2373–2382.

Adams, G. M. W., Wright, R. L., and Jarvik, J. W. (1985). *J. Cell Biol.* **100,** 955–964.

Adams, C. R., Stamer, K. A., Miller, J. K., McNally, J. G., Kirk, M. M., and Kirk, D. L. (1990). *Curr. Genet.* **18,** 141–153.

Aitchison, W. A., and Brown, D. L. (1986). *Cell Motil. Cytoskeleton* **6,** 122–127.

Amati, B. N., Goldschmidt-Clermont, M., Wallace, C. J. A., and Rochaix, J.-D. (1988). *J. Mol. Evol.* **28,** 151–160.

Baker, R. T., and Board, P. G. (1991). *Nucleic Acids Res.* **19,** 1935–1940.

Balshüsemann, D., and Jaenicke, L. (1990a). *Eur. J. Biochem.* **181,** 231–237.

Balshüsemann, D., and Jaenicke, L. (1990b). *FEBS Lett.* **264,** 56–58.

Bandziulis, R. J., and Rosenbaum, J. L. (1988). *Mol. Gen. Genet.* **214,** 204–212.

Barbacid, M. (1987). *Annu. Rev. Biochem.* **56,** 779–827.

Beckmann, M., and Hegemann, P. (1991). *Biochemistry* **30,** 3692–3697.

Bell, G. (1985). *In* "The Origin and Evolution of Sex" (H. Halversen, ed.), pp. 221–256. Alan R. Liss, New York.

Bell, G., and Koufopanou, V. (1986). *In* "Oxford Surveys in Evolutionary Biology" (R. Dawkins and M. Ridley, eds.), Vol. 3, pp. 83–131. Oxford Univ. Press, Oxford.

Birchem, R. (1978). Ph.D. Thesis, Univ. of Georgia, Athens.

Birchem, R., and Kochert, G. (1979). *Protoplasma* **100,** 1–12.

Birnstiel, M. L., Busslinger, M., and Strub, K. (1985). *Cell* **41,** 349–359.

Bisalputra, T., and Stein, J. R. (1966). *Can. J. Bot.* **44,** 1697–1710.

Bourne, H. R. (1988). *Cell* **53,** 669–671.

Bourne, H. R., Sanders, D. A., and McCormick, F. (1990). *Nature (London)* **348,** 125–132.

Bourne, H. R., Sanders, D. A., and McCormick, F. (1991). *Nature (London)* **349,** 117–127.

Bucci, C., Frunzio, R., Chiariotti, L., Brown, A. L., Rechler, M. M., and Bruni, C. B. (1988). *Nucleic Acids Res.* **16,** 9979–9993.

Buchanen, M. J., and Snell, W. J. (1988). *Exp. Cell Res.* **179,** 181–193.

Buchheim, M. A., and Chapman, R. L. (1991). *BioSystems* **25,** 85–100.

Buchheim, M. A., Turmel, M., Zimmer, E. A., and Chapman, R. L. (1990). *J. Phycol.* **26,** 689–699.

Burr, F. A., and McCracken, M. D. (1973). *J. Phycol.* **9,** 345–346.

Caboche, M., and Rouzé, P. (1990). *Trends Genet.* **6,** 187–192.

Callis, J., Pollmann, L., Wettern, M., Shanklin, J., and Viersta, R. (1989). *Nucleic Acids Res.* **17,** 8377.

Cassab, G. I., and Varner, J. E. (1988). *Annu. Rev. Plant Physiol. Plant Mol. Biol.* **39,** 321–353.

Cavalier-Smith, T. (1974). *J. Cell Sci.* **16,** 529–556.

Cavalier-Smith, T. (1978). *J. Cell Sci.* **34,** 247–278.

Cavalier-Smith, T. (1991). *Trends Genet.* **7,** 145–148.

Chaboute, M. E., Chaubet, N., Clement, B., Gigot, C., and Philipps, G. (1988). *Gene* **71,** 217–223.

Chant, J., and Herskowitz, I. (1991). *Cell* **65,** 1203–1212.

Chant, J., Corrado, K., Pringle, J. R., and Herskowitz, I. (1991). *Cell* **65,** 1213–1224.

Chardin, P. (1988). *Biochimie* **70,** 865–868.

Chavrier, P., Vingron, M., Sander, C., Simons, K., and Zerial, M. (1990). *Mol. Cell. Biol.* **10,** 6578–6585.

Cleveland, D. W., and Sullivan, K. F. (1985). *Annu. Rev. Biochem.* **54,** 331–365.

Coggin, S. J., and Kochert, G. (1986). *J. Phycol.* **22,** 370–381.

Cole, R. D. (1984). *Anal. Biochem.* **136,** 24–36.

Coleman, A. W. (1979). *In* "Biochemistry and Physiology of Protozoa" (M. Levandosky, S. H. Hutner, and L. Provasoli, eds.), 2nd Ed., Vol. 1, pp. 307–340. Academic Press, New York.

Colling, C., Gilles, R., Cramer, M., Nass, N., Moka, R., and Jaenicke, L. (1988). *Second Messengers Phosphoproteins* **12,** 123–133.

Conner, T. W., Thompson, M. D., and Silflow, C. D. (1989). *Gene* **84,** 345–358.

Coss, R. A. (1974). *J. Cell Biol.* **63,** 325–329.

Crawford, N. M., Smith, M., Bellissimo, D., and Davis, R. W. (1988). *Proc. Natl. Acad. Sci. U.S.A.* **85,** 5006–5010.

Cresnar, B., Mages, W., Müller, K., Salbaum, J. M., and Schmitt, R. (1990). *Curr. Genet.* **18,** 337–346.

Crick, F. H. C. (1966). *J. Mol. Biol.* **19,** 548–555.

Crow, W. B. (1918). *New Phytol.* **17,** 151–158.

Darnell, J. E. (1978). *Science* **202,** 1257–1260.

Dauwalder, M., Whaley, W. G., and Starr, R. C. (1980). *J. Ultrastruct. Res.* **70,** 318–335.

Deason, T. R., and Darden, W. H., Jr. (1971). *In* "Contributions in Phycology" (B. C. Parker and R. M. Brown, Jr., eds.), pp. 67–79. Allen Press, Lawrence, Kansas.

Desnizky, A. G. (1984). *Tsitolgiya* **26,** 269–274.

Desnizky, A. G. (1985a). *Tsitolgiya* **27,** 227–229.

Desnizky, A. G. (1985b). *Tsitolgiya* **27,** 921–927.

Desnizky, A. G. (1986). *Tsitolgiya* **28,** 545–551.

Desnizky, A. G. (1987). *Tsitolgiya* **29,** 448–453.

Desnizky, A. G. (1990). *Bot. Zh. (Leningrad)* **75,** 181–186.

Detmers, P. A., Carboni, J. M., and Condeelis, J. (1985). *Cell Motil.* **5,** 415–430.

Diefenthal, T. (1989). Ph.D. Thesis, Univ. Köln, Köln, Germany.

Dietmaier, W., Fabry, S., Weig, I., and Schmitt, R. (1992). In preparation.

Doenecke, D. (1988). *In* "Architecture of Eukaryotic Genes" (G. Kahl, ed.), pp. 123–141. Verlag Chemie, Weinheim.

Domozych, D. S. (1987). *Protoplasma* **136,** 170–182.

Doolittle, W. F. (1978). *Nature (London)* **272,** 581–582.

Doonan, J. H., and Grief, C. (1987). *Cell Motil. Cytoskeleton* **7,** 381–392.

Dover, G. A. (1982). *Nature (London)* **299,** 111–117.

Eick, S., Nicolai, M., Mumberg, D., and Doenecke, D. (1989). *Eur. J. Cell Biol.* **49,** 110–115.

Ertl, H. (1989). Ph.D. Thesis, Univ. Regensburg, Regensburg, Germany.

Ertl, H., Mengele, R., Wenzl, S., Engel, J., and Sumper, M. (1989). *J. Cell Biol.* **109,** 3493–3501.

Ettl, H. (1976). *Beih. Nova Hedw.* **49,** 1122–1135.

Fabry, S., Huber, H., Jacobsen, A., Palme, K., and Schmitt, R. (1992a). In preparation.

Fabry, S., Nass, N., Huber, H., Palme, K., Jaenicke, L., and Schmitt, R. (1992b). *Gene,* in press.

Fernandez, E., Schnell, R., Ranum, L. W. P., Hussey, S. C., Silflow, C. D., and Lefebvre, P. A. (1989). *Proc. Natl. Acad. Sci. U.S.A.* **86,** 6449–6453.

Fidel, S., Doonan, J. H., and Morris, N. R. (1988). *Gene* **70,** 283–293.

Finley, D., and Varshavsky, A. (1985). *Trends Biochem. Sci.* **10,** 343–346.

Floyd, G. (1978). *J. Phycol.* **14,** 440–445.

Foster, K. W., Saranak, J., Patel, N., Zarilli, G., Okabe, M., Kline, T., and Nakanishi, K. (1984). *Nature (London)* **311,** 756–759.

Frederickson, R. M., Micheau, M. R., Iwamoto, A., and Miyamoto, N. G. (1989). *Nucleic Acids Res.* **17,** 253–270.

Fritsch, F. E. (1935). "The Structure and Reproduction of Algae," Vol. 1. Cambridge Univ. Press, Cambridge.

Fulton, A. B. (1978). *Dev. Biol.* **61,** 224–236.

Gaffal, K. P. (1988). *Protoplasma* **143,** 118–129.

Gallwitz, D., and Sures, I. (1980). *Proc. Natl. Acad. Sci. U.S.A.* **77,** 2546–2550.

Gallwitz, D., Donath, C., and Sander, C. (1983). *Nature (London)* **306,** 704–707.

Gantt, J. S., and Key, J. L. (1987). *Eur. J. Biochem.* **166,** 119–125.

Gerlach, W. L., and Bedbrook, J. K. (1979). *Nucleic Acids Res.* **7,** 1869–1885.
Gigot, C. (1988). *In* "Architecture of Eukaryotic Genes" (G. Kahl, ed.), pp. 229–242. Verlag Chemie, Weinheim.
Gilbert, S. (1991). "Developmental Biology," 3rd Ed., pp. 273–280. Sinauer, Sunderland, Massachusetts.
Gilbert, W., Marchionni, M., and McKnight, G. (1986). *Cell* **46,** 151–154.
Gilles, R., Gilles, C., and Jaenicke, L. (1983). *Naturwissenschaften* **70,** 571–572.
Gilles, R., Gilles, C., and Jaenicke, L. (1984). *Z. Naturforsch., C* **39,** 584–592.
Gilles, R., Moka, R., and Jaenicke, L. (1985). *FEBS Lett.* **184,** 309–312.
Gilles, R., Balshüsemann, D., and Jaenicke, L. (1987). *In* "Algal Development (Molecular and Cellular Aspects)" (W. Weissner, D. G. Robinson, and R. C. Starr, eds.), pp. 50–57. Springer-Verlag, Berlin.
Goodenough, U. W., and Heuser, J. E. (1985). *J. Cell Biol.* **101,** 1550–1568.
Goodenough, U. W., and Heuser, J. E. (1988). *J. Cell Sci.* **90,** 717–733.
Green, K. J. (1982). Ph.D. Thesis, Washington Univ., St. Louis.
Green, K. J., and Kirk, D. L. (1981). *J. Cell Biol.* **91,** 743–755.
Green, K. J., Viamontes, G. I., and Kirk, D. L. (1981). *J. Cell Biol.* **91,** 756–769.
Gruber, H., Goetinck, S., Kirk, D. L., and Schmitt, R. (1992). *Gene,* in press.
Gruel, B. T., and Floyd, G. L. (1985). *J. Phycol.* **21,** 358–371.
Günther, R., Bause, E., and Jaenicke, L. (1987). *FEBS Lett.* **221,** 293–298.
Gunderson, J. H., Elwood, H., Ingold, A., Kindle, K., and Sogin, M. L. (1987). *Proc. Natl. Acad. Sci. U.S.A.* **84,** 5823–5827.
Hagen, G., and Kochert, G. (1980). *Exp. Cell Res.* **127,** 451–457.
Harper, J. D. I., and John, P. C. L. (1986). *Protoplasma* **131,** 118–130.
Harper, J. F., and Mages, W. (1988). *Mol. Gen. Genet.* **213,** 315–324.
Harper, J. F., Huson, K. S., and Kirk, D. L. (1987). *Genes Dev.* **1,** 573–584.
Harris, E. H. (1989). "The *Chlamydomonas* Sourcebook: A Comprehensive Guide to Biology and Laboratory Use." Academic Press, San Diego.
Hartz, H., and Hegemann, P. (1991). *Nature (London)* **351,** 489–491.
Haubruck, H., Engelke, U., Mertins, P., and Gallwitz, D. (1990). *EMBO J.* **9,** 1957–1962.
Hengst, L., Lehmeier, T., and Gallwitz, D. (1990). *EMBO J.* **9,** 1949–1955.
Hentschel, C. C., and Birnstiel, M. L. (1981). *Cell* **25,** 301–331.
Hershko, A., and Ciechanover, A. (1986). *Prog. Nucleic Acid Res. Mol. Biol.* **33,** 19–56.
Hills, G. J., Gurney-Smith, M., and Roberts, K. (1973). *J. Ultrastruct. Res.* **43,** 179–182.
Hirono, M., Endoh, H., Okada, N., Numata, O., and Watanabe, Y. (1987). *J. Mol. Biol.* **194,** 181–192.
Holmes, J. A., and Dutcher, S. K. (1989). *J. Cell Sci.* **94,** 273–285.
Holst, O., Christoffel, V., Fründ, R., Moll, H., and Sumper, M. (1989). *Eur. J. Biochem.* **181,** 345–350.
Hoops, H. J. (1984). *J. Phycol.* **20,** 20–27.
Hoops, H. J., and Floyd, G. L. (1982). *Br. Phycol. J.* **17,** 297–310.
Huang, B., Ramanis, Z., Dutcher, S. K., and Luck, D. J. L. (1982). *Cell* **29,** 745–753.
Hurley, C. K., and Stout, J. T. (1980). *Biochemistry* **19,** 410–416.
Huskey, R. J., and Griffin, B. E. (1979). *Dev. Biol.* **72,** 226–235.
Huskey, R. J., Griffin, B. E., Cecil, P. O., and Callahan, A. M. (1979a). *Genetics* **91,** 229–244.
Huskey, R. J., Semenkovich, C. F., Griffin, B. E., Cecil, P. O., Callahan, A. M., Chace, K. V., and Kirk, D. L. (1979b). *Mol. Gen. Genet.* **169,** 157–161.
Ikemura, T. (1985). *Mol. Biol. Evol.* **2,** 13–34.
Ivanchenko, M., Georgieva, E., Uschewa, A., and Avramova, Z. (1987). *Eur. J. Biochem.* **162,** 339–344.
Jaenicke, L. (1991). *Prog. Bot.* **52,** 138–189.

Jaenicke, L., and Gilles, R. (1985). *Differentiation* **29,** 199–206.
Jaenicke, L., and Waffenschmidt, S. (1979). *FEBS Lett.* **107,** 250–253.
Jaenicke, L., and Waffenschmidt, S. (1981). *Ber. Dtsch. Bot. Ges.* **94,** 375–386.
Jaenicke, L., Kuhne, W., Spessert, R., Wahle, U., and Waffenschmidt, S. (1987). *Eur. J. Biochem.* **170,** 485–491.
Jentsch, S., McGrath, J. P., and Varshavsky, A. (1987). *Nature (London)* **329,** 131–134.
Johnson, D. I., and Pringle, J. R. (1990). *J. Cell Biol.* **111,** 143–152.
Johnson, U. G., and Porter, K. R. (1968). *J. Cell Biol.* **38,** 403–425.
Kamiya, R., and Witman, G. B. (1984). *J. Cell Biol.* **98,** 97–107.
Kater, J. M. (1929). *Univ. Calif. Publ. Zool.* **33,** 125–168.
Kindle, K. (1990). *Proc. Natl. Acad. Sci. U.S.A.* **87,** 1228–1232.
Kinghorn, J. R., and Campbell, E. (1989). *In* "Molecular and Genetic Aspects of Nitrate Assimilation" (J. L. Wray and J. R. Kinghorn, eds.) pp. 385–403. Oxford Sci. Publ., Oxford.
Kirk, D. L. (1987). *In* "Translational Regulation of Gene Expression" (J. Ilan, ed.), pp. 229–243. Plenum, New York.
Kirk, D. L. (1988). *Trends Genet.* **4,** 32–36.
Kirk, D. L. (1990). *In* "Experimental Phycology. 1: Cell Walls and Surfaces, Reproduction, Photosynthesis" (W. Wiessner, J. D. Robinson, and R. C. Starr, eds.), pp. 81–94. Springer-Verlag, Berlin.
Kirk, D. L., and Harper, J. F. (1986). *Int. Rev. Cytol.* **99,** 217–293.
Kirk, D. L., and Kirk, M. M. (1986). *Science* **231,** 51–54.
Kirk, D. L. (1975). "Biology Today," 2nd Ed. Random House, New York.
Kirk, D. L., Birchem, R., and King, N. (1985). *J. Cell Sci.* **80,** 207–231.
Kirk, D. L., Baran, G. J., Harper, J. F., Huskey, R. J., Huson, K. S., and Zagris, N. (1987). *Cell* **48,** 11–24.
Kirk, D. L., Kaufman, M. R., Keeling, R. M., and Stamer, K. A. (1991a). *Development, Suppl.* No. 1, 67–82.
Kirk, D. L., Kirk, M. M., Stamer, K. A., and Larson, A. (1991b). *In* "The Unity of Evolutionary Biology" (E. C. Dudley, ed.). pp. 568–581. Dioscorides Press, Portland, Oregon.
Kirk, M. M., Ransick, A. J., McRae, S., and Kirk, D. L. (1992). In preparation.
Kochert, G. (1968). *J. Protozool.* **15,** 438–452.
Kochert, G. (1973). *In* "Developmental Regulation: Aspects of Cell Differentiation" (S. J. Coward, ed.), pp. 155–167. Academic Press, New York.
Kochert, G. (1975). *Dev. Biol., Suppl.* No. 8, 55–90.
Kochert, G. (1981). *In* "Sexual Interactions in Eukaryotic Microbes" (D. H. O'Day and P. A. Horgan, eds.), pp. 73–93. Academic Press, New York.
Kochert, G., and Olsen, L. W. (1970). *Arch. Mikrobiol.* **74,** 19–30.
Kochert, G., and Yates, I. (1970). *Dev. Biol.* **23,** 128–135.
Korolkov, S. N., Garnovskaya, M. N., Basov, A. S., Chuaev, A. S., and Dumler, I. L. (1990). *FEBS Lett.* **270,** 132–134.
Koufopanou, V. (1990). Ph.D. Thesis, McGill Univ., Montreal.
Kuhsel, M. G., Strickland, R., and Palmere, J. D. (1990). *Science* **250,** 1570–1573.
Lang, N. J. (1963). *Am. J. Bot.* **50,** 280–300.
Larson, A., Kirk, M. M., and Kirk, D. L. (1992). *Mol. Biol. Evol.* **9,** 85–105.
LeDizet, M., and Piperno, G. (1986). *J. Cell Biol.* **103,** 13–22.
Lindauer, A., Müller, K., and Schmitt, R. (1992). *Gene.* Submitted for publication.
Mages, H.-W., Tschochner, H., and Sumper, M. (1988). *FEBS Lett.* **234,** 407–410.
Mages, W. (1990). Ph.D. Thesis, Univ. Regensburg, Regensburg, Germany.
Mages, W., Salbaum, J. M., Harper, J. F., and Schmitt, R. (1988). *Mol. Gen. Genet.* **213,** 449–458.

Marchant, H. J. (1977). *Protoplasma* **93,** 325–339.
Matsuda, Y., Saito, T., Yamaguchi, T., and Kawase, H. (1985). *J. Biol. Chem.* **260,** 6373–6377.
Matsui, M., Sasamoto, S., Kunieda, T., Nomura, N., and Ishizaki, R. (1989). *Gene* **76,** 313–319.
McCracken, M. D., and Barcelona, W. J. (1976). *J. Histochem. Cytochem.* **24,** 668–673.
Melkonian, M. (1984a). *In* "Systematics of the Green Algae" (D. E. G. Irvine and D. M. John, eds.), pp. 73–120. Academic Press, New York.
Melkonian, M. (1984b). *In* "Compartments in Algal Cells and Their Interaction" (W. Wiessner, D. Robinson, and R. C. Starr, eds.), pp. 96–108. Springer-Verlag, Berlin.
Melkonian, M., Reize, I. B., and Presig, H. R. (1987). *In* "Algal Development (Molecular and Cellular Aspects)" (W. Weissner, D. G. Robinson, and R. C. Starr, eds.), pp. 102–113. Springer-Verlag, Berlin.
Mesquita, J. M., and Fátima Santos, M. (1984). *Cytologia* **49,** 229–241.
Metzner, J. (1945). *Torrey Bot. Club Bull.* **72,** 86–136.
Miyake, S., and Yamamoto, M. (1990). *EMBO J.* **9,** 1417–1422.
Mohr, E., Trieschmann, L., and Grossbach, U. (1989). *Proc. Natl. Acad. Sci. U.S.A.* **86,** 9308–9312.
Moka, R. (1985). Diplom-Thesis, Univ. Köln, Köln, Germany. Cited in Jaenicke and Gilles (1985).
Moka, R. (1988). Ph.D. Thesis, Univ. Köln, Köln, Germany. Cited in Jaenicke (1991).
Monterio, M. J., and Cox, R. A. (1987). *J. Mol. Biol.* **193,** 427–438.
Montolin, L., Puigdomenech, P., and Rigau, J. (1990). *Gene* **94,** 201–207.
Müller, K., and Schmitt, R. (1988). *Nucleic Acids Res.* **16,** 4121–4136.
Müller, K., Lindauer, A., Brüderlein, M., and Schmitt, R. (1990). *Gene* **93,** 167–175.
Müller, K., Klose, R., Lindauer, A., Mages, W., and Schmitt, R. (1992). In preparation.
Nozaki, H. (1986). *Jpn. J. Phycol.* **34,** 232–247.
Nozaki, H. (1988). *Phycologia* **27,** 209–220.
Ochman, H., and Wilson, A. C. (1987). *J. Mol. Evol.* **26,** 74–86.
Pall, M. (1975). *In* "Developmental Biology: Pattern Formation, Gene Regulation" (D. McMahon and C. F. Fox, eds.), pp. 148–156. Academic Press, New York.
Palme, K., Diefenthal, T., Vingron, M., Sander, C., and Schell, J. (1992). *Proc. Natl. Acad. Sci. U.S.A.* **89,** 787–791.
Pickett-Heaps, J. D. (1970). *Planta* **90,** 174–190.
Pickett-Heaps, J. D. (1975). "Green Algae: Structure, Reproduction and Evolution in Selected Genera." Sinauer, Sunderland, Massachusetts.
Pickett-Heaps, J. D., and Marchant, H. J. (1972). *Cytobios* **6,** 225–264.
Pommerville, J. C., and Kochert, G. D. (1981). *Eur. J. Cell Biol.* **24,** 236–243.
Pommerville, J. C., and Kochert, G. D. (1982). *Exp. Cell Res.* **140,** 39–45.
Powers, S., Gonzales, E., Christensen, T., Cubert, J., and Broek, D. (1991). *Cell* **65,** 1225–1231.
Proudfoot, N. (1982). *Nature* (*London*) **298,** 516.
Ransick, A. J. (1988). Ph.D. Thesis, Univ. of Texas, Austin.
Ransick, A. (1991). *Dev. Biol.* **143,** 185–198.
Ransick, A. (1992). *In* "Evolutionary Conservation of Developmental Mechanisms" (A. Spradling, ed.). Wiley-Liss, New York. In press.
Rausch, H. (1988). Ph.D. Thesis, Univ. Regensburg, Regensburg, Germany.
Rausch, H., Larsen, N., and Schmitt, R. (1989). *J. Mol. Evol.***29,** 255–265.
Raven, P. H., Evert, R. F., and Eichorn, S. E. (1986). "Biology of Plants," 4th Ed. Worth, New York.
Razafimahatratra, P., Chaubet, N., Philipps, G., and Gigot, C. (1991). *Nucleic Acids Res.* **19,** 1491–1496.

Redman, K. L., and Rechensteiner, M. (1989). *Nature (London)* **338,** 438–440.
Reufer, U., and Nultsch, W. (1991). *Cell Motil. Cytoskeleton* **18,** 269–278.
Rexach, M. F., and Schekman, R. W. (1991). *J. Cell Biol.* **114,** 219–229.
Roberts, K. (1974). *Philos. Trans. R. Soc. London, Ser. B* **268,** 129–146.
Roberts, K., Gurney-Smith, M., and Hills, G. J. (1972). *J. Ultrastuct. Res.* **40,** 599–613.
Rogers, J. (1989). *Trends Genet.* **5,** 213–216.
Salisbury, J. L., Baron, A. T., and Sanders, M. (1988). *J. Cell Biol.* **107,** 635–641.
Sanicola, M., Ward, S., Childs, G., and Emmons, S. W. (1990). *J. Mol. Biol.* **212,** 259–268.
Scagel, R. F., Bandon, R. J., Rouse, G. E., Schofield, W. B., Stein, J. R., and Tayor, T. M. C. (1969). "Plant Diversity: An Evolutionary Approach." Wadsworth, Belmont, California.
Schiedlmeier, B., and Schmitt, R. (1992). *Curr. Genet.* Submitted for publication.
Schlipfenbacher, R., Wenzl, S., Lottspeich, F., and Sumper, M. (1986). *FEBS Lett.* **209,** 57–62.
Schlösser, U. G. (1976). *Ber. Dtsch. Bot. Ges.* **89,** 1–56.
Schmitt, R., and Kirk, D. L. (1992). *In* "The Cytoskeleton of the Algae" (D. Menzel, ed.), pp. 369–392. CRC Press, Boca Raton, Florida.
Segaar, P. J. (1990). *Acta Bot. Neerl.* **39,** 29–42.
Segaar, P. J. (1991). Ph.D. Thesis, Rijksuniv. Leiden, Leiden.
Segaar, P. J., and Gerritsen, A. F. (1989). *Crypt. Bot.* **1,** 249–274.
Segaar, P. J., Gerritsen, A. F., and De Bakker, M. A. G. (1989). *Nova Hedw.* **49,** 1–23.
Sharp, P. M., and Li, W.-H. (1987). *J. Mol. Evol.* **25,** 58–64.
Siegelman, M., Bond, M. W., Gallatin, W. M., St. John, T., Smith, H. T., Fried, V. A., and Weissman, I. L. (1986). *Science* **231,** 823–829.
Silflow, C. D., Chisholm, R. L., Conner, T. W., and Ranum, L. P. W. (1985). *Mol. Cell. Biol.* **5,** 2389–2398.
Starr, R. C. (1969). *Arch. Protistenkd.* **111,** 204–222.
Starr, R. C. (1970). *Dev. Biol., Suppl.* No. 4, 59–100.
Starr, R. C. (1975). *Arch. Protistenkd.* **117,** 187–191.
Starr, R. C. (1980). *In* "Phytoflagellates" (E. Cox, ed.), pp. 147–164. Elsevier, New York.
Starr, R. C., and Jaenicke, L. (1974). *Proc. Natl. Acad. Sci. U.S.A.* **71,** 1050–1054.
Starr, R. C., and Jaenicke, L. (1988). *Sex Plant Reprod.* **1,** 28–31.
Starr, R. C., and Jaenicke, L. (1989). *In* "Algae as Experimental Systems" (A. W. Coleman, L. J. Goff, and J. R. Stein-Taylor, eds.), pp. 135–147. Alan R. Liss, New York.
Stewart, K. D., and Mattox, K. R. (1975). *Bot. Rev.* **41,** 104–135.
Tam, L.-W., and Kirk, D. L. (1991a). *Dev. Biol.* **145,** 51–66.
Tam, L.-W., and Kirk, D. L. (1991b). *Development* **112,** 571–580.
Tam, L.-W., Stamer, K. A., and Kirk, D. L. (1991). *Dev. Biol.* **145,** 67–76.
Tamanoi, F. (1988). *Biochim. Biophys. Acta* **948,** 1–15.
Triemer, R. E., and Brown, R. M., Jr. (1974). *J. Phycol.* **10,** 419–433.
Tschochner, H., Lottspeich, F., and Sumper, M. (1987). *EMBO J.* **6,** 2203–2207.
Valencia, A., Chardin, P., Wittinghofer, A., and Sander, C. (1991). *Biochemistry* **30,** 4637–4648.
Vande Berg, W. J., and Starr, R. C. (1971). *Arch. Protistenkd.* **113,** 195–219.
Viamontes, G. I. (1978). Ph.D. Thesis, Washington Univ., St. Louis.
Viamontes, G. I., Fochtmann, J. L., and Kirk, D. L. (1979). *Cell* **17,** 537–550.
von Holt, C., and Brandt, W. F. (1986). *FEBS Lett.* **194,** 282–286.
Waffenschmidt, S., Knittler, M., and Jaenicke, L. (1990). *Sex Plant Reprod.* **3,** 1–6.
Weisshaar, B., Gilles, R., Moka, R., and Jaenicke, L. (1984). *Z. Naturforsch., C* **39,** 1159–1162.
Wells, D. (1986). *Nucleic Acids Res.* **14,** r119–r149.
Wells, D., and Kedes, L. (1985). *Proc. Natl. Acad. Sci. U.S.A.* **82,** 2834–2838.

Wells, D., Bains, W., and Kedes, L. (1986). *J. Mol. Evol.* **23,** 224–241.
Wenzl, S., and Sumper, M. (1982). *FEBS Lett.* **143,** 311–315.
Wenzl, S., and Sumper, M. (1986a). *Cell* **46,** 633–639.
Wenzl, S., and Sumper, M. (1986b). *Dev. Biol.* **115,** 119–128.
Wenzl, S., and Sumper, M. (1987). *In* "Algal Development (Molecular and Cellular Aspects)" (W. Weissner, D. G. Robinson, and R. C. Starr, eds.), pp. 58–65. Springer-Verlag, Berlin.
Wenzl, S., Thym, D., and Sumper, M. (1984). *EMBO J.* **3,** 739–744.
West, M. H. P., and Bonner, W. M. (1980). *Nucleic Acids Res.* **8,** 4671–4680.
Wettern, M., Parag, H. A., Pollmann, L., Ohad, I., and Kulka, G. (1990). *Eur. J. Biochem.* **191,** 571–576.
Wichmann, H., Disela, C., Haubruck, H., and Gallwitz, D. (1989). *Nucleic Acids Res.* **17,** 6737–6738.
Wiebauer, K., Herrero, J.-J., and Filipowicz, W. (1988). *Mol. Cell. Biol.* **8,** 2042–2051.
Wildemann, A. G. (1988). *Nucleic Acids Res.* **16,** 2553–2564.
Wray, J. L., and Kinghorn, J. R., eds. (1989). "Molecular and Genetic Aspects of Nitrate Assimilation." Oxford Sci Publ., Oxford.
Wright, R. L., Salisbury, J., and Jarvik, J. W. (1985). *J. Cell Biol.* **101,** 1903–1912.
Wu, M., Allis, C. D., Richman, R., Cook, R. G., and Gorovsky, M. A. (1986). *Proc. Natl. Acad. Sci. U.S.A.* **83,** 8674–8678.
Xu, M., Kathe, S. D., Goodrich-Blair, H., Nierzwicki-Bauere, S. A., and Shub, D. A. (1990). *Science* **250,** 1568–1570.
Yates, I., and Kochert, G. (1976). *Cytobios* **15,** 7–21.
Youngblom, J., Schloss, J. A., and Silflow, C. D. (1984). *Mol. Cell. Biol.* **4,** 2686–2696.
Zimmerman, W. (1921). *Jahrb. Wiss. Bot.* **60,** 256–294.

Actin Matrix of Dendritic Spines, Synaptic Plasticity, and Long-Term Potentiation

Eva Fifková and Marisela Morales
Department of Psychology, Center for Neuroscience, University of Colorado, Boulder, Colorado 80309

I. Introduction

The function of dendritic spines has been the subject of much speculation ever since they were discovered. This review is by no means intended to be a comprehensive review of the literature on dendritic spines, but rather to bring into context the composition of the spine cytoplasm with the synaptic activity generated by the spine. We have repeatedly stressed that the actin-based cytoskeleton could to a large extent determine the plastic properties of dendritic spines. The link between the signal-transmitting systems of the membrane and actin-regulatory proteins in the cytoplasm justifies this hypothesis.

II. General Description of Spines

Dendritic spines are appendage-like outgrowths of dendrites that were first observed in the light microscope by Ramón y Cajal (1891) and in the electron microscope by Gray (1959). They contain an enlarged terminal part (the spine head) and a slender stalk connecting the head with the parent dendrite. Whereas the stalk is always narrow, the head varies a great deal in shape and size (Fig. 1). In the cortex, hippocampus, and the dentate fascia, there is sometimes a small protrusion emanating from the spine head into the axon terminal contacting the dendritic spine. These protrusions are called spinules, and they originate within the midst of the synaptic apposition (Tarrant and Routtenberg, 1977). The spine synapse is conspicuous by its postsynaptic density (PSD), which gives to the contact an asymmetric appearance in the electron microscope. The

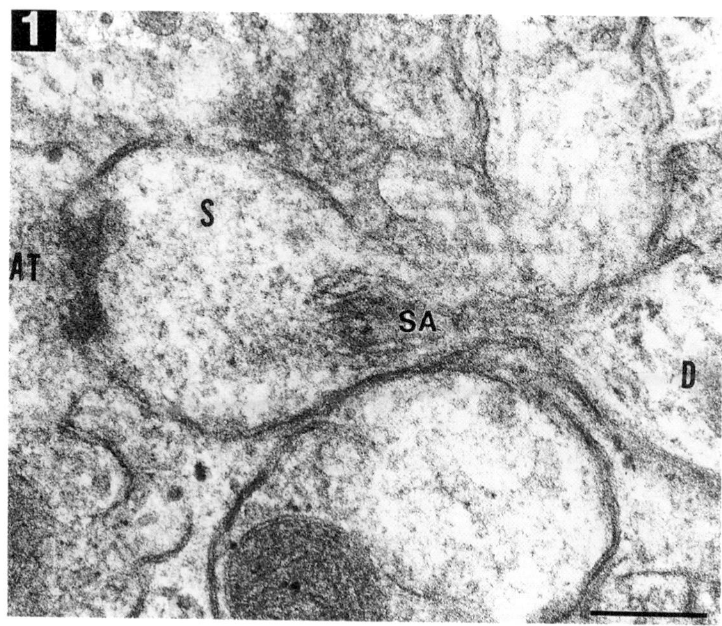

FIG. 1 Dendritic spine (S) associated with its parent dendrite (D) contacted by an axon terminal (AT). This micrograph was prepared from a 0.12-μm thick section and was taken with the high-voltage electron microscope. It shows the narrow spine stalk filled with membranes of the spine apparatus (SA). Bar, 0.25μm.

contact-making axon terminal has a uniform population of round, clear synaptic vesicles. These are morphological features characterizing an excitatory synapse. The spine may have a second synapse positioned either on the head near its junction with the stalk or on the stalk itself (Fifková *et al.*, 1990, 1992). Axon terminals of these synapses contain pleomorphic synaptic vesicles and form symmetrical junctional specializations. In the dentate fascia they contain gamma-aminobutyric acid (GABA) (Fifková *et al.*, 1990, 1992). Whereas the prevailing population of spines is postsynaptic, dendritic spines of the olfactory bulb granule cells may also be presynaptic, containing synaptic vesicles and making reciprocal synapses with dendrites of mitral cells.

Dendritic spines are found on many types of neurons (for review see Peters *et al.*, 1976). Although spines are mostly associated with dendrites, they were also observed on perikarya and on the axon initial segment. Within a defined brain region, there may be spiny types of neurons, whereas other neuronal types may be spine-free. Some brain pathways terminate solely on dendritic spines, whereas other pathways seem to avoid them. In some brain regions, such as the cerebral cortex, hippocampus, and the dentate fascia, an inverse relation has been observed between the dimensions of the spine and its parental dendrite: large spines reside generally on small dentrites and vice versa. Although there are so far no systematic data available on the density of dendritic spines in various types of spiny neurons, in the cortical pyramids it has been shown that the density may vary across the dendritic tree, being much higher on the apical than on basilar dendrites or on oblique branches emanating from the apical dendrites. The proximal dendritic segments are usually spine-free. It has been estimated that spines account for 43% of the combined surface area of the perikaryon and the dendritic tree.

III. Spines and Synaptic Plasticity

The function of dendritic spines has been the subject of much speculation ever since they were discovered (Crick, 1982). It has been repeatedly shown that dendritic spines are endowed with a considerable degree of plasticity because they are readily modified by stimulation and changes in the environment. Because in vertebrates the majority of brain synapses is on dendritic spines, much of the synaptic plasticity is carried by the spines. Synaptic plasticity allows for adaptive changes in the brain in response to past events and experiences; therefore, it is one of the most important properties of the brain. Thus, understanding the molecular basis of synaptic plasticity became one of the central issues of neuroscience.

The actin-based cytoskeleton is a dynamic component of living cells with major structural and contractile properties involved in essential cellular processes. Recent literature indicates that neuronal activity induces dynamic interactions between the plasma membrane and the cytoskeleton. Cytoskeletal proteins are endowed with considerable versatility. Moreover, they interact with membrane proteins and membrand lipids and are, therefore, the most likely candidates for mediating synaptic plasticity. Signals transduced by the plasma membrane activate a cascade of biochemical events in the cell membrane, which are translated into changes of the cytoskeleton. Although such changes occur immediately in response to the stimulus, they may, in addition, trigger a chain of events outlasting the initial stimulus, which could ultimately lead to long-lasting modifications in the cytoplasm. For example, stimulation-induced changes in actin networks were shown to occur with a short latency and to persist over extended periods of time (Stossel, 1982). In view of these observations,

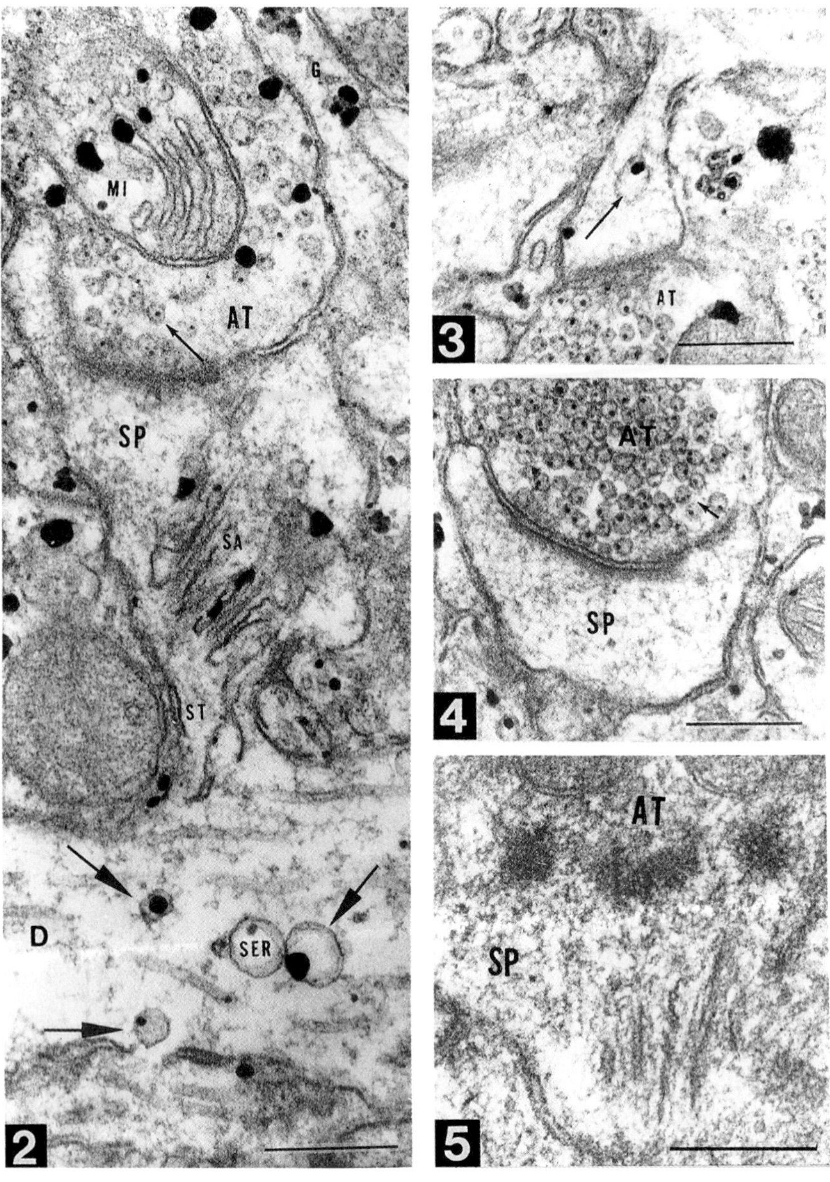

FIG. 2 Dendritic spine (SP) attached with a stalk (ST) to a dendrite (D) contains a spine apparatus (SA). Ca^{2+}-antimonate deposits are seen in the dendritic smooth endoplasmic reticulum (SER, large arrows) and in the spine apparatus (SA) in the head of the dendritic spine (SP). The spine stalk (ST) contains a spine apparatus sac. Precipitates of varying dimensions are seen in the mitochondrion (MI), the axon terminal (AT), and synaptic vesicles (small arrows). A glial process (G) contains aggregated precipitates. Bar, 0.25μm.

identification, characterization, and distribution of cytoskeletal and contractile proteins in dendritic spines are the first steps toward understanding the molecular basis of synaptic plasticity.

IV. Spine Organelles

The internal structure of postsynaptic spines is conspicuous by the absence of cytoplasmic organelles with the exception of the spine apparatus (SA), smooth endoplasmic reticulum (SER), and occasional coated vesicles or ribosomes. Spines are noted for the absence of mitochondria. The spine apparatus was first described by Gray (1959, 1982). It consists of parallel membranous sacs that alternate with plates of dense material, which are contacted by dendritic microtubules during development (Westrum *et al.*, 1980). SA may be seen along the entire length of the stalk, partly protruding into the spine head (Fig. 2). The SA membranous sacs are in continuity with the SER of the parent dendrite. In some spines the spine apparatus may occupy a large proportion of the spine stalk cytoplasm, whereas other spines may possess only simple loops of the SER. As to whether the spine apparatus or SER is inherent to the spine may be determined from serial thin sections. With this method it has been shown that in the hippocampus some spines lack either of these organelles, whereas in the striatum all the spines had one or the other organelle. The degree of complexity of the spine apparatus seems to be the function of the dimensions of spines rather than to be region-specific (Harris and Stevens, 1989). Although the function of the SA and SER is not yet fully understood, it may play an important role in Ca^{2+} regulation. Calcium has been demonstrated in these organelles (Fifková *et al.*, 1982a, 1983; Burgoyne *et al.*, 1983) by a Ca^{2+}-precipitation technique (Figs. 2–5). Furthermore, a receptor for inositol 1,4,5-trisphosphate (IP_3), which has a Ca^{2+}-mobilizing effect, was found

FIG. 3 Ca^{2+}-antimonate deposits in the SER of the spine (small arrow). Axon terminal (AT) with vesicles containing precipitates. Bar, 0.25μm.

FIG. 4 Axon terminal (AT) with synaptic vesicles containing small deposits is making a synaptic contact with a spine head (SP). An arrow points to a vesicle with two deposits. No deposits are found in the spine head in the absence of SA or SER. Bar, 0.25μm.

FIG. 5 Spine head with a spine apparatus (SP) makes contact with an axon terminal (AT). This is a control preparation that was treated with EGTA prior to staining. The absence of precipitates following this treatment is explained by the chelating effect of EGTA on Ca^{2+}. Bar, 0.25μm. (From Fifková *et al.*, 1983. Reproduced with permission.)

in the SA and SER membranes (Satoh *et al.*, 1990). This issue is discussed in greater detail later.

V. Composition of the Spine Cytoplasm

The spine may be viewed as a specific domain of the neuron from the electrical as well as cytochemical aspect (Steward and Levy, 1982). The absence of cytoplasmic organelles from the spine allows for a full use of the actin cytomatrix, which is here particularly dense. Out of the entire neuron, the dendritic spines contain the highest density of actin filaments, which are arranged in such a way as to determine the characteristic shape of the spine (Fifková and Delay, 1982). The high density of the actin cytomatrix suggested very low resting levels of free Ca^{2+} (Fifková, 1985a,b; Markham and Fifková, 1986). This suggestion was confirmed experimentally (Segal *et al.*, 1990). In freeze-fractured preparations the spine cytoplasm was described by Landis and Reese (1983).

A. Actin

Although brain actin was identified biochemically in the early 1970s (Berl *et al.*, 1973), actin filaments were recognized rather recently. The main reason for this delay is the vulnerability of actin filaments to electron-microscopy fixatives, particularly to OsO_4. In the earlier electron-microscope literature (Peters *et al.*, 1976), cytoplasmic "floccular material" was described in spines, which makes them stand out in the neuropile of conventionally fixed tissue. The floccular material actually represents disintegrated actin filaments. Because the severity of actin filament damage is the function of the concentration, pH, and temperature of the OsO_4 solution, and the length of exposure, it is possible, by controlling these parameters (Pollard, 1986), to reduce the adverse impact of OsO_4 and to preserve the actin filaments in the brain tissue (Fifková and Delay, 1982).

Actin is a highly conserved molecule, which in nonmuscle cells is present in a monomeric (G-actin) and polymeric form (F-actin) in equal proportions. Actin filament assembly is a self-sustaining process controlled by physicochemical conditions of the cytoplasm and by actin-regulatory proteins. Through these proteins actin filaments are linked to the plasma membrane and to cytoplasmic organelles. In addition, actin-binding proteins also cross-link actin into networks or bundles that constitute the cytoplasmic actin gel (Condeelis, 1983; Stossel, 1983).

Due to the fast and easy interconversion of G- and F-actin, the actin network is a dynamic structure. This property ensures transient assemblies of the network by external stimuli inducing localized changes in ionic composition, pH, and temperature (Craig and Pollard, 1982). Versatility is conferred on actin networks by actin-regulatory proteins, cytoplasmic Ca^{2+}, and phosphoinositides (Condeelis, 1983; Stossel, 1983, 1989). This may be of particular significance in neurons where free Ca^{2+} levels fluctuate with neuronal activity. Various stimuli can rapidly alter the actin cytomatrix, as was demonstrated in stimulated platelets (Escolar *et al.,* 1986; Nakata and Hirokawa, 1987), and the induced changes may persist over varying periods of time (Stossel, 1982). Consequently, the flexible actin-dependent activities facilitate speedy reactions in response to changing functional requirements of the cell (Stossel, 1982). Thus, the actin-based cytomatrix is uniquely suited to underlie the mechanism of various plastic reactions in neurons in general and in dendritic spines in particular.

We have used two methods to identify actin in spines (Fifková and Delay, 1982; Morales and Fifková, 1989a,b). One method uses the affinity of actin filaments for the S-1 subfragment of myosin (S-1 fragment; Ishikawa *et al.,* 1969; LeBeux and Willemont, 1975; Katsumaru *et al.,* 1982) and the other benefits from the high resolution of immunogold electron microscopy (Cohen *et al.,* 1985; Morales and Fifková, 1989a). Actin filaments are polar structures, and the polarity is revealed with the S-1 fragment (Ishikawa *et al.,* 1969). This fragment binds to individual actin molecules under an angle of 45 degrees so that the superimposed S-1 fragments along the filament give it the appearance of an arrowhead pattern (Figs. 6–10). Because actin monomers have greater affinity for the barbed end than for the pointed end, the filament grows three times faster at the former than at the latter end (Pollard *et al.,* 1976). The immunogold electron microscopy reveals details in the associations of actin filaments with the plasma membrane, PSD, and membranes of cytoplasmic organelles (Fig. 11). Moreover, it permitted us to study co-localization of actin with other proteins. Thus, the S-1 fragment technique and immunolabeling complement each other in revealing different aspects of the properties of the actin cytomatrix in the spine.

With these two methods we have demonstrated that the bulbous spine head contains a dense actin network, whereas the slender spine stalk contains a fascicle of actin filaments oriented in parallel with the long axis of the stalk (Fig. 6). That the actin cytomatrix controls the spine shape comes from observations showing that the spine retains its shape even when the plasma membrane is disrupted (Markham and Fifková, 1986). Within the spine actin filaments are associated with the PSD, the

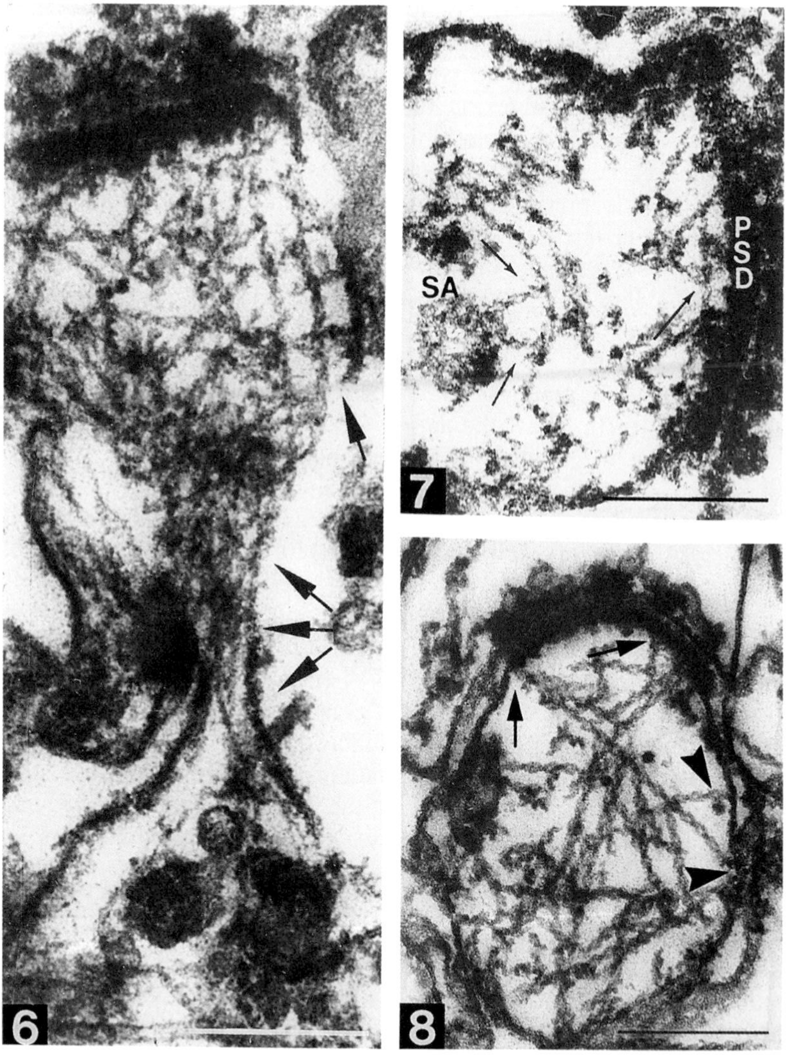

FIG. 6 Dendritic spine. Actin filaments are visualized with the S-1 fragment of myosin following permeation of the plasma membrane with saponin. Prominent are the actin filament network in the spine head (single large arrow) and longitudinally oriented filaments in the spine stalk (three large arrows). Bar, 0.25μm. (From Fifková, 1985a. Reproduced with permission.)

FIG. 7 Spine head with postsynaptic density (PSD). The actin network fills the spine head. Two short filaments (marked with arrows) connect the spine apparatus (SA) with a long filament. An actin filament associated laterally with the PSD is marked with a single arrow. Treatment with the S-1 fragment. Bar, 0.25μm.

FIG. 8 Spine head with a network of actin filaments. Simple arrows indicate end-on association of actin filaments with the postsynaptic density. Filled arrows indicate end-on association of the actin filament with the spine plasma membrane. Bar, 0.25μm.

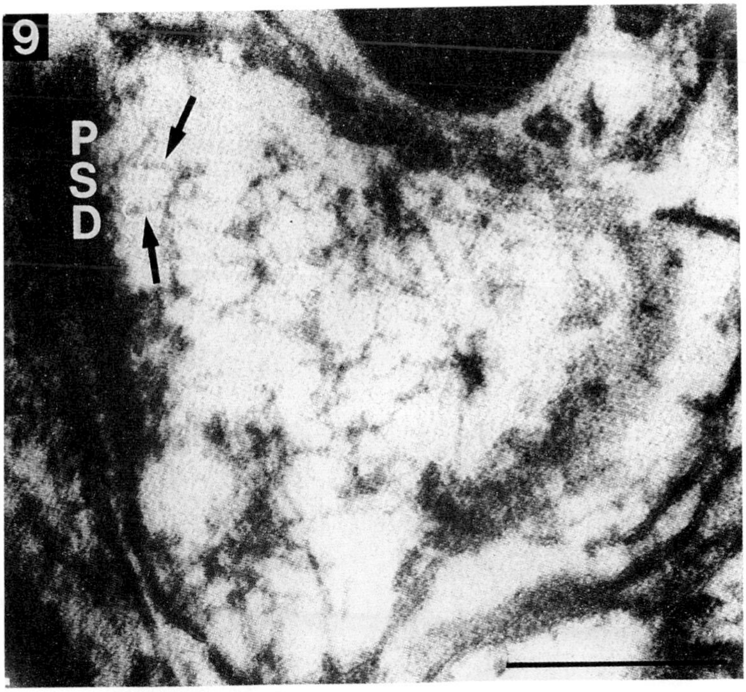

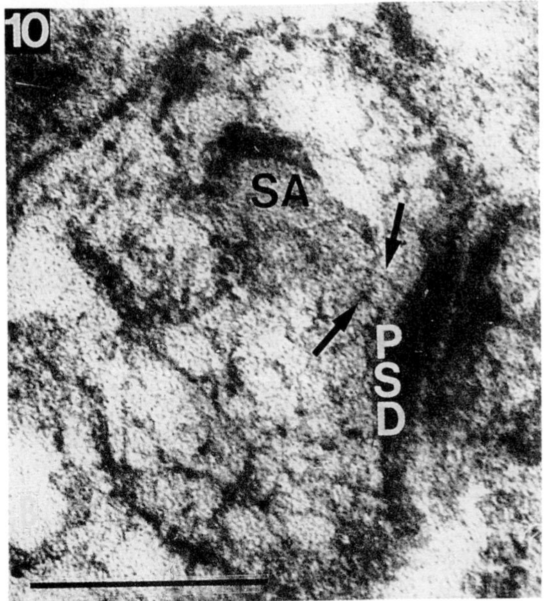

FIG. 9 Spine from a control block. Tissue was treated with saponin solution without the S-1 fragment of myosin. The actin filaments appear thin; however, their organization is similar to that of Figs. 6–8. There are filaments associated with the PSD (arrows). Bar, 0.25μm.

FIG. 10 Spine head from a control block contains untreated actin network. The filament (arrows) connecting postsynaptic density (PSD) with the spine apparatus (SA) is prominent. Same procedure as in Fig. 9. Bar, 0.25μm.

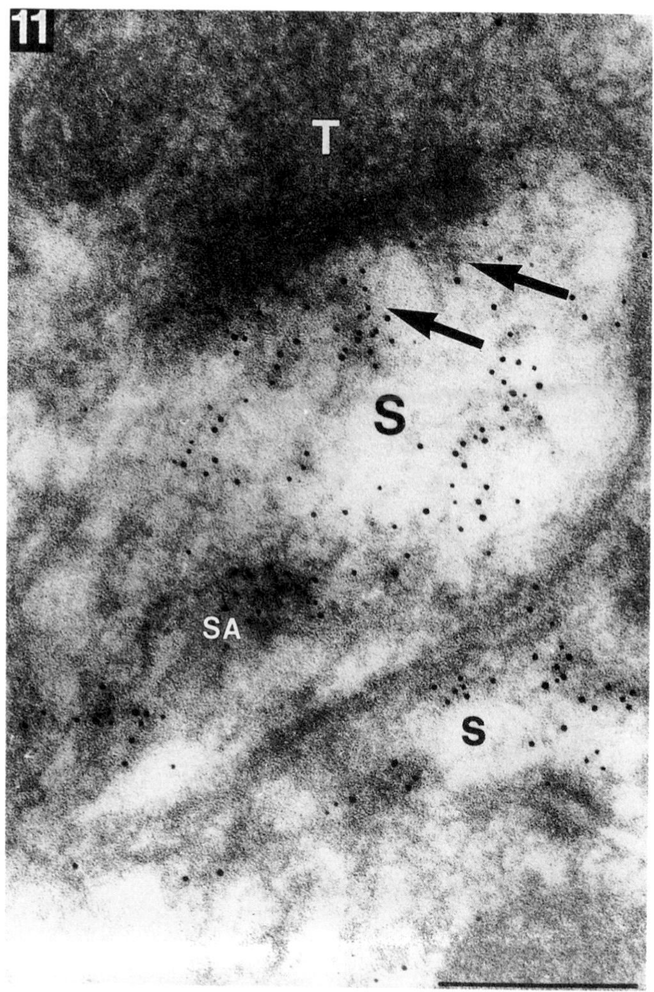

FIG. 11 Actin filaments in a spine (S) labeled with a monoclonal antiactin antibody (Amersham, Arlington Heights, Illinois) in a postembedding procedure. The primary antibody was visualized by a secondary antibody that was bound to 5nm colloidal gold. There is an association of labeled actin filament with postsynaptic density (arrows) and the label on the spine apparatus (SA). Labeling of the axon terminal (T) is considerably lower. Bar, 0.25 μm. (From Morales and Fifková, 1989a. Reproduced with permission.)

plasma membrane, and with endomembranes of SA and the spine's SER. Actin filaments extend from the plasma membrane to these organelles so that SA and SER appear to be suspended within the actin network (Figs. 6–10).

B. Role of Ca^{2+} in Organization of Actin Cytomatrix

The concentration of the free cytosolic Ca^{2+} is critically important in controlling the spatial organization of the actin cytomatrix. Therefore, fluctuation of the free cytoplasmic Ca^{2+} that occurs during intense synaptic activity is likely to affect consistency of the cytoplasm with subsequent changes in the morphology of synaptic sites and possible rearrangement of synaptic organelles. These may be the most important functions of neuronal actin. They appear to be unique to this protein because they are not shared with other cytoskeletal elements, such as microtubules or neurofilaments. The density of actin networks is inversely related to the concentration of free cytoplasmic Ca^{2+}. The lower the level of Ca^{2+}, the denser the meshwork of the actin cytomatrix, the stiffer the actin gel. The stiffness of the gel may impact the distribution of cytoplasmic organelles. Large organelles get excluded from regions with high density of the actin network, whereas smaller organelles may get locked within the actin matrix. The extent of actin gelation also modulates contractile activity of the cytomatrix. A minimal gel structure of the matrix is necessary to transmit contractile forces; however, an increased gel structure inhibits the rate of contraction (Janson and Taylor, 1990). In resting conditions free Ca^{2+} concentration is thought to be the lowest under the plasma membrane due to the activity of Ca^{2+} channels and pumps. The low resting Ca^{2+} concentration in spines (Segal *et al.*, 1990) could account for the strikingly high density of the actin cytomatrix. The low diffusibility of Ca^{2+} in the cytoplasm (Hodgkin and Keynes, 1957) suggests that changes in the Ca^{2+} concentration in the neuron could be highly localized (Tsien and Tsien, 1990). This in turn could translate into localized changes of the actin cytomatrix. In spines the slender stalk filled with SA or SER membranes coupled with the low diffusibility of the cytoplasm could help to stabilize Ca^{2+} levels in the spine (Horwitz, 1984; Harris and Stevens, 1989) by restricting Ca^{2+} movements.

The phosphoinositide cascade plays a central role in transduction of incoming signals. Hydrolysis of phosphatidylinositol 4,5-bisphosphate (PIP_2) by receptor-activated phospholipase C generates two intracellular messengers, inositol 1,4,5-trisphosphate (IP_3) and diacylglycerol (DAG). IP_3 has a Ca^{2+}-mobilizing effect that is mediated by binding to intracellular receptors, IP_3R. These receptors were shown to be localized in membranes of the SA or SER (Satoh *et al.*, 1990), which agrees with previous experiments, suggesting a Ca^{2+}sequestering role for SA and SER (Fifková *et al.*, 1982a, 1983; Burgoyne *et al.*, 1983). The highly cooperative opening of Ca^{2+} channels by nanomolar concentrations of IP_3 enables the cell to detect and amplify very small increases in the concentration of this messenger. Given the time scale of less than a second for opening and closing the Ca^{2+} channels, large changes in the concentration of Ca_{2+} may be rapidly achieved (Meyer *et al.*, 1988) in the spine cytoplasm. In addition, Ca^{2+}-dependent ATPase activity was observed in the membranes of SA and SER (Kriho and Cohen, 1988). These findings bring further evidence for the Ca^{2+}-regulatory role of SA and SER. In dendritic spines the presence of IP_3R and Ca^{2+} dependent ATPase activity in the membranes of the SA and SER together with detection of Ca^{2+} in the lumen of these organelles (Figs. 2–5) (Fifková *et al.*, 1982a, 1983; Burgoyne *et al.*, 1983) strongly suggest that these organelles play a crucial role during signal transduction.

C. Brain Actin-Associated Proteins

The actin-associated proteins constitute a group of proteins that binds to actin monomers or actin filaments to regulate polymerization or to attach filaments to other cytoskeletal elements or membranes. According to their activities, actin-associated proteins may be divided into several groups. We follow the divisions of several review articles (Schliwa, 1981; Craig and Pollard, 1982; Weeds, 1982; Stossel *et al.*, 1985). Only those actin-associated proteins that have been purified from the brain are discussed.

1. Actin Depolymerization and Polymerization Inhibiting Proteins

These proteins complex with actin monomers so as to prevent them entering the polymerization cycle. They may account for a large proportion of the monomeric actin (50%) in the cytoplasm (Flock *et al.*, 1981). Several low-molecular mass actin-depolymerizing proteins have been isolated from brain, chick actin-depolymerizing factor (ADF), destrin, cofilin, brain inhibitory protein (BIP), and profilin.

The chick ADF and destrin are capable of severing F-actin and binding actin monomers. *In vitro* studies have indicated that these activities are pH- and Ca^{2+}-independent (Bamburg *et al.*, 1980; Nishida *et al.*, 1984a, 1985; Giuliano *et al.*, 1988). Both proteins have a molecular mass of 19 kDa; ADF has been purified from chick brain and destrin from porcine brain. cDNA clones coding for chick ADF (Adams *et al.*, 1990) and porcine destrin (Moriyama *et al.*, 1990) have been obtained. Comparison of the deduced amino acid sequence (165 amino acids) obtained from both cDNAs indicated that porcine destrin differs from chick brain ADF by only eight amino acid substitutions.

Cofilin (21 kDa) has similar functions as destrin, but the interaction of cofilin with actin is pH-sensitive (Nishida *et al.*, 1984b). The action of cofilin at pH higher than 7.3 is similar to that of destrin; they both bind G-actin in a 1:1 molar ratio and quickly depolymerize F-actin in a stoichiometrical manner. However, at near-neutral pH, cofilin binds to F-actin in a 1:1 molar ratio of cofilin to actin in the filament and induces only limited depolymerization of F-actin. Contrary to that, destrin is capable of complete F-actin depolymerization. Matsuzaki *et al.* (1988) have isolated cDNA clones coding for porcine cofilin; the deduced amino acid sequence of cofilin is similar to that of destrin (71% identical). Thus, destrin and cofilin are related proteins that are coded by different genes. The distribution of these proteins in the brain is unknown. However, because the amino acid sequence of cofilin and destrin is known, it will be possible to produce specific antibodies directed against each individual protein.

The BIP has been isolated from rat brain (Berl *et al.*, 1983). It has a molecular mass of 21 kDa. This protein prevents polymerization and induces depolymerization of actin filaments. The relationship between BIP and cofilin or destrin has not been investigated.

Profilin binds to monomeric actin and does not interact with F-actin directly. Profilin and actin monomer form a 1:1 complex that cannot polymerize. It has been shown that brain profilin inhibits the polymerization of brain actin more strongly than polymerization of muscle actin. This suggests that brain profilin is capable of discriminating between β- and γ-actins, present in brains, and α-actin, present in muscle (Nishida *et al.*, 1984c). The binding of profilin to actin monomers is a Ca^{2+}-independent process. It has been shown that profilin is tightly bound to PIP_2, which is involved in the membrane-signaling path. When PIP_2 is broken down, profilin is released and is free to interact with actin and influence polymerization of actin filaments (Lassing and Lindberg, 1985, 1988).

2. Actin Modulators

These proteins regulate the rigidity of actin gels. They induce solation of actin gels by breaking the links formed by actin cross-linkers. They are also described as actin-capping or actin–end-blocking proteins. The end binding of these proteins to actin filaments seems to be mediated by at least two mechanisms: one Ca^{2+} dependent and the other Ca^{2+} independent. Both types of capping proteins have been isolated from brain.

The brain Ca^{2+}-independent capping protein is a dimer consisting of 36 and 31 kDa peptides (Kiliman and Isenberg, 1982). The binding of this dimer to the barbed end of actin filaments is a Ca^{2+}-independent event, which causes a shortening of the filaments. Because, however, the filaments are not completely disassembled, they may serve as nuclei for regrowth of the filament to full length. The regrowth process requires release of the capping protein at the barbed end.

Two brain Ca^{2+}-dependent capping proteins (Cap 90 and gelsolin) have been recognized and distinguished according to their effect on actin filaments. The actin filament capping protein, Cap 90, has a molecular mass of 90 kDa. It binds to the fast-growing end of actin filaments in a Ca^{2+}-dependent manner. Cap 90 does not sever actin filaments (Isenberg *et al.*, 1983).

Gelsolin also has a molecular mass of 90 kDa (Petrucci *et al.*, 1983; Verkhovskij *et al.*, 1984), although variation of the molecular mass may occur (Nishida *et al.*, 1981; Petrucci *et al.*, 1983). The variation may be a result of proteolytic activity during the isolation of gelsolin (Petrucci *et al.*, 1983). The interactions of gelsolin and actin are multiple, as gelsolin breaks the noncovalent bonds between actin monomers within actin filament and thus severs the filament. It also binds to the fast-growing (barbed) end of the filament, blocking monomer exchange from that end, thus raising the critical concentration for actin polymerization. This process results in depolymerization of the filaments. Lastly, gelsolin also nucleates actin assembly. These three effects of gelsolin on actin depend on Ca^{2+} concentrations. Recent observations indicate that phosphoinositides dissociate gelsolin from actin filaments and thus inhibit the severing activity of gelsolin (Janmey and Stossel, 1987, 1989).

3. Actin Cross-Linking Proteins

These proteins mediate lateral associations and branching polymerization of actin filaments. They have two binding sites for actin so that two filaments may be joined together under an angle that will determine the angle at which actin filaments are cross-linked (Stossel, 1983). These proteins stiffen the actin gels and hence increase the resistance of the cytoplasm. Myosin, microtubule-associated protein 2 (MAP 2), and spectrin are actin filament cross-linking proteins that have been observed in spines and are discussed in the following paragraphs.

Brain myosin shares a number of properties with other muscle and nonmuscle myosins: asymmetrical shape of the molecule, molecular mass of about 450 kDa, ATPase activity, reversible interaction with actin, and its ability to form filaments (Burridge and Bray, 1975; Kuczmarski and Rosenbaum, 1979a,b; Hobbs and Fredriksen, 1980; Barylko and Sobieszek, 1983; Barylko *et al.*, 1986). Brain myosin is formed of two heavy chains of molecular mass about 210 kDa and light chains of molecular mass from 17 to 23 kDa (Fig. 12). The 20 kDa myosin light chain (MLC) can

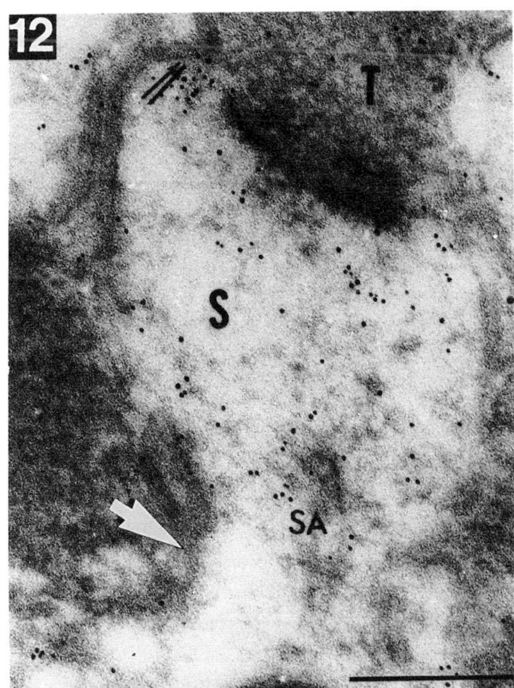

FIG. 12 Localization of myosin in the spine head (S) and stalk (large arrow). Myosin was labeled in a postembedding procedure with a polyclonal antihuman platelet myosin antibody (Dr. J. Scholey) and visualized in the same way as in Fig. 11. The label (5 nm colloidal gold) is associated with actin filaments, is in the cytoplasm, on membranes of the spine apparatus (SA), and on the plasma membrane (double arrows). T, axon terminal. Bar, 0.25μm. (From Morales and Fifková, 1989a. Reproduced with permission.)

be phosphorylated by Ca^{2+}/calmodulin-dependent MLC kinase (MLCK; Barylko *et al.*, 1986) and by Ca^{2+}/calmodulin-dependent protein kinase II (CAMK II; Tanaka *et al.*, 1986). The phosphorylation of MLC is required for the myosin ATPase activity activated by actin.

We have studied the distribution of myosin in the brain by using an immunocytochemical approach. Myosin was detected with a polyclonal antihuman platelet myosin antibody that recognized a single polypeptide that corresponds to brain myosin in immunoblots. In preparations of isolated brain myosin filaments, which displayed characteristic interactions between the tail and head of the molecule, the antimyosin antibody recognized the tail portion of the molecule. Immunolabeling revealed the reactivity prevailingly in dendritic spines. Myosin was observed in both compartments of the spine, in the head and the stalk. The immunolabel was distributed alongside the actin filaments as well as in the cytplasm. Little labeling was seen in the PSD. However, the gold probe was regularly associated with the spine plasma membrane and the spine endomembranes SA and SER (Morales and Fifková, 1989a) (Fig. 12).

In dendritic spine, the relative amounts of actin exceed those of myosin, which agrees with the ratio observed in other nonmuscle cells (1:100). This could be due to the fact that actin subserves both cytoskeletal and motile functions, whereas myosin is a mechanochemical transducer (Pollard *et nal.*, 1976). In nonmuscle cells this function may require a relatively low concentration of the protein (Taylor and Condeelis, 1985). Because of its mechanochemical activity, myosin is most often considered as a contraction- and movement-generating protein (Stossel *et al.*, 1985). However, myosin *in vitro* was also shown to be a potent cross-linking factor of actin filaments (Brotschi *et al.*, 1978). Brain myosin distinguishes from other nonmuscle myosins in that the filaments *in vitro* are more stable than filaments of other nonmuscle myosins (Barylko *et al.*, 1986), which enables brain myosin to form various three-dimensional configurations *in vitro* (Matsumura *et al.*, 1985; Morales and Fifková, 1989a). The functional significance of this property is at present unclear; it could, however, add stability to the spine actin matrix.

MAP 2 is a phosphoprotein of 280 kDa molecular mass (Burns and Islam, 1984), which binds microtubules and actin filaments (Sattilaro *et al.*, 1981; Sattilaro, 1986; Griffith and Pollard, 1982) (Figs. 13 and 14). Immunocytochemical studies have demonstrated the presence of MAP 2 in dendrites (Cáceres *et al.*, 1983, 1984; Bernhardt and Matus, 1984; DeCamilli *et al.*, 1984; Binder *et al.*, 1986; Morales and Fifková, 1989b).

The identification of MAP 2 in dendritic spines has been a matter of debate; we have utilized an immunogold approach to demonstrate that MAP 2 is indeed a component of the dendritic spines (Morales and Fifková, 1989b). MAP 2 was detected with a monoclonal anti-MAP 2 antibody, previously characterized by Binder *et al.* (1986). Western blots of the whole brain extracts and purified MAP 2 reacted with this antibody demonstrated that peptides corresponding to MAP 2a and b were specifically labeled. In the tissue MAP 2 was identified prevailingly in dendritic spines and dendrites. In spines the immunolabel was primarily associated with actin filaments, PSD, and the spine endomembranes SA and SER. In some instances the area adjacent to PSD displayed a distinct pattern of crossed-over actin filaments, which were also immunolabeled for MAP 2. Occasionally, MAP 2 immunoreactivity was associated with the spine plasma membrane (Morales and Fifková, 1989b).

MAP 2 is generally considered a microtubule-binding protein; however, *in vitro* studies have shown that it also binds actin filaments (Griffith and Pollard, 1978, 1982; Nishida *et al.*, 1981; Sattilaro *et al.*, 1981). This

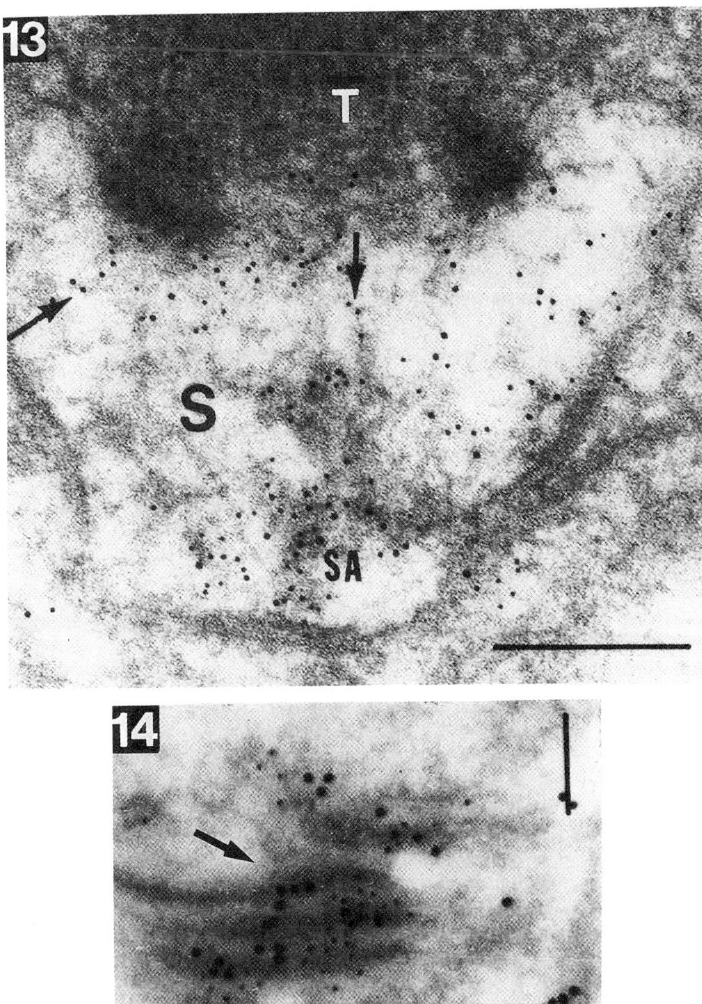

FIG. 13 MAP 2 in a spine head (S) reacted with a monoclonal anti-MAP 2 antibody, IgG, clone 14 (Drs. L. I. Binder and A. Frankfurter) and visualized with a secondary antibody bound to 5 nm colloidal gold. There is a label of MAP 2 mainly on actin filaments (arrows) and membranes of the spine apparatus (SA). T, axon terminal. Bar, 0.25μm.

FIG. 14 Co-localization of MAP 2 (5 nm colloidal gold) and actin (10 nm colloidal gold) on the spine apparatus (arrow). Bar, 0.125 μm. (From Morales and Fifková, 1989b. Reproduced with permission.)

interaction is strongly inhibited by tubulin (Sattilaro, 1986). During early development tubulin expression follows that of actin and MAP 2 in dendrites (Bernhardt and Matus, 1982). This suggests that, in the absence of tubulin, MAP 2 may regulate the gel-sol transition of actin (Sattilaro, 1986). Similar actin–MAP 2 interactions could also take place in dendritic spines because adult spines, too, are devoid of tubulin (Westrum *et al.*, 1980; Morales and Fifková, 1991). In the brain in resting conditions, MAP 2 is highly phosphorylated on multiple sites of the molecule (Sloboda *et al.*, 1975; Vallee, 1980; Schulman, 1984; Goldenring *et al*, 1985; Yamamoto *et al.*, 1985; Akiyama *et al.*, 1986; Yamauchi and Fujisawa, 1988). The number and combination of phosphorylation sites on the MAP 2 molecule may alter the biological function of this protein (Aoki and Siekevitz, 1988). The level of phosphorylation of MAP 2 is inversely related to its actin cross-linking: phosphorylated MAP 2 has a reduced ability to cross-link actin filaments (Nishida *et al.*, 1981; Selden and Pollard, 1983; Pollard *et al.*, 1984). However, phosphorylation does not inhibit MAP 2 binding to actin filaments so that both phosphorylated and unphosphorylated MAP 2 may use actin as an anchoring substrate (Sattilaro, 1986). The aforementioned *in vitro* studies have demonstrated interactions between MAP 2 and actin. The observed co-localization of MAP 2 with actin filaments in dendritic spines (Morales and Fifková, 1989b) indicates that interactions between MAP 2 and actin take place *in vivo* (Figs. 13–16).

Spectrin molecules were originally discovered under the plasma membrane of erythrocytes (Fig. 17). However, later on it has been shown that they are also associated with plasma membranes of nonerythrocytes, including neurons. Brain spectrin is a fibrous protein with subunits of 240 kDa (α subunit) and 235 kDa (β subunit). The molecule of spectrin is formed as a tetrameric complex: $(\alpha\beta)_2$ (for review see Goodman *et al.*, 1986). Zagon *et al.* (1986), using an immunoperoxidase method, demonstrated the presence of spectrin in dendritic spines. Within the spine immunolabel was associated with PSD.

We had used an immunogold technique to further investigate the distribution of spectrin in the dendritic spines. To detect brain spectrin, we have used a monoclonal antibody against human erythrocyte α- and β-spectrin. This antibody recognized rat brain and erthrocyte spectrin. When we analyzed the distribution of immunolabel in the dendritic spines, we found a distinct association with the cytoplasmic aspect of the plasma membrane, PSD, SA, and spine SER. A similar pattern of labeling could be seen in the spines of all regions studied.

Spectrin is the major protein of the submembrane skeleton that links membrane receptors directly or indirectly with cytoplasmic actin filaments. Spectrin by binding calmodulin and activating myosin ATPase forms, with submembrane actin and myosin, a membrane-associated force-generating complex (Connolly, 1984; Connolly and Graham, 1985;

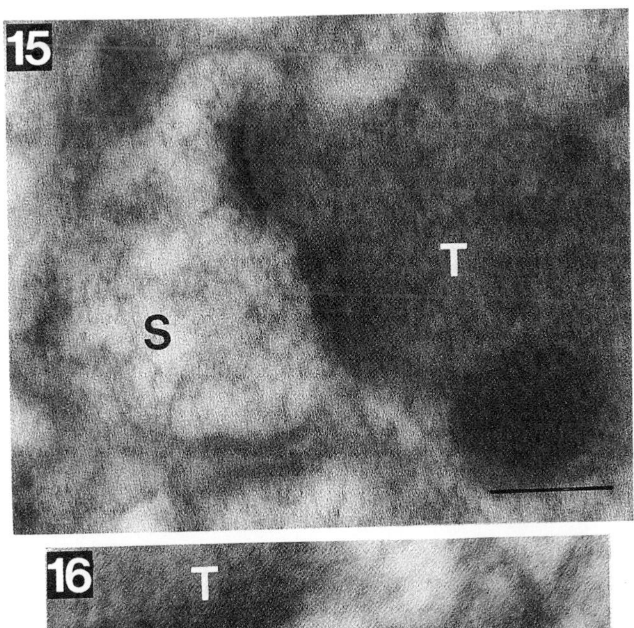

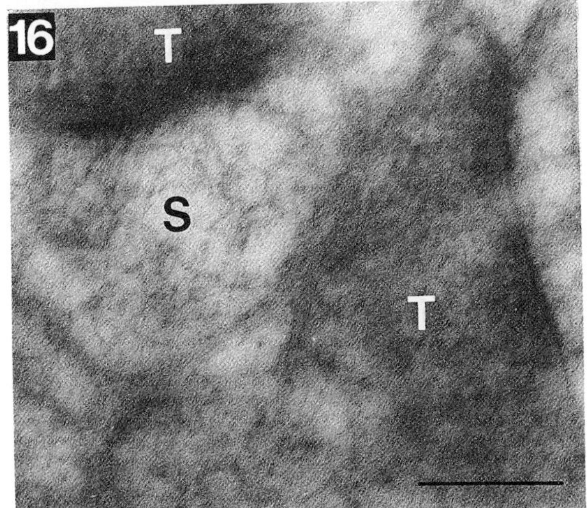

FIG. 15 Controls. When the primary anti-MAP 2 antibody was substituted with buffer (Fig. 16), the immunolabel was completely abolished. S, dendritic spines; T, axon terminals. Bar, 0.25μm. Similar controls were also performed for actin, myosin, and spectrin.

FIG. 16 Controls. When the primary anti-MAP 2 antibody was preadsorbed with an excess of antigen, the immunolabel was completely abolished. S, dendritic spines; T, axon terminals. Bar, 0.25μm. Similar controls were also performed for actin, myosin, and spectrin. (From Morales and Fifková, 1989b. Reproduced with permission.)

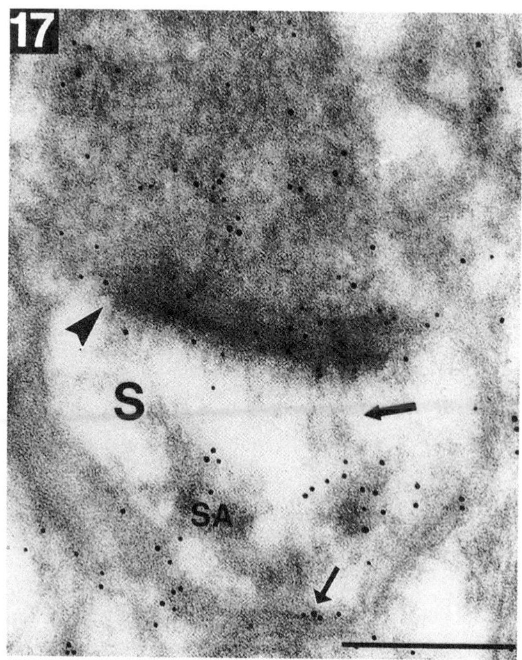

FIG. 17 Localization spectrin label in the spine (S). Spectrin was labeled with a monoclonal antierythrocyte spectrin antibody (MCA 448, Bioproducts for Science, Indianapolis, Indiana) in a procedure similar to that of actin and myosin using 5 nm colloidal gold probe. Spectrin label is on the postsynaptic density (large arrowhead), the spine apparatus (SA), and the plasma membrane (small arrow). There is sparse immunostaining on actin filaments (arrow). Bar, 0.25μm.

Peng and Phelan, 1984). Such a complex is uniquely suited to stabilize receptors in synaptic membranes and to mobilize and translocate them during a process that leads to changes in the length of synaptic appositions in response to stimulation or other interventions. The observed co-distribution of spectrin with actin, myosin, and MAP 2 on the membranes of the SA and SER suggests that force-generating complexes similar to those of the plasma membrane could also operate under these endomembranes. If so, they could regulate Ca^{2+}-transport-related IP_3 receptors that are associated with the SA and SER (Satoh *et al.*, 1990). Moreover, subplasmalemmal spectrin associates indirectly cytoplasmic actin filaments with the plasma membrane. Given that similar molecular constituents are under the plasma membrane and under SA and SER endomembranes, the endomembranes, too, could display actin-anchoring capacity. Our experiments strongly support this possibility. Hence, SA or SER may be confined to a particular position within the spine actin matrix spanning from the plasma membrane to the spine apparatus. This position could change during reorganization induced in the actin cytomatrix by stimulation. Because the spine stalk is such an extremely narrow passage connecting the spine head and the dendrite (Fig. 1), it represents a locus of increased electrical resistance (Rall and Rinzel, 1971a,b) and a locus restricting free exchange of molecules and ions with the parent dendrite (Horwitz, 1984; Harris and Stevens, 1989). Both parameters could be modified if redistribution of the SA were to occur after synaptic activation. Collectively, these factors could affect synaptic potentials generated by the spine.

In summary, we have established the existence and distribution of actin filaments, myosin, MAP 2, and spectrin in dendritic spines. The behavior of these proteins in the brain *in vivo* is at present unknown. As was already mentioned, the actin network and its interaction with other cellular components is regulated by the action of actin-associated proteins. Furthermore, the interactions of actin-associated proteins with actin are also regulated, and this regulation is mediated by several factors: ionic concentration, pH, presence of another actin-associated protein, phosphoinositides, Ca^{2+} concentration, and phosphorylation. We will next review some of the factors and conditions that may influence the arrangement of actin in network dendritic spines.

Caldesmon cross-links actin filaments; however, at micromolar concentrations of Ca^{2+}, it binds to calmodulin, which inhibits its actin cross-linking activity (Kakiuchi and Sobue, 1983). Alpha actinin binds and cross-links actin filaments into a side-by-side configuration; the brain alpha-actinin seems to be Ca^{2+}-sensitive (Duhaiman and Bamburg, 1984). Bundling of actin filaments is also the function of low molecular weight proteins, such as *tau* (Griffith and Pollard, 1982; Selden and Pollard, 1983) and a 53 kDa protein (Maekawa *et al.,* 1983, 1988). *In vitro* the nerve growth factor (NGF) forms well-ordered actin bundles. It is interesting that only the biologically active form of NGF is endowed with this property (Castellani and O'Brien, 1981).

VI. Physiological Significance of Actin-Associated Proteins

The question of physiological relevance of actin-associated proteins has been repeatedly raised. Specific effects of actin-associated proteins were gathered from *in vitro* experiments where the reaction of actin with one or two associated proteins is tested. *In vivo,* however, actin may be exposed simultaneously to a number of factors that may cooperate or compete with each other, thus affecting the actin network in a way that may not always be predicted from the *in vitro* studies (Geiger, 1982). Therefore, caution has to be exercised when extrapolating results from *in vitro* to *in vivo* conditions.

The first insight into the physiological significance of cytoskeletal proteins in the brain was gained through the work of Aoki and Siekevitz (1985). These authors have shown that relative levels of phosphorylation change the biological function of MAP 2. They have postulated that a phosphorylated MAP 2 during critical period keeps the cytoskeletal architecture of the visual cortex primed to change its organization in response to light-evoked neuronal activity. A decline in synaptic plasticity marks the end of the critical period, and it is accompanied by progressive dephosphorylation of MAP 2 (Aoki and Siekevitz, 1985). A rapid 70% decrease in the phosphorylation of MAP 2 can be induced by NMDA application to hippocampal slices (Halpain and Greengard, 1990). However, the importance of this change for hippocampal physiology is not clear at present.

VII. Long-Term Potentiation: A Model of Synaptic Plasticity

Experiments on cytoskeletal proteins were undertaken to understand the mechanism of morphometric changes in dendritic spines that we have observed concurrently with long-term potentiation (LTP) in the dentate fascia (VanHarreveld and Fifková, 1975; Fifková and VanHarreveld, 1977; Fifková and Anderson, 1981; Fifková *et al.*, 1982b; Fifková, 1985c). LTP, first described by Bliss and Lømo (1973) and Bliss and Gardner-Medwin (1973), is currently considered to be the best physiological model of activity-dependent synaptic plasticity. Moreover, there is growing evidence that LTP may actually represent the cellular substrate for some types of learning and memory (Morris *et al.*, 1986; Sarvey, 1988a,b). Because the main afferent path, the perforant path (pp), to the dentate fascia terminates solely on dendritic spines in restricted zones of the dentate molecular layer (Nafstadt, 1967; Fifková, 1975), quantitative ultrastructural studies in this region are greatly facilitated. Stimulated spines show long-lasting enlargement of the spine head: cross-sectional area and perimeter (Fifková and VanHarreveld, 1977; Fifková *et al.*, 1982b), widening and shortening of the spine stalk (Fifková and Anderson, 1981), and increased length of synaptic appositions (Fifková, 1985c). Similar changes associated with LTP were subsequently described in other laboratories (Moshkov *et al.*, 1977, 1980; Petukhov and Popov, 1986; Desmond and Levy, 1983, 1986a,b; Wenzel *et al.*, 1985; Andersen *et al.*, 1988a,b). Morphometric changes were also observed in dendritic spines of a variety of species where they were associated with stimulation-induced enhancement of synaptic activity (Bradley and Horn, 1979; Brown and Horn, 1979; Coss and Globus, 1978; Coss *et al.*, 1980; Brandon and Coss, 1982; Burgess and Coss, 1982, 1983). Thus, the morphometric changes in stimulated spines may represent a more general mechanism of synaptic plasticity. The theoretical significance of these observations is that for the first time they brought attention to a modulatory effect of experimentally induced changes in spine geometry on synaptic activity. Because structural changes in spines are most likely preceded by molecular changes, clarification of the molecular events underlying these processes in the spines during increased activity may provide an extremely valuable tool in studies leading to an understanding of higher brain functions.

A. Postsynaptic Mechanism of LTP

Although the presynaptic and postsynaptic side are both involved in the mechanism of LTP (for reviews see Wickens, 1988; Madison *et al.*, 1991), here we consider only the postsynaptic mechanism.

1. Calcium

Depolarization of the postsynaptic membrane (Gustafsson and Wigström, 1986; Gustafsson *et al.*, 1987) and increased Ca^{2+} entry into dendritic spines via NMDA receptors, induced by repetitive activation of excitatory hippocampal synapses that release glutamate (Bliss *et al.*, 1988), were shown to be necessary to induce LTP and sufficient to potentiate synaptic transmission (Kauer *et al.*, 1988; Malenka *et al.*, 1988). Increased intraspinal Ca^{2+} seems to be critical for the induction of LTP, as well as for its maintenance. A specific NMDA receptor-dependent influx of Ca^{2+} into stimulated spines was observed using fura-2 injection (Müller and Connor, 1991). As has been already pointed out, the long-term increase in internal Ca^{2+} induced by glutamate may be due not only to the activation of the NMDA receptor but also to activation IP_3 receptors in the membranes of SA and SER. The latter possibility is supported by experiments in which dantrolene, a drug that blocks Ca^{2+} release from the sarcoplasmic reticulum (Mody *et al.*, 1988), blocks also LTP, even 2 hours after it was induced (Obenaus *et al.*, 1988).

The exclusive role of NMDA receptors in the mechanism of LTP has been recently challenged when it was shown that NMDA receptor antagonists do not block LTP induced in the termination site of the lateral pp in the dentate molecular layer (Bramham *et al.*, 1990), where there are few NMDA receptors (Monaghan and Cotman, 1985; Cotman *et al.*, 1987). Moreover, topical application of NMDA or glutamate will produce only short-lasting potentiation (Collingridge *et al.*, 1983; Kauer *et al.*, 1988), despite the Ca^{2+} influx through the NMDA ionophore. Adding to this challenge are observations indicating that LTP can be prevented by pretreatment with pertussis toxin (PTX) (Goh and Pennefather, 1989, 1990) to which NMDA receptors are not sensitive. Taken together, these observations suggest an additional mechanism for LTP. This mechanism could be mediated through the PTX-sensitive ACPD receptor. Glutamate binding to the ACDP receptor activates G-proteins and phospholipase C and thus triggers a phosphoinositide cascade in the membrane, which leads to the release of internal Ca^{2+}, production of DAG, and activation of the PKC. This notion is supported by the observation suggesting that high frequency stimulation that is inducing LTP is also inducing a sustained increase in phosphoinositide hydrolysis (Bonner *et al.*, 1991) and that LTP can be blocked by phospholipase inhibitors (Okada *et al.*, 1989).

2. Protein Kinases

Increased levels of Ca^{2+} will activate Ca^{2+}-dependent protein kinases, which are capable of mediating different long-lasting modifications in the cell (Schwartz and Greenberg, 1987; Linden and Routtenberg, 1989). Because number of protein kinase blockers were shown to suppress LTP, kinase activity induced by the increase of Ca^{2+} in the dendritic spine is thought to be critically involved in the mechanism of LTP (Malinow *et al.*, 1988). Two kinases, PKC and CAMK II, are considered to be prime candidates in the generation and maintenance of LTP (Lisman, 1985; Miller and Kennedy, 1986; Reymann *et al.*, 1988a,b; Malenka *et al.*, 1989). CAMK II and PKC are both components of the PSD (Wolf *et al.*, 1986; Kennedy *et al.*, 1983) and both are also present in dendritic spines (CAMK II; Ouimet *et al.*, 1984; PKC; Kose *et al.*, 1990). Similarly, calmodulin, which is necessary for CAMK II activation, is also found in spines (Cáceres *et al.*, 1983). CAMK II has the interesting property that following Ca^{2+} activation it phosphorylates not only its substrates but also itself at specific residues. After completion of the autophosphorylation, the kinase switches to a new state in which it has substantial catalytic activity even in the absence of Ca^{2+}. This means that CAMK II may retain information about a previous activating Ca^{2+} signal long after the Ca^{2+} concentration has returned to resting levels (Kennedy, 1989). This property of CAMK II may be important for LTP. Similarly, a novel, persistently active cytosolic form of PKC found in the hippocampus may be of importance because elevated activity of this PKC is associated with the maintenance of LTP

(Klann *et al.*, 1991). In addition to CAMK II and PKC, the presence of tyrosine- and serine-threonine kinases was shown to be required postsynaptically for LTP induction (O'Dell *et al.*, 1991a). The physiological relevance of these kinases can be revealed only through the action of their respective substrate proteins.

3. Substrates

Substrate proteins of CAMK II and PKC are among other actin–cross-linking proteins: myosin, MAP 2, and spectrin. Myosin light chain (MLC), the regulatory protein of myosin, is a substrate for CAMK II (Tanaka *et al.*, 1986) and PKC (Klann *et al.*, 1991); MAP 2 is a substrate for CAMK II (Yamauchi and Fujisawa, 1988) and for PKC (Nishizuka, 1986; Kikkawa *et al.*, 1988); spectrin is a substrate for PKC (Gregorio and Repasky, 1990). The three actin cross-linking proteins are also substrates for other kinases [MLC for MLC kinase (MLCK), MAP 2 for cAMP kinase (Vallee *et al.*, 1981), spectrin for MLCK]. Involvement of the latter group of kinases in LTP has so far not been studied. In addition to protein kinases, protein phosphatases, such as calcineurin, were shown to be activated by the Ca^{2+}-calmodulin complex during NMDA stimulation causing dephosphorylation of MAP 2 (Halpain and Greengard, 1990). In sum, activation of these enzymes and the ensuing phosphorylation or dephosphorylation of myosin, MAP 2, and spectrin could affect binding properties of these proteins with actin and thus change the dynamics of the actin matrix in the spine. By modifying organization of the actin network and its contractile activity, these actin-associated proteins could contribute to the morphometric changes observed in stimulated spines (Fifková and Morales, 1989; Morales and Fifková, 1989a,b). So far, the sequence of events that may occur in the actin matrix during LTP is not known; however, given comparable conditions in activated nonneural cells, there seems to be little doubt that the actin network of the spine will be modified during LTP.

B. Organization of the Actin Cytomatrix

1. Role of Phosphorylation

Increased concentration of free cytoplasmic Ca^{2+} in spines may affect levels of phosphorylation of the actin-associated proteins through the activation of kinases and phosphatases. Myosin is one of the proteins that regulates assembly of the actin network. CAMK II, which is present in spines, and MLC kinase are capable of phosphorylating the MLC. This phosphorylation is required for actin–myosin interactions and for actin-induced activation of myosin ATPase. The conformation of myosin molecules and their assembly into filaments changes with phosphorylation. Phosphorylated myosin filaments can cross-link and organize actin filaments, whereas a folded dephosphorylated myosin molecule does not interact with actin and thus is not able to maintain actin cytoskeletal structure. This has been shown in experiments in which dephosphorylation of the regulatory MLC_{20} by Type I protein phosphatase caused a complete disassembly of actin filaments (Fernandez *et al.*, 1990). MLC_{20} is phosphorylated by Ca^{2+}-calmodulin-activated MLCK. When the synthesis of MLCK is blocked (Shoemaker *et al.*, 1990) or its activity suppressed by specific antibodies (Lamb *et al.*, 1988), a cell rounding has occurred, indicating that the cell shape supported by the actin cytoskeleton has been altered by myosin dephosphorylation. In instances in which the ACDP receptors induce hydrolysis of PIP_2 in the spine membrane, products of this hydrolysis, IP_3 and DAG, could also alter the actin network. IP_3 mobilizes Ca^{2+} from the SA or spine SER and DAG activates PKC, which is also present in spines. *In vitro* studies have shown that PKC and CAMK II, by phosphorylating MAP 2, inhibit MAP 2-actin cross-linking activity. In addition to myosin phosphorylation, contraction of the actin matrix can be induced also by solation. Stiff actin gel resists contractile forces of myosin, but the solating gel cannot (Janson *et al.*, 1991). Thus, the structure and organization of the gel network may affect myosin's ability to cause contraction.

2. Role of Actin Polymerization–Depolymerization

Two actin-associated proteins, which affect actin filament assembly, profilin and gelsolin, were shown to be regulated by membrane phospholipids. Stossel (1989) has used this observation for a model that would explain the regulation of filament organization by transmembrane signals. When PIP_2 is hydrolyzed, profilin is dissociated from this membrane phospholipid (Lassing and Lindberg, 1985, 1988). Profilin is then free to bind to actin monomers, thus preventing actin filament polymerization. The increase of intracellular Ca^{2+}, mediated through IP_3 receptors in the endomembranes of SA, activates gelsolin, which shortens actin filaments by severing and blocking the fast-growing end. These activities promote disassembly of actin filaments. When PIP_2 is restored in the membrane and Ca^{2+} levels in the cytoplasm return to prestimulation levels, gelsolin is dissociated from the ends of F-actin, thereby generating free filament ends that act as nuclei for rapid assembly of monomeric actin into filaments (Cunningham *et al.*, 1991). Thus, both profilin and gelsolin may represent links through which receptor-activated membrane phosphoinositides regulate modifications of the actin cytomatrix. Direct evidence showing that actin polymerization alone can generate morphological changes was provided by experiments in which, in the absence of motor molecules, phospholipid vesicles loaded with G-actin became deformed into irregular shapes when actin was polymerized into filaments (Cortese *et al.,* 1989). These examples drawn from nonneural tissue clearly indicate that various conditions of stimulation can induce changes in the actin matrix with subsequent morphological changes in the cell. Thus, by analogy, similar changes may be expected to occur in stimulated dendritic spines. The importance of morphological changes of dendritic spines for LTP is, however, a debated issue and will be addressed in Section VII,C.

An alternate mechanism has been proposed for the postsynaptic mechanism of LTP, which is based on the assumption that membrane-associated, calcium-activated neutral protease calpain hydrolyzes in spines spectrin, the main protein of the submembrane cytoskeleton. This was thought to result in remodeling of the postsynaptic membrane so that previously latent glutamate receptors are exposed and the membrane is sensitized to the subsequent stimuli (Lynch and Baudry, 1984). However, the glutamate binding measured in these experiments actually represents glutamate transport into vesicles of the membrane preparations (Pin *et al.,* 1984). Thus, there is no evidence that calcium-dependent proteolysis unmasks latent glutamate receptors (Mellgren, 1987). Moreover, calpain was reported to be absent from spines in the hippocampus (Hamabuko *et al.,* 1986). It has been shown that the effect of calpain on spectrin can be blocked by leupeptin. Because induction of LTP can be disrupted by chronic administration of leupeptin, this was taken as a support for the calpain–spectrin hypothesis of LTP (Staubli *et al.,* 1988). However, leupeptin may affect axon terminals. The main function of calpain in axon terminals is to keep the neurofilaments dissociated. Chronic suppression of calpain activity by leupeptin will fill up the terminals with neurofilaments a state that resembles fibrillar degeneration (Roots, 1983). Leupeptin-affected terminals could still respond to baseline stimulation (Staubli *et al.,* 1988); however, they might not be capable to respond to the tetanic stimulus necessary to induce LTP. Thus, the leupeptin induced block of LTP (Staubli *et al.,* 1988) suggests a block at the presynaptic rather than postsynaptic side. Therefore, it would appear that the calpain theory of LTP requires a thorough revision.

Recently, three articles have appeared in close succession, all of which are implicating nitric oxide (NO) in the mechanism of LTP (Böhme *et al.,* 1991; O'Dell *et al.,* 1991b; Schuman and Madison, 1991). This mechanism is based on an analogy with the mechanism operating in the vascular system. NO has been originally described as a diffusible messenger synthesized in endothelial cells (endothelium-derived relaxing factor; Palmer *et al.,* 1987; for review see Furchgott and Vanhoutte, 1989) which activates soluble guanylate cyclase. This results in an increase of cGMP in vascular smooth muscles, and this increase precedes the relaxation of the vascular musculature. NO constitutes a widespread transduction mechanism for soluble guanylate cyclase (Knowles *et al.*, 1989), which catalyzes the formation of cGMP from GTP. Cyclic GMP induces smooth muscle relaxation by dephosphorylating the MLC and by phosphorylating several other low-molecular weight proteins (Rapoport *et al.*, 1983). Because NO is produced in endothelial cells and because the smooth muscle fibers of the vessel wall are confined to the overlying pericytes, NO acts as an intercellular messenger reaching its target by diffusion. This is evidenced by the relaxation-blocking effect of hemoglobin, which binds avidly NO. This deactivation occurs in the extracellular space as the large hemoglobin molecule does not enter the cell (Martin *et al.*, 1985; Palmer *et al.*, 1987).

In the nervous system glutamate was shown to increase cGMP levels (Drummond, 1983) in a reaction that also involves cell-to-cell interactions (Garthwaite and Garthwaite, 1987; Garthwaite *et al.*, 1988). In stimulated neurons, glutamate induces NO formation through NMDA receptors in a calcium-(Garthwaite *et al.*, 1988) and calmodulin-(Bredt and Snyder, 1990) dependent manner. NO activity then accounts for the cGMP (Garthwaite *et al.*, 1988) and ADP ribosyltransferase responses (O'Dell *et al.*, 1991b) associated with NMDA receptor activation. Garthwaite *et al.* (1988) suggested that in the nervous system NO released from the stimulated cells may act as a retrograde messenger that would be involved in synaptic modifications and in the mechanism of LTP. Subsequently, this suggestion became validated when NO synthetase (NOS) was isolated and localized in neurons of several brain regions (Bredt *et al.*, 1990, 1991).

NOS activity in the postsynaptic neuron was established as a critical factor in the generation of LTP. Postsynaptic application of an NOS inhibitor N^{G}-methyl-L-arginine (L-Me-Arg) blocks the reaction, whereas its enantiomer D-Me-Arg, which does not inhibit NOS activity, also does not suppress LTP (Schuman and Madison, 1991). However, according to O'Dell *et al.* (1991b) and Schuman and Madison (1991), NO is not acting in the postsynaptic neuron. Rather, in analogy with the vascular system, NO is thought to diffuse out of the cell and to reach through the extracellular space axon terminals of the stimulated neuron where it activates the guanylate cyclase and ADP ribosyltransferase (Garthwaite *et al.*, 1988; O'Dell *et al.*, 1991b). These two enzymes are thought to be involved in an increase in transmitter release, which by some is considered to be the mechanism of LTP (Bliss *et al.*, 1986; Errington *et al.*, 1987; Bekkers and Stevens, 1990; Malinow and Tsien, 1990). In order to prove that, advantage has been taken of hemoglobin-mediated inactivation of NO. The hemoglobin-induced block of LTP (O'Dell *et al.*, 1991b; Schuman and Madison, 1991) was taken as evidence that NO was prevented from reaching the presynaptic terminals, thus confirming the presynaptic mechanism of LTP (O'Dell *et al.*, 1991b; Schuman and Madison, 1991). However, there is an alternate interpretation of the blocking effect of hemoglobin on LTP. Hemoglobin, by binding NO in the extracellular space, may increase the concentration gradient of NO across the cell membrane and so increase the net rate of diffusion, thus depleting the stimulated postsynaptic neuron of NO (Martin *et al.*, 1985). The ultimate target of NO action is the MLC, which becomes dephosphorylated by cGMP. The dephosphorylated MLC, which in the vascular musculature causes relaxation of muscle tension, could in neurons contribute to the relaxation of tension of the actin matrix with consequent morphometric spine changes. Depletion of the postsynaptic neuron of NO would reduce the chance that MLC will be dephosphorylated and the actin matrix affected. In line with the postsynaptic mechanism of NO, action is the observed block of LTP by NOS inhibitors in the postsynaptic neuron (Schuman and Madison, 1991). In addition, the essential difference in the organization of the blood vessel wall and the neuropile makes it unlikely that NO can act as a specific retrograde messenger in the nervous system. In blood vessels endothelial cells are adjacent to the smooth muscle-carrying pericytes; therefore, NO can directly diffuse from the site of origin to the target cells. However, in the nervous system the organization of individual elements is less structured and NO can freely diffuse into the extracellular space. Therefore, it is rather unlikely that NO would retrogradely reach only those axon terminals that contact the activated neuron. Rather, all profiles equidistant to the NO-producing neuron could be affected. Thus, the data presented in the literature do not unequivocally favor the presynaptic action of NO. NO could equally well induce the production of cGMP in the postsynaptic neuron and cause dephosphorylation of the MLC with the consequent relaxation of the actin matrix in dendritic spines.

C. Changes in Dendritic Spines

Several types of morphological changes were observed in association with LTP. One group of investigators found in the dentate fascia and CA_3 an enlargement of dendritic spines and increase in the synaptic length (VanHarreveld and Fifková, 1975; Fifková and VanHarreveld, 1977; Fifková and Anderson, 1981; Fifková *et al.*, 1982a,b; Fifková, 1985c; Moshkov *et al.*, 1977, 1980; Desmond and Levy, 1983, 1986a,b; Petukhov and Popov, 1986; Wenzel *et al.*, 1985; Andersen *et al.*, 1988a,b), whereas the other group observed in CA_1 an increased density of synaptic contacts (Lee *et al.*, 1980; Chang and Greenough, 1984). Lee *et al.* (1980) have observed, in addition, a "rounding up" of dendritic spines without an increase of the cross-sectional area. Given that the "rounding up" of spines actually represents a change from an elliptical to a circular shape (as seen in the electron micrographs), a 33% increase in the cross-sectional area should result. In the absence of such an increase, a shrinkage of the spine head must have occurred. Given morphological changes that occur concurrently with LTP, it is now important to consider the extent to which the observed alteration of spine morphology could account for the increase in synaptic efficacy during LTP.

D. Modeling of Spine Function

Because of their small size, spines have remained inaccessible to direct electrophysiological investigations. Therefore, for now the functional analysis of spines must be based on biophysical models. It has been recognized by Chang (1952) that the electrical resistance of the thin spine stalk could attenuate the synaptic response generated by the spine head in the postsynaptic neuron. Rall and Rinzel (1971a,b) introduced the idea that values of the spine stalk resistance could change the efficacy of an axospinous synapse. If a passive membrane (i.e., without voltage-sensitive ion channels) is assumed for the spine head, then the voltage at the spine base (i.e., in the parent dendrite) will be half of that at the spine head provided that the spine stalk resistance equals that of the dendritic branch resistance. If an active membrane (i.e., with voltage-sensitive ion channels) for the spine head is assumed, an action potential generated at the spine head could result in amplification of the synaptic input at the parent dendrite (Jack *et al.,* 1975; Miller *et al.,* 1985; Perkel and Perkel, 1985). Thus, within an optimal range of resistance values, the spine stalk has emerged as an important locus for changes in synaptic efficacy (Segev and Rall, 1988; Rall and Segev, 1988).

Excitable spines are capable of mutual interactions. The synaptic response of an active spine increases considerably the amplitude of the excitatory postsynaptic potential (EPSP) generated. This may subsequently increase the amount of current that flows into the parent dendritic branch. The current that spreads from the stimulated spine may activate sequentially neighboring excitable spines by bringing them to their respective thresholds of activation, thus generating a saltatory propagation of an action potential in the dendritic branch (Shepherd *et al.,* 1985). Because excitable spines generate such large potentials, they constitute probably only a subpopulation of the total spine population in the region. They will be most efficient when located in distal parts of the dendritic tree (Shepherd and Brayton, 1987). Triggering of the response in the parent dendrite and the spine-to-spine impulse communication is critically dependent on morphological and physiological parameters of the spine stalk, namely, its dimensions and electrical resistance, respectively (Rall, 1970, 1974, 1978; Rall and Rinzel, 1971a,b; Koch and Poggio, 1983; Wilson, 1984; Coss and Perkel, 1984; Shepherd *et al.*, 1985; Shepherd and Brayton, 1987). Changes in the stalk resistance could result from decreasing the stalk diameter, increasing the stalk length, or partially occluding the stalk with membranes of the SA (Miller *et al.*, 1985).

The extent to which the charge transferred from the spine head to the parent dentrite will be attenuated by the spine stalk will also be determined by the magnitude of the synaptic conductance. It has been calculated that if the snyaptic conductance were 5 nS, then 33% of the spine stalks (in the CA_1 region) are sufficiently thin to reduce the charge to the parent dendrite by more than 10% (Harris and Stevens, 1989). It has been argued that when current injection is used to model synaptic activation rather than a change in synaptic conductance, no current attenuation is seen in spines (Kawato and Tsukahara, 1983; Turner and Schwartzkroin, 1983). Because, however, this manipulation does not take into account the reversal potential of the synapse, it cannot be considered as an appropriate treatment (Koch and Poggio, 1985).

The spine stalk, besides providing an electrical resistance to the current flow, also provides a diffusional resistance to the flow of ions and molecules. During synaptic activation of a spine, a large electrochemical gradient is generated that would attract charged molecules into the spine head from the dendrite (Horwitz, 1984) and control the diffusion of molecules in and out of the spine (Shepherd, 1979). By restricting the flow of molecules, Ca^{2+} in particular, the spine stalk may effectively isolate the spine head and thus provide a localized environment in which reactions specific to a particular synapse would occur (Holmes, 1990). Therefore, a large transient increase of Ca^{2+} might be attained in the spine head, according to the model of Gamble and Koch (1987) and Zador *et al.* (1990). However, this increase will be restricted to spines with long, thin stalks (Holmes, 1990). Thus, in addition to the electrical resistance, its diffusional resistance may also be a key function of the spine stalk (Holmes, 1990). As has been already pointed out, the conclusions discussed in preceding paragraphs were arrived at by using modeling procedures. The only parameter so far experimentally verified pertains to levels of Ca^{2+} in dendritic spines.

E. GABA-ergic Input to Dendritic Spines

Glutamate has been so far the only transmitter considered in the mechanism of LTP in the hippocampus and dentate fascia. In the dentate fascia

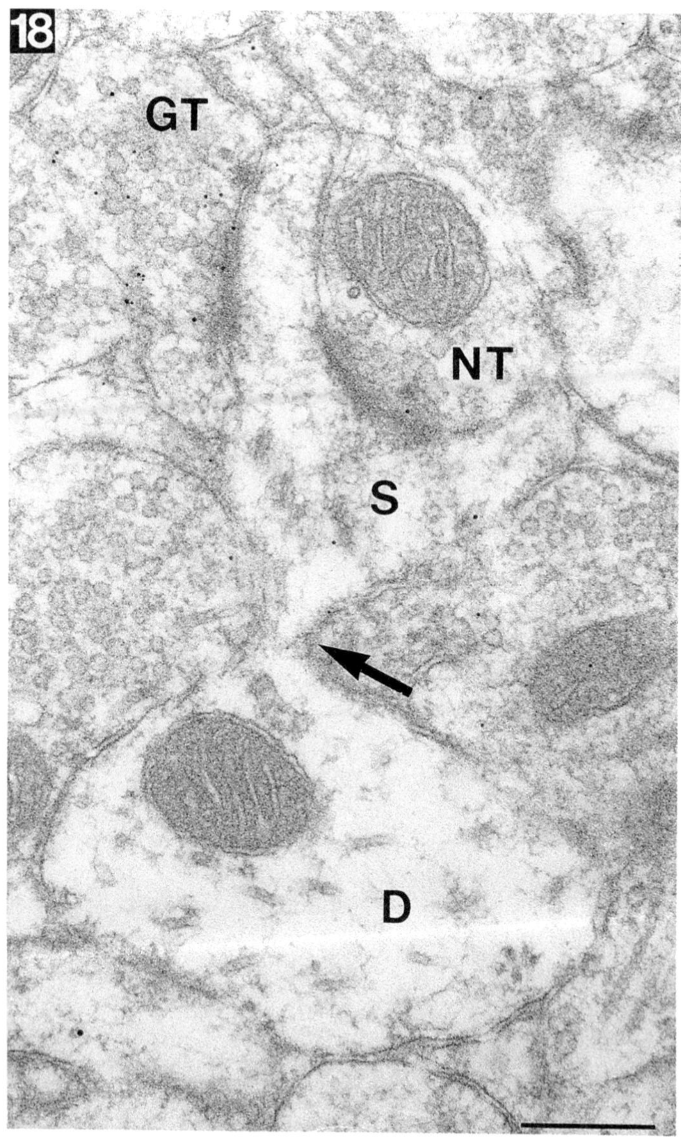

FIG. 18 Cup-shaped dendritic spine (S) in the distal third of the dentate molecular layer is associated with parent dendrite (D) by a stalk (arrow). It is contacted by an asymmetrical non-GABA-ergic terminal (NT), which is positioned in the "cup" formed by the spine. A GABA-ergic terminal (GT) forms a symmetrical contact on the side of the spine head. The anti-GABA antibodies (Chemicon, Temecula, California) are visualized by secondary antibody bound to 5 nm colloidal gold. Bar, 0.25μm. (From Fifková *et al.*, 1992. Reproduced with permission.)

[in the distal two-thirds of the dentate molecular layer (DML)], however, we have recently observed that on some spines GABA-ergic symmetrical synapses (presumed inhibitory) are co-localized with the asymmetrical excitatory glutamatergic synapses. Invariably, the asymmetrical synapse occupies the center of the spine head, whereas the GABA-ergic synapse is either on the lateral aspect of the spine head or on the spine stalk (Fifková *et al.*, 1990, 1992) (Figs. 18–20). The function of inhibitory contacts on dendritic spines is at present not fully clarified; however, it could be related to the excitability properties of the spine plasma membrane. As has been already mentioned, Segev and Rall (1988) have postulated that spines could have either passive or active membrane properties. The importance of this distinction follows from the fact that active spines could amplify their synaptic input to dendrites, whereas passive spines cannot (Segev and Rall, 1988). Moreover, active spines are postulated to represent a rather small proportion of the total spine population of a neuron with prevalent distribution in the distal segments of the dendritic tree. The latter two properties coincide with properties of those spines that carry GABA-ergic contacts in the DML. Thus, GABA synapses could be contacting prevailingly active spines (Fifková *et al.*, 1990, 1992). In this context it is important that induction of LTP is greatly facilitated by blocking the GABA by antagonists, such as bicuculline or picrotoxin (Wigström and Gustafsson, 1983, 1985; Douglas and Dempster, 1987; Tomasulo *et al.*, 1990). Thus, the GABA input to spines in the DML may regulate the induction of LTP.

VIII. Conclusion

This review was aimed at outlining the importance of the spine cytoskeleton and its modifications for the mechanism of synaptic plasticity in general and LTP in particular. So far we can only speculate on the role of the actin cytoskeleton in these phenomena as experimental evidence is slowly forthcoming. However, several recent observations point to the plausibility of our hypothesis. Central to this hypothesis is modification of the actin network in dendritic spines. Two mechanisms, mutually not exclusive, may participate in this modification. One implicates modification in cross-linking of actin filaments and the other depolymerization–repolymerization of actin. Both mechanisms require an increase of intraspinal Ca^{2+} that is provided by extracellular and intracellular sources via the NMDA ionophore and IP_3 receptors (of SA or SER), respectively. The depolymerization–repolymerization of the actin is linked to the action of actin-regulatory proteins, profilin and gelsolin, which are associated with the signal-

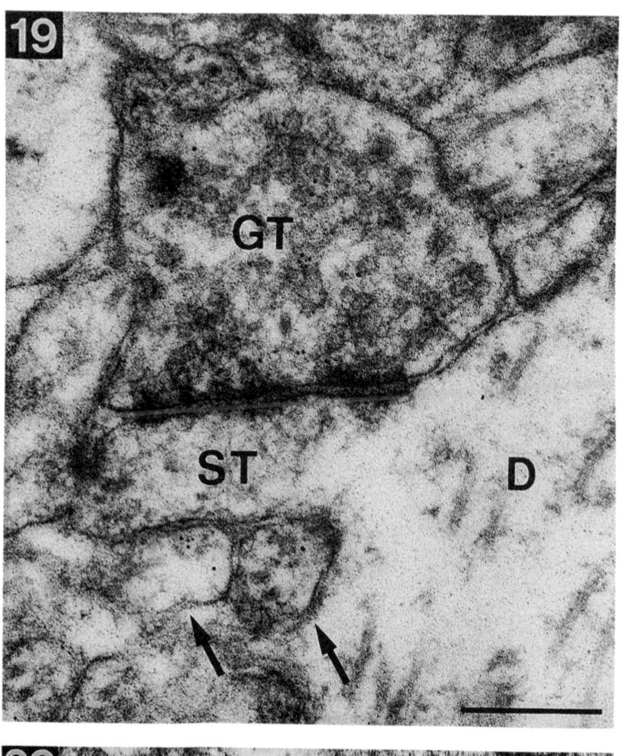
19
GT
ST
D

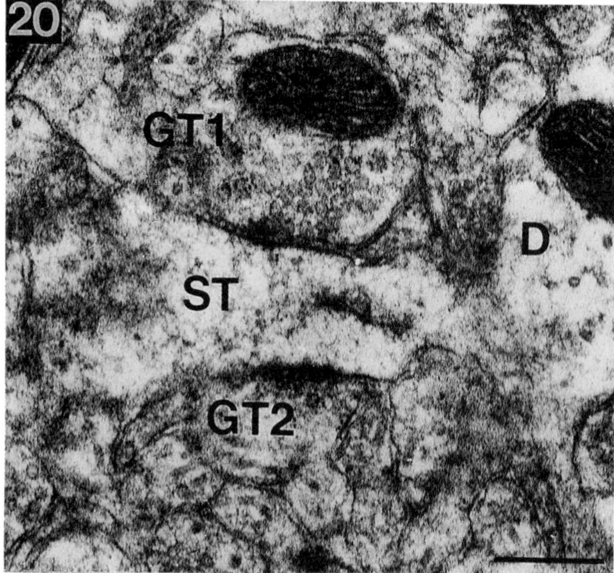
20
GT1
D
ST
GT2

transducing phosphoinositides (PIP_2). The cross-linking of actin filaments is the function of actin cross-linking proteins, three of which were studied in detail in dendritic spines (myosin, MAP 2, and spectrin). These cross-linkers are phosphorylated through their respective kinases (PKC and CAMK II), which were shown to be involved in the maintenance of LTP. The effect of phosphorylation of myosin, MAP 2, and spectrin in spines, or for that matter in the brain, is not known. However, if it were similar to the effect of phosphorylation *in vitro,* it would change the binding activity of these proteins to actin. If so, it could induce rearrangements of the actin network and movements of the cytoplasm within the spine itself and between the spine and its parent dendrite. These modifications could result in morphometric changes, which were observed by us and others in spines during increased synaptic activity and LTP.

The concerted action of actin-associaed proteins on the spine actin cytomatrix may endow dendritic spines with dynamic properties necessary for synaptic plasticity. Research along these lines should contribute to the understanding of the molecular events that determine the structural changes in dendritic spines and to the understanding of the effect these structural modifications could have on the function of the neuron.

Acknowledgments

This investigation was supported in part by NIA grant AG04804, NIAAA grant AA06196, and by NIMH grant MH41834 to Eva Fifková. The authors express their sincere thanks to Drs. L. I. Binder, University of Alabama and A. Frankfurter, University of Virginia, for the generous gift of a monoclonal anti-MAP 2 antibody (IgG, clone AP 14), to Dr. J. Scholey, National Jewish Center for Immunology and Respiratory Medicine in Denver, Colorado, for the generous gift of the polyclonal human antiplatelet myosin antibodies, and B. Peterson for typing the manuscript.

References

Adams, M. E., Minamide, L. S., Duester, G., and Bamburg, J. R. (1990). *Biochemistry* **29,** 7414–7420.

FIG. 19 GABA-ergic terminal (GT) in the distal third of the dentate molecular layer is making a symmetrical contact on a spine stalk (ST) emerging from a dendrite (D). Two unmyelinated GABA-ergic axons are indicated with arrows. Bar, 0.25μm.

FIG. 20 Two GABA-ergic terminals (GT_1 and GT_2) are making symmetrical contacts on a spine stalk (ST) emerging from dendrite (D). Bar, 0.25 μm. (From Fifková *et al.*, 1992. Reproduced with permission.)

Akiyama, T., Nishida, E., Ishida, J., Saji, N., Ogawara, N., Hoshi, M., Miyata, Y., and Sakai, H. (1986). *J. Biol. Chem.* **261,** 15648–15651.

Andersen, P., Blackstad, T., Hulleberg, G., Vaaland, J. L., and Trommald, M. (1988a). *Proc Physiol. Soc. S* p. 3.

Andersen, P., Blackstad, T., Hulleberg, G., Trommald, M., and Vaaland, J. L. (1988b). *Proc. Physiol. Soc. P.C.* p. 50.

Aoki, C., and Siekevitz, P. (1985). *J. Neurosci.* **5,** 2465–2483.

Aoki, C., and Siekevitz, P. (1988). *Sci. Am.* **259,** 56–64.

Bamburg, J. R., Harris, H. E., and Weeds, A. G. (1980). *FEBS Lett.* **121,** 178–182.

Barylko, B., and Sobieszek, A. (1983). *EMBO J.* **2,** 369–374.

Barylko, B., Tooth, P., and Kendrick-Jones, J. (1986). *Eur. J. Biochem.* **158,** 271–282.

Bekkers, J. M., and Stevens, C. F. (1990). *Nature (London)* **346,** 724–729.

Berl, S., Puszkin, S., and Nicklas, W. J. (1973). *Science* **179,** 441–446.

Berl, S., Chou, M., and Mytilineau, C. (1983). *J. Neurochem.* **40,** 1397–1405.

Bernhardt, R., and Matus, A. (1982). *J. Cell Biol.* **92,** 589–593.

Bernhardt, R., and Matus, A. (1984). *J. Comp. Neurol.* **226,** 203–221.

Binder, L. I., Frankfurter, A., and Rebhun, L. I. (1986). *Ann. N.Y. Acad. Sci.* **466,** 145–166.

Bliss, T. V. P., and Lømo, T. (1973). *J. Physiol. (London)* **232,** 331–356.

Bliss, T. V. P., and Gardner-Medwin, A. R. (1973). *J. Physiol. (London)* **232,** 357–374.

Bliss, T. V. P., Douglas, R. M., Errington, M. L., and Lynch, M. A. (1986). *J. Physiol. (London)* **377,** 391–408.

Bliss, T. V. P., Clements, M. P., Errington, M. L., Lynch, M. A., and Williams, J. H. (1988). *Soc. Neurosci. Abstr.* **14,** Part 1, 564.

Böhme, G. A., Bon, C., Stutzmann, J.-M., Doble, A., and Blanchard, J.-C. (1991). *Eur. J. Pharmacol.* **199,** 379–381.

Bonner, M. J., Burgard, E. C., and Sarvey, J. M. (1991). *Soc. Neurosci. Abstr.* **17,** 383.

Bradley, P., and Horn, G. (1979). *Brain Res.* **162,** 148–153.

Bramham, C. R., Milgram, N. W., and Srebro, B. (1990). *Soc. Neurosci. Abstr.* **16,** 981.

Brandon, J. G., and Coss, R. G. (1982). *Brain Res.* **252,** 51–61.

Bredt, D. S., and Snyder, S. H. (1990). *Proc. Natl. Acad. Sci. U.S.A.* **87,** 682–685.

Bredt, D. S., Hwang, P. M., and Snyder, S. H. (1990). *Nature (London)* **347,** 768–770.

Bredt, D. S., Hwang, P. M., Glatt, C. E., Lowenstein, C., Reed, R. R., and Snyder, S. H. (1991). *Nature (London)* **351,** 714–718.

Brown, M. W., and Horn, G. (1979). *Brain Res.* **162,** 142–147.

Brotschi, E. A., Hartwig, J. H., and Stossel, T. P. (1978). *J. Biol. Chem.* **253,** 8988–8993.

Burgess, J. W., and Coss, R. G. (1982). *Dev. Psychobiol.* **15,** 461–470.

Burgess, J. W., and Coss, R. G. (1983). *Brain Dev.* **266,** 217–233.

Burgoyne, R. D., Gray E. G., and Barron, J. (1983). *J. Anat.* **136,** 634–635.

Burns, R. G., and Islam, K. (1984). *Eur. J. Biochem.* **141,** 599–608.

Burridge, K., and Bray, D. (1975). *J. Mol. Biol.* **99,** 1–13.

Cáceres, B., Bender, P., Snavely, L., Rebhun, L. I., and Steward, O. (1983). *Neuroscience* **10,** 449–461.

Cáceres, A., Binder, L. I., Payne, M. R., Bender, P., Rebhun, L. I., and Steward, O. (1984). *J. Neurosci.* **4,** 394–410.

Castellani, L., and O'Brien, E. (1981). *J. Mol. Biol.* **147,** 205–213.

Chang, F. L. F., and Greenough, W. T. (1984). *Brain Res.* **309,** 35–46.

Chang, H. T. (1952). *Cold Spring Harbor Symp. Quant. Biol.* **17,** 189–202.

Cohen, R. S., Chung, S. K., and Pfaff, D. W. (1985). *Cell. Mol. Neurobiol.* **5,** 271–284.

Collingridge, G. L., Kehl, S. J., and McLennan, H. (1983). *J. Physiol. (London)* **334,** 33–46.

Condeelis, J. (1983). *In* "Spatial Organization of Eukaryotic Cells. Modern Cell Biology" (J. R. McIntosh, ed.), Vol. 2, pp. 225–240. Alan R. Liss, New York.

Connolly, J. A. (1984). *J. Cell Biol.* **99,** 148–154.
Connolly, J. A., and Graham, A. J. (1985). *Eur. J. Cell Biol.* **37,** 191–195.
Cortese, J. D., Schwab, B., Frieden, C., and Elson, E. L. (1989). *Proc. Natl. Acad. Sci. U.S.A.* **86,** 5773–5777.
Coss, R. G., and Globus, A. (1978). *Science* **200,** 787–790.
Coss, R. G., and Perkel, D. H. (1985). *Behav. Neurol.* **44,** 151–185.
Coss, R. G., Brandon, J. G., and Globus, A. (1980). *Brain Res.* **192,** 49–59.
Cotman, C. W., Monaghan, D. T., Ottersen, O. P., and Storm-Mathisen, J. (1987). *Trends Neurosci.* **10,** 273–280.
Craig, S. W., and Pollard, T. D. (1982). *Trends Biochem. Sci.* **7,** 88–92.
Crick, F. (1982). *Trends Neurosci.* **5,** 44–46.
Cunningham, C. C., Stossel, T. P., and Kwiatkowski, D. J. (1991). *Science* **251,** 1233–1236.
DeCamilli, P., Miller, P. E., Levitt, P., Walter, U., and Greengard, P. (1984). *Neuroscience* **11,** 819–846.
Desmond, N. L., and Levy, W. B. (1983). *Brain Res.* **265,** 21–30.
Desmond, N. L., and Levy, W. B. (1986a). *J. Comp. Neurol.* **253,** 466–475.
Desmond, N. L., and Levy, W. B. (1986b). *J. Comp. Neurol.* **253,** 476–482.
Douglas, R. M., and Dempster, A. (1987). *Soc. Neurosci. Abstr.* **13,** 1664.
Drummond, G. I. (1983). *Adv. Cyclic Nucleotide Res.* **15,** 373–494.
Duhaiman, A. S., and Bamburg, J. R. (1984). *Biochemistry* **23,** 1600–1608.
Errington, M. L., Lynch, M. A., and Bliss, T. V. P. (1987). *Neuroscience* **20,** 279–284.
Escolar, C., Krumwide, M., and White, J. G. (1986). *Am. J. Pathol.* **123,** 86–94.
Fernandez, A., Brautigan, D. L., Mumby, M., and Lamb, N. J. C. (1990). *J. Cell Biol.* **111,** 103–112.
Fifková, E. (1975). *Brain Res.* **96,** 169–175.
Fifková, E. (1985a). *Brain Res. Rev.* **9,** 187–215.
Fifková, E. (1985b). *Cell Mol. Neurobiol.* **5,** 47–63.
Fifková, E. (1985c). *In* "Recent Achievements in Restorative Neurology" (M. Dimitrijevic and J. C. Eccles, eds.), pp. 263–271. Karger, Basel.
Fifková, E., and Anderson, C. L. (1981). *Exp. Neurol.* **74,** 621–627.
Fifková, E., and Delay, R. J. (1982). *J. Cell Biol.* **95,** 345–350.
Fifková, E., and Morales, M. (1989). *Ann. N.Y. Acad. Sci.* **568,** 131–137.
Fifková, E., and VanHarreveld, A. (1977). *J. Neurocytol.* **6,** 211–230.
Fifková, E., Markham, J. A., and Delay, R. J. (1982a). *Soc. Neurosci. Abstr.* **8,** 279.
Fifková, E., Anderson, C. L., Young, S. J., and VanHarreveld, A. (1982b). *J. Neurocytol.* **11,** 183–210.
Fifková, E., Markham, J. A., and Delay, R. J. (1983). *Brain Res.* **266,** 163–168.
Fifková, E., Eason, H., and Schaner, P. (1990). *Soc. Neurosci. Abstr.* **16,** 695.
Fifková, E., Eason, H., and Schaner, P. (1992). *Brain Res.* **577,** 331–336.
Flock, A., Cheung, H. C., Block, B., and Utter, G. (1981). *J. Neurocytol.* **10,** 133–147.
Furchgott, R. F., and Vanhoutte, P. M. (1989). *FASEB J.* **3,** 2007–2018.
Gamble, E., and Koch, C. (1987). *Science* **236,** 1311–1315.
Garthwaite, J., and Garthwaite, G. (1987). *J. Neurochem.* **48,** 29–39.
Garthwaite, J., Charles, S. L., and Chess-Williams, R. (1988). *Nature* (*London*) **336,** 385–388.
Geiger, G. (1982). *Trends Biochem. Sci.* **7,** 388–389.
Giuliano, K. A., Khatib, F. A., Hayden, S. M., Daoud, E. W. R., Adams, M. E., Amorese, D. A., Bernstein, B. W., and Bamburg, J. R. (1988). *Biochemistry* **27,** 8931–8938.
Goh, J. W., and Pennefather, P. S. (1989). *Science* **244,** 980–983.
Goh, J. W., and Pennefather, P. S. (1990). *Brain Res.* **511,** 345–348.
Goldenring, J. R., Vallano, M. L., and DeLorenzo, R. J. (1985). *J. Neurochem.* **45,** 900–905.
Goodman, S. R., Riederer, B. M., and Zagon, S. (1986). *BioEssays* **5,** 25–29.

Gray, E. G. (1959). *J. Anat.* **93,** 420–433.
Gray, E. G. (1982). *Trends Neurosci.***5,** 5–6.
Gregorio, C. C., and Repasky, E. A. (1990). *J. Cell Biol.* **111,** 424a.
Griffith, L. M., and Pollard, T. D. (1978). *J. Cell Biol.* **78,** 958–965.
Griffith, L. M., and Pollard, T. D. (1982). *J. Biol. Chem.* **257,** 9143–9151.
Gustafsson, B., and Wigström, H. (1986). *J. Neurosci.* **6,** 1575– 1582.
Gustafsson, B., Wigström, H., Abraham, W. C., and Huang, Y.-Y. (1987). *J. Neurosci.* **7,** 774–780.
Halpain, S., and Greengard, P. (1990). *Neuron* **5,** 237–246.
Hamabuko, T., Kannagi, R., Murachi, T., and Matus, A. (1986). *J. Neurosci.* **6,** 3103–3111.
Harris, K. M., and Stevens, J. K. (1989). *J. Neurosci.* **9,** 2982–2997.
Hobbs, D. S., and Fredriksen, D. W. (1980). *Biophys. J.* **32,** 705–718.
Hodgkin, A. L., and Keynes, R. D. (1957). *J. Physiol.* (*London*) **138,** 253–281.
Holmes, W. R. (1990). *Brain Res.* **519,** 338–342.
Horwitz, B. (1984). *Neuroscience* **12,** 887–905.
Isenberg, G., Ohnheiser, R., and Maruta, H. (1983). *FEBS Lett.* **163,** 225–229.
Ishikawa, H., Bischoff, R., and Holtzer, H. (1969). *J. Cell Biol.* **43,** 312–328.
Jack, J. J. B., Noble, D., and Tsien, R. W. (1975). "Electric Current Flow in Excitable Cells." Oxford Univ. Press, London.
Janmey, P. A., and Stossel, T. P. (1987). *Nature* (*London*) **325,** 362–364.
Janmey, P. A., and Stossel, T. P. (1989). *J. Biol. Chem.* **264,** 4825–4831.
Janson, L. W., and Taylor, D. L. (1990). *J. Cell Biol.* **111,** 289a.
Janson, L. W., Kolega, J., and Taylor, D. L. (1991). *J. Cell Biol.* **114,** 1005–1015.
Kakiuchi, S., and Sobue, K. (1983). *Trends Biochem. Sci.* **81,** 59–62.
Katsumaru, H., Murakami, F., and Tsukahara, N. (1982). *Biomed. Res.* **3,** 337–340.
Kauer, J. A., Malenka, R. C., and Nicoll, R. A. (1988). *Nature* (*London*) **334,** 250–252.
Kawato, M., and Tsukahara, N. (1983). *J. Theor. Biol.* **103,** 507–522.
Kennedy, M. B. (1989). *Ann. N.Y. Acad. Sci.* **568,** 193–197.
Kennedy, M. B., Bennett, M. K., and Erondu, N. E. (1983). *Proc. Natl. Acad. Sci. U.S.A.* **80,** 7357–7361.
Kikkawa, U., Ogita, K., Go, M., Nomura, H., Kitano, T., Hashimoto, T., Ase, K., Sekigushi, K., Koumoto, J., Nishizuka, Y., Saito, N., and Tanaka, C. (1988). *In* "Advances in Second Messenger and Phosphoprotein Research" (R. Adelstein, C. Klee, and M. Rodbell, eds.), pp. 67–74. Raven, New York.
Kiliman, M. W., and Isenberg, G. (1982). *EMBO J.* **1,** 889–894.
Klann, E., Chen, S. J., and Sweatt, J. D. (1991). *Soc. Neurosci. Abstr.* **17,** 382.
Knowles, R. G., Palacios, M., Palmer, R. M. J., and Moncada, S. (1989). *Proc. Natl. Acad. Sci. U.S.A.* **86,** 5159–5162.
Koch, C., and Poggio, T. (1983). *Proc. R. Soc. London Ser. B* **218,** 455–477.
Koch, C., and Poggio, T. (1985). *J. Theor. Biol.* **113,** 225–229.
Kose, A., Ito, A., Saito, N., and Tanaka, C. (1990). *Brain Res.* **518,** 209–217.
Kriho, V., and Cohen, R. S. (1988). *Soc. Neurosci. Abstr.* **14,** 106.
Kuczmarski, E. R., and Rosenbaum, J. L. (1979a). *J. Cell Biol.* **80,** 341–355.
Kuczmarski, E. R., and Rosenbaum, J. L. (1979b). *J. Cell Biol.* **80,** 356–371.
Lamb, N. J. C., Fernandez, A., Conti, M. A., Adelstein, R., Glass, D. B., Welch W. J., and Feramisco, J. R. (1988). *J. Cell Biol.* **106,** 1955–1971.
Landis, D. M. D., and Reese, T. S. (1983). *J. Cell Biol.* **97,** 1169–1178.
Lassing, I., and Lindberg, U. (1985). *Nature* (*London*) **314,** 472–474.
Lassing, I., and Lindberg, U. (1988). *J. Cell. Biochem.* **37,** 255–267.
LeBeux, Y. L., and Willemont, J. (1975). *Cell Tissue Res.* **160,** 1–36.
Lee, K. S., Schottler, F., Oliver, M., and Lynch, G. (1980). *J. Neurophysiol.* **44,** 247–258.

Linden, D. J., and Routtenberg, A. (1989). *Brain Res. Rev.* **14,** 279–296.
Lisman, J. E. (1985). *Proc. Natl. Acad. Sci. U.S.A.* **82,** 3055–3057.
Lynch, G., and Baudry, M. (1984). *Science* **224,** 1057–1063.
Madison, D. V., Malenka, R. C., and Nicoll, R. A. (1991). *Annu. Rev. Neurosci.* **14,** 379–397.
Maekawa, S., Endo, S., and Sakai, H. (1983). *J. Biochem. (Tokyo)* **94,** 1329–1337.
Maekawa, S., Ohta, K., and Sakai, H. (1988). *Cell Struct. Funct.* **13,** 373–385.
Malenka, R. C., Kauer, J. A., Zucker, R. S., and Nicoll, R. A. (1988). *Science* **242,** 81–84.
Malenka, R. C., Kauer, J. A., Perkel, D. J., Mauk, M. D., Kelly, P. T., Nicoll, R. A., and Waxham, M. N. (1989). *Nature (London)* **340,** 554–557.
Malinow, R., and Tsein, R. W. (1990). *Nature (London)* **346,** 177–180.
Malinow, R., Madison, D. V., and Tsien, R. W. (1988). *Nature (London)* **335,** 820–824.
Markham, J. A., and Fifková, E. (1986). *Dev. Brain Res.* **27,** 263–269.
Martin, W., Villani, G. M., Jothianandan, D., and Furchgott, R. F. (1985). *J. Pharmacol. Exp. Ther.* **232,** 708–716.
Matsumura, S., Kumon, A., and Chiba, T. (1985). *J. Biol Chem.* **260,** 1959–1966.
Matsuzaki, F., Matsumoto, S., Yahara, I., Yonezawa, N., Nishida, E., and Sakai, H. (1988). *J. Biol. Chem.* **263,** 11564–11568.
Mellgren, R. L. (1987). *FASEB J.* **1,** 110–115.
Meyer, T., Holowka, D., and Stryer, L. (1988). *Science* **240,** 653–656.
Miller, J. P., Rall, W., and Rinzel, J. (1985). *Brain Res.* **325,** 325–330.
Miller, S. G., and Kennedy, M. B. (1986). *Cell* **44,** 861–870.
Mody, T., MacDonald, J. F., and Baimbridge, K. G. (1988). *Soc. Neurosci. Abstr.* **14,** Part 1, 94.
Monaghan, D. T., and Cotman, C. W. (1985). *J. Neurosci,* **5,** 2909–2919.
Morales, M., and Fifková, E. (1989a). *J. Comp. Neurosci.* **279,** 666–674.
Morales, M., and Fifková, E. (1989b). *Cell Tissue Res.* **256,** 447–456.
Morales, M., and Fifková, E. (1991). *Cell Tissue Res.* **265,** 415–423.
Moriyama, K., Nishida, E., Yonezawa, N., Sakai, H., Matsumoto, S., Tida, K., and Yahara, T. (1990). *J. Biol. Chem.* **265,** 5768–5773.
Morris, R. G. M., Anderson, E., Lynch, G., and Baudry, M. (1986). *Nature (London)* **319,** 774–776.
Moshkov, D. A., Petrovskaia, L. L., and Bragin, A. G. (1977). *Dokl. Akad. Nauk USSR* **237,** 1525–1528.
Moshkov, D. A., Petrovskaia, L. L., and Bragin, A. G. (1980). *Tsitologiya* **22,** 20–26.
Müller, W., and Connor, J. A. (1991). *Soc Neurosci. Abst.* **17,** Part 1, 1.
Nafstadt, P. M. J. (1967). *Z. Zellforsch. Mikrosk. Anat.* **76,** 532–542.
Nakata, T., and Hirokawa, N. (1987). *J. Cell Biol.* **105,** 1771–1780.
Nishida, E., Kuwaki, T., and Sakai, H. (1981). *J. Biochem. (Tokyo)* **90,** 575–578.
Nishida, E., Maekawa, S., Muneyuki, E., and Sakai, H. (1984a). *J. Biochem. (Tokyo)* **95,** 378–398.
Nishida, E., Maekawa, S., and Sakai, H. (1984b). *Biochemistry* **23,** 5307–5313.
Nishida, E., Maekawa, S., and Sakai, H. (1984c). *J. Biochem (Tokyo)* **95,** 399–404.
Nishida, E., Muneyuki, E., Maekawa, S., Ohta, Y., and Sakai, H. (1985). *Biochemistry* **24,** 6624–6630.
Nishizuka, Y. (1986). *Science* **233,** 305–311.
Obenaus, A., Mody, T., and Baimbridge, K. G. (1988). *Soc. Neurosci. Abstr.* **14,** 567.
O'Dell, T. J., Kandel, E. R., and Grant, S. G. N. (1991a). *Soc. Neurosci. Abstr.* **17,** 383.
O'Dell, T. J., Hawkins, R. D., Kandel, E. R., and Arancio, O. (1991b). *Proc. Natl. Acad. Sci U.S.A.* **88,** 11285–11289.
Okada, D., Yamagishiki, S., and Sugiyama, H. (1989). *Neurosci. Lett.* **100,** 141–146.
Ouimet, C. C., McGuinness, T. L., and Greengard, P. (1984). *Proc. Natl. Acad. Sci. U.S.A.* **81,** 5604–5608.

Palmer, R. M. J., Ferrige, A. G., and Moncada, S. (1987). *Nature (London)* **327,** 524–526.
Peng, H. B., and Phelan, K. A. (1984). *J. Cell Biol.* **99,** 344–349.
Perkel, D. H., and Perkel, D. J. (1985). *Brain Res.* **325,** 331–335.
Peters, A., Palay, S. L., and Webster, H. deF. (1976). "The Fine Structure of the Nervous System." Saunders, Philadelphia.
Petrucci, T. C., Thomas, C., and Bray, D. (1983). *J. Neurochem.* **40,** 1507–1516.
Petukhov, V. V., and Popov, V. I. (1986). *Neuroscience* **18,** 823–835.
Pin, J.-P., Bockaert, J., and Recasesn, M. (1984). *FEBS Lett.* **175,** 31–36.
Pollard, T. D. (1986). *J. Cell. Biochem.* **31,** 87–95.
Pollard, T. D., Fujiwara, K., Niederman, R., and Maupin-Szamier, P. (1976). *In* "Cell Motility," Book B (R. Goldman, T. D. Pollard, and J. Rosenbaum, eds.), pp. 689–724. Cold Spring Harbor Lab., Cold Spring Harbor, New York.
Pollard, T. D., Selden, S. C., and Maupin, P. (1984). *J. Cell Biol.* **99,** 33s–37s.
Rall, W. (1970). *In* "Excitatory Synaptic Mechanisms," (P. Anderson and J. K. S. Jensen, eds.). pp. 175–187. Universitets Forlaget, Oslo.
Rall, W. (1974). *In* "Cellular Mechanisms Subserving Changes in Neuronal Activity" (C. Woody, K. Brown, T. Crow, and J. Knispel, eds.). pp. 13–21 Brain Inf. Serv., UCLA, Los Angeles.
Rall, W. (1978). *In* "Studies in Neurophysiology" (A. K. McIntyre and K. Porter, eds.). pp. 203–209. Cambridge Univ. Press, Cambridge.
Rall, W., and Rinzel, J. (1971a). *Int. Congr. Physiol. Sci., 25th* **9,** 466.
Rall, W., and Rinzel, J. (1971b). *Soc. Neurosci. Abstr., Annu. 1st Meet.,* p. 64.
Rall, W., and Segev, I. (1988). *Neurol. Neurobiol.* **37,** 263–282.
Ramón y Cajal, S. (1891). *Cellule* **7,** 124–176.
Rapoport, R. M., Draznin, M. B., and Murad, F. (1983). *Nature (London)* **306,** 174–176.
Reymann, K. G., Brodemann, R., Kase, H., and Matthies, H. (1988a). *Brain Res.* **461,** 388–392.
Reymann, K. G., Frey, U., Jork, R., and Matthies, H. (1988b). *Brain Res.* **440,** 305–314.
Roots, B. I. (1983). *Science* **221,** 971–972.
Sarvey, J. M. (1988a). *In* "Sensitization in the Nervous System" (P. W. Kalivas and C. D. Barnes, eds.), pp. 47–78. Telford, Caldwell, New Jersey.
Sarvey, J. M. (1988b). *In* "Biophysics to Behavior" (P. Landfield and S. Deadwyler, eds.), pp. 329–353. Alan R. Liss, New York.
Satoh, T., Ross, C. A., Villa, A., Supattapone, S., Pozzan, T., Snyder, S. H., and Meldolesi, J. (1990). *J. Cell Biol.* **111,** 615–624.
Sattilaro, R. F. (1986). *Biochemistry* **25,** 2003–2009.
Sattilaro, R. F., Dentler, W. L., and LeCluyse, E. L. (1981). *J. Cell Biol.* **90,** 467–473.
Schliwa, M. (1981). *Cell* **25,** 587–590.
Schulman, H. (1984). *J. Cell Biol.* **99,** 11–19.
Schuman, E. M., and Madison, D. V. (1991). *Science* **254,** 1503–1506.
Schwartz, J. H., and Greenberg, S. M. (1987). *Annu. Rev. Neurosci.* **10,** 459–476.
Segal, M., Guthrie, P., and Kater, S. B. (1990). *Soc. Neurosci. Abstr.* **16,** 468.
Segev, I., and Rall, W. (1988). *J. Neurophysiol.* **60,** 499–523.
Selden, S. C., and Pollard, T. D. (1983). *J. Biol. Chem.* **258,** 7064–7071.
Shepherd, G. M. (1979). "The Synaptic Organization of the Brain," p. 364. Oxford Univ. Press, New York.
Shepherd, G. M., and Brayton, R. K. (1987). *Neuroscience* **21,** 151–165.
Shepherd, G. M., Brayton, R. K., Miller, J. P., Segev, I., Rinzel, J., and Rall, W. (1985). *Proc. Natl. Acad. Sci. U.S.A.* **82,** 2192–2195.
Shoemaker, M. O., Lau, W., Shattuck, R. L., Kwiatkowski, A. P., Matrisian, P. E., Guerra-Santos, L., Wilson, E., Lukas, T. J., VanEldik, L. J., and Watterson, D. M. (1990). *J. Cell Biol.* **111,** 1107–1125.

Sloboda, R. D., Rudolph, S. A., Rosenbaum, J. L., and Greengard, P. (1975). *Proc. Natl. Acad. Sci. U.S.A.* **72,** 177–181.

Staubli, U. J., Larson, J., Thibault, O., Baudry, M., and Lynch, G. (1988). *Brain Res.* **444,** 153–158.

Steward, O., and Levy, W. B. (1982). *J. Neurosci.* **2,** 284–291.

Stossel, T. P. (1982). *Philos. Trans. R. Soc. London, Ser. B* **299,** 275–289.

Stossel, T. P. (1983). *In* "Spatial Organization of Eukaryotic Cells. Modern Cell Biology" (J. R. McIntosh, ed.), Vol. 2, pp. 203–223. Alan R. Liss, New York.

Stossel, T. P. (1989). *J. Biol. Chem.* **264,** 18261–18264.

Stossel, T. P., Chapponier, C., Ezzell, R. M., Hartwig, J. H., Janmey, P. A., Kwiatkowski, D. J., Lind, S. E., Smith, D. B., Southwick, F. S., Yin, H. L., and Zaner, K. S. (1985). *Annu. Rev. Cell Biol.* **1,** 353–402.

Tanaka, E., Fukunaga, K., Yamamoto, H., Iwasa, T., and Miyamoto, E. (1986). *J. Neurochem.* **47,** 254–262.

Tarrant, S. B., and Routtenberg, A. (1977). *Tissue Cell* **9,** 461–473.

Taylor, D., and Condeelis, J. (1985). *Int. Rev. Cytol.* **56,** 57–144.

Tomasulo, R. A., Levy, W. B., and Steward, O. (1990). *Soc. Neurosci Abstr.* **16,** 873.

Tsien, R. W., and Tsien, R. Y. (1990). *Annu. Rev. Cell Biol.* **6,** 715–760.

Turner, D. A., and Schwartzkroin, P. A. (1983). *J. Neurosci.* **3,** 2381–2394.

Vallee, R. B. (1980). *Proc. Natl. Acad. Sci. U.S.A.* **77,** 3206–3210.

Vallee, R. B., DiBartolomeis, M. J., and Theurkauf, W. E. (1981). *J. Cell Biol.* **90,** 568–576.

VanHarreveld, A., and Fifková, E. (1975). *Exp. Neurol.* **49,** 736–749.

Verkhovskij, A. B., Surgucheva, I. G., and Gelfand, V. I. (1984). *Biochem. Biophys. Res. Commun.* **123,** 596–603.

Weeds, A. (1982). *Nature (London)* **296,** 811–816.

Wenzel, J., Schmidt, C., Duwe, G., Skrebitz, W. G., and Kudrjats, I. (1985). *J. Hirnforsch.* **26,** 573–583.

Westrum, L. E., Jones, D. H., Gray, E. G., and Barron, J. (1980). *Cell Tissue Res.* **208,** 171–181.

Wickens, J. (1988). *Prog. Neurobiol.* **31,** 507–528.

Wigström, H., and Gustafsson, B. (1983). *Nature (London)* **301,** 603–604.

Wigström, H., and Gustafsson, B. (1985). *Acta Physiol. Scand.* **125,** 159–172.

Wilson, C. J. (1984). *J. Neurosci.* **4,** 281–297.

Wilson, C. J., Groves, P. M., Kitai, S. T., and Linder, J. C. (1983). *J. Neurosci.* **3,** 383–398.

Wolf, M., Burgess, S., Misra, U. K., and Sahyoun, N. (1986). *Biochem. Biophys. Res. Commun.* **140,** 691–698.

Yamamoto, H., Fukunaga, K., Goto, S., Tanaka, E., and Miyamoto, E. (1985). *J. Neurochem.* **44,** 759–768.

Yamauchi, T., and Fujisawa, H. (1988). *Biochim. Biophys. Acta* **968,** 77–85.

Zador, A., Koch, C., and Brown, T. H. (1990). *Proc. Natl. Acad. Sci. U.S.A.* **87,** 6718–6722.

Zagon, I. S., Higbee, R., Riederer, B. M., and Goodman, S. R. (1986). *J. Neurosci.* **6,** 2977–2986.

Role of Signal Transduction Systems in Cell Proliferation in Yeast

Isao Uno
Life Science Research Center, Nippon Steel Corporation, Kawasaki, Japan

I. Introduction

In *Saccharomyces cerevisiae* intracellular cAMP mediates environmental signals that regulate cellular metabolism and cell proliferation. The studies on the cAMP-requiring mutants and their suppressors in the yeast revealed that cAMP-dependent protein phosphorylation is involved in the G1 phase of the cell cycle, stimulation of the phosphoinositide pathway, and the postmeiotic stage of sporulation, and that inhibition of cAMP-dependent protein phosphorylation is required to go into the G0 stage and induce meiotic division.

The experiments of signal transduction systems lead to work using yeast as a model of eukaryotic cell regarding the work of cell cycle and signal transduction. Phosphorylation of cellular proteins is required in these processes, and the nature of these proteins phosphorylated by protein kinases is important to the understanding of the role of signal transduction for growth and differentiation in yeast.

In yeast we know of the existence of signal transduction systems, including cAMP, inositol phospholipid, calcium, and tyrosine kinase. The occurrence of cAMP has been reported to be widespread among prokaryotic and eukaryotic organisms (Robison *et al.*, 1971; Botsford, 1981). Cyclic AMP stimulates transcription of various operons of prokaryotic cells by binding with a protein (CAP). In eukaryotes cAMP may not be usually involved in the direct stimulation of transcription but plays a role as second messenger in a wide variety of functions. It has been well documented that cAMP in eukaryotes mainly acts through activation of protein kinases that, in turn, phosphorylate and thereby activate cellular proteins critical for phenotypic expression (Robison *et al.*, 1971). The existence of enzymes involved in the cAMP–cascade system, namely, adenylate cyclase, phosphodiesterases, and cAMP-dependent protein kinase, was indicated in yeast (Uno, 1988). cAMP-gated channel proteins were identified as cAMP-binding protein in mammalian cells. However, phosphatidylinositol (PI), Ca^{2+}, and tyrosine kinase signal transduction systems were found in yeast, as in mammalian cells.

The information bearing on the involvement of signal transduction in the control of the cell division cycle has been suggested earlier by a number of studies with mammalian cells (Friedman *et al.*, 1976; Boynton and Whitfield, 1983). The studies with mammalian cell lines were clearly valuable in helping to identify the role of signal transduction systems in cell proliferation and transformation (Gottesman, 1980). To elucidate further the molecular aspects of the role of the signal transduction system in cell division and differentiation, various mutants of yeast, which were altered in the signal transduction system, have been studied in the expectation that the powerful genetic and molecular approaches possible in yeast will enable us to understand the functions of signal transduction systems in eukaryotes. Specific biochemical changes caused by the mutation can then be correlated with altered cellular functions in yeast, such as cell division and cellular differentiation. In this review of the signal transduction system in yeast, major progress made in the genetic analysis of the mutants altered in the signal transduction system and in the studies of the function of second messenger in yeast cells will be discussed.

II. Cyclic AMP Cascade

A. Isolation of Mutants Related to cAMP Cascade

The cAMP-requiring mutants were able to be isolated from strains utilizing cAMP (Matsumoto *et al.*, 1982) with the genetic background of P-28-24C. Although *amp1* and *cam* mutations enhance uptake of exogenous cAMP, P-28-24C derivatives can utilize the exogenous cAMP without *amp1* and *cam* mutations because they carry mutations or genetic variations called *cas1* (Boutelet *et al.*, 1985) or *rca1* (Boy-Marcotte *et al.*, 1987). The *cyr1* mutants, which were isolated as cAMP-requiring mutant, carried a lesion in the structural gene for adenylate cyclase (Matsumoto *et al.*, 1982). Adenylate cyclase gene was cloned by isolation of DNA fragment complemented *cyr1* mutation (Masson *et al.*, 1984; Casperson *et al.*, 1985) and sequenced (Kataoka *et al.*, 1985a Masson *et al.*, 1986).

Several approaches were carried out to identify the functional domain of *CYR1* gene product, including molecular and genetic analyses (Uno *et al.*, 1985b, 1987; Field *et al.*, 1987; Marshall *et al.*, 1988; Colicelli *et al.*, 1990; Field *et al.*, 1990a). The catalytic domain was localized in the near

C-terminal region. The analysis of the *CYR1* gene by deletion and insertion mutagenesis to localize regions required for activation by the Ras2 protein revealed that leucine-rich repeats and carboxy-terminus are required for interaction of yeast adenylate cyclase with Ras proteins (Suzuki *et al.*, 1990). These results did not coincide with the data described by Uno *et al.* (1987), indicating that a region between a leucine-rich repeat and the catalytic domain is associated with the regulatory function of the Ras products, indicating that the results may be artifact. Also, COOH-terminal residues (Yamawaki-Kataoka *et al.*, 1989) are required for proper Ras-dependent regulation of adenylate cyclase. The immunochemical analysis suggests that the N-terminal region is an inhibitory domain mediating the effect of the Ras proteins (Heideman *et al.*, 1990); further, this model is supported by experiments with genetically truncated Cyr1 proteins.

Further, adenylate cyclase genes were cloned and sequenced from two kinds of fungi, *S. pombe* (Yamawaki-Kataoka *et al.*, 1989; Young *et al.*, 1989) and *N. crassa* (Kore-eda *et al.*, 1991), and bovine brain (Krupinski *et al.*, 1989). The amino acid sequence of three fungal adenylate cyclases is highly conserved, especially catalytic domain and leucine-rich repeats, and their domain structure was also conserved. The amino acid sequence of their catalytic domain is highly homologous to that of bovine brain, but the composition of their functional domains is different from that of bovine brain because transmembrane structure was not found in fungal adenylate cyclases. In yeast adenylate cyclase is a peripheral membrane protein (Matsumoto *et al.*, 1984), suggesting the existence of a protein-anchoring adenylate cyclase to the membrane (Heideman *et al.*, 1990; Mitts *et al.*, 1990). The adenylate cyclase activity of *S. pombe* is localized to the plasma membrane of the cell. The activity is not stimulated by guanyl nucleotides as *S. pombe* adenylate cyclase when its COOH-terminus is deleted (Engelberg *et al.*, 1990), suggesting that the COOH-terminus may be important for response to *RAS* gene products. These results may suggest the localization of each functional domain of adenylate cyclase, but the determination of its tertiary structure has yet to be elucidated.

B. Regulation of Adenylate Cyclase

1. *RAS* Genes

Evidence on the regulation of adenylate cyclase has been accumulated as follows. The *RAS* genes are a highly conserved family of genes, first discovered as the oncogenes of rat sarcoma virus (for reviews see Chardin, 1988; Gibbs and Marshall, 1989). Mammalian *ras* genes encoding altered proteins are found in many tumor cells and are capable of the morphological and tumorigenic transformation of mammalian cells. Yeast cells contain two closely related but distinct *RAS* genes, *RAS1* and *RAS2* (Defeo-Jones *et al.*, 1983; Papageorge *et al.*, 1984; Powers *et al.*, 1984). Neither *RAS1* nor *RAS2* is by itself an essential gene (Kataoka *et al*,. 1984), but *ras1 ras2* mutant cells are arrested by G1 phase like the *cyr1* mutants. The *bcy1* mutation, which suppresses the *cyr1* mutation (Matsumoto *et al.*, 1982), suppresses lethality in *ras1 ras2* yeast (Toda *et al.*, 1985). Compared with wild-type strains, intracellular cAMP levels are significantly depressed in *ras2* strains and virtually undetectable in *ras1 ras2 bcy1* strains. Membranes from *ras1 ras2, bcy1* lack the GTP-stimulated adenylate cyclase activity present in membranes from wild-type cells. Mixing membranes from *ras1, ras2* yeast with membranes from adenylate cyclase-deficient mutant (*cyr1*) reconstituted a GTP-dependent adenylate cyclase. A yeast strain containing *RAS2*$^{\mathrm{Val19}}$, a *RAS2* allele with a mutation that is analogous to the oncogenic mammalian *ras* genes (Kataoka *et al.*, 1984), has increased levels of an apparently GTP-independent adenylate cyclase activity and of intracellular cAMP. The yeast adenylate cyclase is regulated by the *RAS2* gene product that is a GTP-binding protein as mammalian adenylate cyclase.

The *ras* mutant cells were able to remain viable if they carry a mammalian *rasH* gene. In addition, yeast–mammalian hybrid genes and a deletion mutant yeast *ras1* gene were shown to induce morphogenetic transformation of mouse NIH3T3 cells when the genes had a point mutation analog to one that increases the transforming activity of mammalian *ras* genes. The results establish the functional relevance of the yeast system to the genetic and biochemistry of cellular transformation induced by mammalian *ras* genes (Defeo-Jones *et al.*, 1985; Kataoka *et al.*, 1985b).

The processing of Ras proteins is required for fatty acid acylation and membrane localization (Fujiyama and Tamanoi, 1986). A mutant defective in yeast gene that is required for *RAS* functions, designated *SUPH* or *DPR1*, was temperature-sensitive for growth and showed sterile phenotype specific to **a** cells (Powers *et al.*, 1986; Fujiyama *et al.*, 1987). The *supH* is allelic to *ste16*, a gene required for the production of the mating pheromone **a**-factor. Both *RAS* and **a**-factor coding sequences terminate with the potential acyltransferase recognition sequence Cys-A-A-X, where A is an aliphatic amino acid. Mutations in *SUPH-STE16* prevent the membrane localization and maturation of Ras protein, as well as the fatty acid acylation of it and other membrane proteins. The designation *RAM* (Ras protein and **a**-factor maturation function) for *SUPH* and *STE16* was proposed. *RAM* may encode an enzyme responsible for the modification and membrane localization of proteins with this C-terminal sequence.

RAS1 and *RAS2* gene products were phosphorylated *in vivo* and *in vitro*, and it was suggested that these proteins may be phosphorylated by cAMP-dependent protein kinase (Resnick and Racker, 1988; Cobitz *et al.*, 1989). However, the physiological function of their phosphorylation is unknown

Ruggieri *et al.* (1989) attempted to identify genes involved in the regulation of cAMP pathway, and *MSI1* was found to encode a putative protein that shows homology to the beta subunit of the mammalian guanine nucleotide-binding regulatory proteins (Gs), which are composed of heterotrimer (alpha, beta, and gamma subunit) (Rodbell, 1980; Gilman, 1984; Katada and Ui, 1982). This suggests the existence of gamma subunit and the regulation of adenylate cyclase by Gs protein as found in mammalian cells.

2. SRV2(CAP1), CDC25, and IRA Genes

There is evidence to show that other genes, *CDC25, SRV2(CAP1), IRA1, IRA2,* and *MSI1*, are related to the regulation of adenylate cyclase activity. Fedor-Chaiken *et al.* (1990) found a gene, *SRV2,* mutations of which alleviate stress sensitivity in strains carrying an activated *RAS* gene. Epistasis analysis suggests that the *SRV2* gene affects accumulation of cAMP in the cell. Direct assays of cAMP accumulation indicate that mutations of the gene diminish the rate of *in vivo* production of cAMP following stimulation by an activated *RAS* allele. Null mutation of *srv2* results in lethality, which cannot be suppressed by mutational activation of the cAMP-dependent protein kinase. The sequence of the gene indicates that it encodes an adenylate cyclase-associated protein, CAP, which is a product of *CAP* gene identified by Field *et al.* (1988, 1990b). These results demonstrate that Srv2 protein is required for *RAS*-activated adenylate cyclase activity but that it participates in other essential cellular functions as well. This effect required the N-terminal domain of CAP (Gerst *et al.*, 1991).

The level of cAMP decreased significantly in the *cdc25* cells when they were shifted to the restrictive temperature, and the *cdc25* arrest was suppressed by the addition of cAMP to the medium (Camonis *et al.*, 1986; Robinson *et al.*, 1987). Further, *cyr2,* originally identified as a cAMP-requiring gene, is an allele of *CDC25* gene, as described by Morishita and Uno (1991). However, the *CDC25* gene product is essential for activation of adenylate cyclase (Broek *et al.*, 1987; Robinson *et al.*, 1987). The biochemical experiments suggested that the *CDC25* gene product may regulate adenylate cyclase by regulating the guanine nucleotide bound to RAS proteins (Daniel and Simchen, 1986; Daniel *et al.*,1987; Engelberg *et al.*, 1990). The *CDC25* gene has been cloned and sequenced (Camonis *et al.*, 1986; Broek *et al.*, 1987). Disruption of the *CDC25* gene was lethal and the lethality could be suppressed by the presence of the activated *RAS2*^{Val19} (Broek *et al.*, 1987). Further mutant alleles of *RAS2* were discovered that dominantly interfered with wild-type *RAS* function (Powers *et al.* 1988; Morishita and Uno 1991). One of them, which was originally identified as *CYR3* mutation (Morishita and Uno, 1991), indicates null phenotype unless other mutations had temperature-sensitive growth. The inhibitory effects of the *RAS2* can be overcome by overexpression of *CDC25* but only in the presence of wild-type *RAS*. These results suggest that these mutant *RAS* genes interfere with the normal interaction of *RAS* and *CDC25* proteins and that this interaction is direct. These indicated that the *CDC25* product regulates adenylate cyclase by regulating the guanine nucleotide bound to RAS proteins.

Damak *et al.* (1991) cloned a DNA fragment that suppresses *cdc25* mutation but could not suppress *ras1, ras2,* or *cdc35* mutations. This fragment contains a 5′-truncated open reading frame that shares 47% identity with the C-terminal part of the *CDC25* gene. They named the entire gene *SDC25,* but this gene is dispensable for cell growth under usual conditions. No noticeable phenotype was found in the deleted strain. Another gene, the *NSP1* gene, has been identified by its ability, when expressed at high levels, to bypass the *CDC25* requirement for growth. Sequence analysis of the cloned *NSP1* locus suggests that the NSP1 product contains 269 amino acids and has a membrane-spanning domain at its carboxy-terminus (Tripp *et al.,* 1989). Phosphoprotein analysis of *NSP1*-suppressed cells indicates that the NSP1 product controls the phosphorylation of two 31 kDa proteins, the phosphorylation and dephosphorylation of which are strongly correlated with cell cycle arrest and proliferation, respectively, and suggests that the NSP1 product is an important downstream element of a *CDC25*- dependent, nutrient-responsive phosphorylation pathway.

Ras1 and Ras2 activity is negatively regulated by two GTPase activating protein (GAP) homologs, *IRA1,* formerly designated as *ppd1,* and *IRA2* (Tanaka *et al.,* 1988, 1989, 1990a,b). *IRA1* and *IRA2* potentially encode extremely large proteins (more than 300 kDa) that are 45% identical to each other. Both Ira proteins share a stretch of about 380 amino acids that is structurally and functionally related to the catalytic domain of the mammalian $p21^{ras}$GAP, and IRA2-encoded peptide that spans this catalytic domain stimulates the GTPase activity of Ras proteins. Interestingly, the negative regulation of Ras proteins mediated by these two different Ira proteins is not redundant because either the loss of *IRA1* function alone or the loss of *IRA2* function alone significantly increases Ras activity. Nonredundant Ras GAPs may also exist in animal cells, in which two distinct proteins with GAP activity for p21 have been discovered: cytoplasmic GAP and the NF1 gene product.

These results indicate that several kinds of gene products regulate adenylate cyclase activity. However, the regulation mechanism of adenylate cyclase is not so clear, and the precise biochemical analysis is required for understanding.

C. Phosphodiesterases

Cyclic AMP level may be regulated by phosphodiesterases, which degrade cAMP to 5′-AMP. A low and high Km cAMP phosphodiesterase was identified and characterized (Londesborough and Suoranta, 1983; Suoranta and Londesborough, 1984). The structural genes of these enzymes were cloned and sequenced. A gene, *PDE2,* has been cloned from the yeast that, when present in high copy, reverses the phenotypic effects of *RAS2*^{Val19}, a mutant form of the *RAS2* gene that renders yeast cells sensitive to heat shock and starvation (Sass *et al.*, 1986). *PDE2* encodes a high-affinity cAMP phosphodiesterase that shares sequence homology with animal cell phosphodiesterases. Nikawa *et al.* (1987b) isolated the *PDE1* gene, which encodes a low-affinity cAMP phosphodiesterase. These two genes represent highly divergent branches in the evolution of phosphodiesterases. High copy number plasmids containing either *PDE1* and *PDE2* can reverse the growth arrest defects of yeast cells carrying the *RAS2*^{Val19} mutation. Disruption of both *PDE* genes results in a phenotype that resembles that induced by the *RAS2*^{Val19} mutation. These results suggest that these enzymes may control intracellular cAMP level.

D. Regulation of Intracellular cAMP Level

The wild-type cells respond to the environmental signals, such as sulfur starvation, heat shock, or temperature shift. Growth of the *cyr1* cells could be arrested at the G1 phase after reaching the stationary phase or sulfur starvation in the presence of at least one normal *RAS* gene but not arrested in the absence of normal *RAS*. The *cyr1* mutant cells were more resistant to the heat treatment compared with *cyr1 ras1 ras2* mutants. These data indicate that these environmental signals are received by the *RAS* products and transferred to adenylate cyclase. The signals may be first received by particular receptors, but the nature of such receptors is not known. Addition of glucose-related fermentable sugars to cells cultivated under derepressed condition triggers a Ras protein-mediated cAMP signal, which induces a protein phosphorylation cascade. The *cdc25* mutant strains were deficient in basal cAMP synthesis and in the glucose-induced cAMP signal. These results indicate that the *CDC25* gene product is required not only for basal cAMP synthesis in yeast but also for specific activation of cAMP synthesis by the signal pathway leasing from glucose to adenylate cyclase (Aelst *et al.,* 1991). However, Thevelein *et al.* (1987a,b) reported that the effect of glucose cannot be explained on the basis of effects known to be caused by the membrane-depolarizing compounds, such as dinitrophenol, which causes increases in the cAMP level. There is no evidence to explain the activation mechanism of adenylate cyclase by the addition of glucose.

Using the technique of centrifugal elutriation, it was demonstrated that during the cell division cycle of yeast there are stage-specific fluctuations in the intracellular concentration of cAMP. Smith *et al.* (1990) indicated that the intracellular concentration of cAMP is at its highest during the division cycle and at its lowest immediately prior to and just after cell separation, and also extracellular cAMP level being 10–100 times higher than intracellular levels. During the cell cycle of *S. cerevisiae* the extracellular level of cAMP does not fluctuate.

Nikawa *et al.* (1987a) reported that manipulation of genes related to the RAS/adenylate cyclase pathway revealed a system for feedback control that can modulate cAMP levels over at least a 10,000-fold range. The feedback control depends upon the activity of the cAMP-dependent protein kinases and requires the presence of the Cdc25 and Ras proteins. The capability for such dramatic control of cAMP levels raises fundamental questions about the normal mechanism of action of the cAMP-signaling system in yeast.

Cyclic AMP can overcome the cell division block caused by thermosensitive mutations in *CYR1 (CDC35)* (Matsumoto *et al.*, 1982), *RAS* (Morishita and Uno, 1991), and *CDC25* genes (Martegani *et al.*, 1986), and this phenotypic suppression by cAMP requires the presence of a mutation in unlinked positions such as *CAS1* (Boutelet *et al.*, 1985) or *RCA1* (Boy-Marcotte *et al.*, 1987). Thus, cAMP seems to be a positive effector, which controls the initiation of a new division cycle.

E. cAMP-Dependent Protein Kinase

In eukaryotic cells cAMP-dependent protein kinases were detected, and two types, Type I and Type II of cAMP-dependent protein kinases, were observed (Taylor *et al.*, 1990). In eukaryotes the effects of cAMP are commonly thought to be due larely to cAMP-dependent protein kinase. Recently, cAMP-dependent ion channels were observed in mammalian cells, and these cAMP receptor proteins do not have protein kinase activity. These cAMP-dependent protein kinases are tetrameric proteins consisting of two regulatory subunits and two catalytic subunits. The regulatory subunits each contain two binding sites for cAMP, which, when occupied, cause the holoenzyme to dissociate two active catalytic subunits, with the regulatory subunits, RI and RII, which are known to be present in mammalian cells.

In yeast only one kind of enzyme, which is similar to the Type II enzyme, was recognized (Takai *et al.*, 1974; Uno *et al.*, 1982, and the structural genes were cloned and sequenced. A suppressor mutation of a cAMP-requiring mutation is a secondary mutation in another locus that reverses the G1 arrest due to the primary mutation and must exert some compensatory functions in the cAMP cascade system. One such suppressor, *bcy1*, was able to suppress *cyr1*, *cdc25* and *ras1 ras2* mutations. The *bcy1* mutant was deficient in the regulatory subunit of cAMP-dependent protein kinase (Matsumoto *et al.*, 1982). Toda *et al.*(1987a) cloned *BCY1* gene, which was found to encode a regulatory subunit of cAMP-dependent protein kinase. However, three genes *(TPK1, TPK2,* and *TPK3),* which are highly homologous, were isolated from yeast and encoded the catalytic subunits of cAMP-dependent protein kinase (Toda *et al.*, 1987b). Gene disruption experiments demonstrated that no two of the three genes are essential by themselves but at least one *TPK* gene is required for a cell to grow normally. A multicopy plasmid carrying the *TPK* gene complemented the cAMP-requiring phenotype of a *cyr1* mutant and the temperature sensitivity of a *ras1 ras2*^{ts} mutant (Toda *et al.*, 1987b). This is consistent with the idea that activation of cAMP-dependent protein kinase can overcome the defects of adenylate cyclase and Ras protein in yeast.

F. Phosphoprotein Phosphatases

In mammalian cells the nature of serine threonine protein phosphatases was controversial. Most of the protein phosphatases in the literature could be explained by four principal catalytic subunits with broad and overlapping substrate specificities, and simple criteria were introduced that could be used to distinguish them (Cohen, 1989). The enzymes were subdivided into two groups (Types I and II) depending on whether they dephosphorylate the beta subunit of phosphorylase kinase specifically and were inhibited by nanomolar concentrations of the small heat- and acid-stable proteins, termed inhibitor 1 and 2 (Type I protein phosphatase), or whether they dephosphorylated the alpha subunit of phosphorylase kinase preferentially and were insensitive to inhibitors (Type II). Type II phosphatases could be subclassified into three distinct enzymes, IIA, IIB, and IIC in a number of ways, but most simply by their dependence on divalent cations. IIB and IIC had absolute requirement for Ca^{2+} and Mg^{2+}, respectively, whereas IIA, like I, was active in the absence of divalent cations. In yeast *dis*$^{2+}$ homolog (Type I protein phosphatase) was cloned by Ohkura *et al.* (1989) from *S. cerevisiae,* but its function is still unknown. A suppressor *(SIT4)* of *HIS4* transcriptional defect encodes a protein with homology to the catalytic subunit of protein phosphatase Type IIA (Arndt *et al.*, 1989). Recently, Liu *et al.* (1991), isolated clones that encode calmodulin-binding proteins and found two closely related genes (*CMP1* and *CMP2*) that encode proteins homologous to the catalytic subunits of phosphoprotein phosphatase Type IIB (calcineurin). The elimination of either or both of these genes had no effect on cell viability, indicating that those genes are not essential for cell growth. As indicated, phosphoprotein phosphatases are not well known in yeast, so that molecular and biochemical analysis was required for understanding.

G. Downstream of cAMP Pathway

There is not much evidence about downstream cAMP pathway. *SCH9* gene was isolated by its ability to complement a *cdc25*^{ts} mutation (Toda *et al.*, 1988), and encodes 90,000-da protein with a carboxy-terminal domain homologous to yeast and mammalian cAMP-dependent protein kinase catalytic subunits. In addition to suppressing loss of *CDC25* function, multicopy plasmids containing *SCH9* gene suppress the growth defects of strains lacking *RAS, CYR1,* and *TPK* genes. Cell lacking *SCH9* grow slowly and have a prolonged G1 phase of the cell cycle. This defect is suppressed by activation of the cAMP effector pathway. *SCH9* gene encodes a protein kinase that may be part of a growth control pathway, which is at least partially redundant within the cAMP pathway.

Garrett and Broach (1989) isolated and characterized revertants capable of growth at the nonpermissive temperature from *ras*^{ts}. Suppressors in one complementation group (designated as *yak1*) are particularly intriguing because they appear to alleviate only the growth defect of the temperature-sensitive *ras* mutants and do not show any of the phenotypes, such as heat shock sensitivity, or starvation sensitivity, associated with increased production of cAMP. The *YAK1* gene has been cloned, and disruptions generated *in vitro* reveal that it is not essential for growth and that its loss confers growth to a strain deleted for *tpk1, tpk2,* and *tpk3,* structural genes for the catalytic subunit of cAMP-dependent protein kinase. Furthermore, the coding region predicts a protein with significant homology to the family of protein kinases, suggesting that loss of cAMP-dependent protein kinase functions can be suppressed by the loss of a second protein kinase (Garrett *et al.*, 1991). These results place *YAK1* downstream from, or on a parallel pathway to, the kinase step in the RAS/cAMP pathway, and suggest the regulation of cellular functions by the protein kinase chain reaction in yeast.

H. Cyclic AMP-Responsive Gene Expression

In mammals cAMP regulates gene expression of several genes through cAMP-dependent protein phosphorylation. Montminy *et al.* (1986) and Comb *et al.* (1986) recognized that cAMP-responsive element (CRE) of rat somatostatin gene and the cAMP-responsive element-binding protein (CREB) were purified from nuclei of PC12 cells (Montminy and Bilezikjian, 1987; Hoeffler *et al.*, 1988). Cyclic AMP stimulates somatostatin gene transcription by phosphorylation of CREB (Gonzalez *et al.*, 1989; Gonzalez and Montminy, 1989).

Mammalian cAMP-responsive element can activate transcription in yeast and bind a yeast factor(s) that resembles the mammalian transcriptional factor ATF, which is involved in the expression of many viral E1a-inducible and cellular cAMP-inducible genes (Jones and Jones, 1989), but it is unknown whether phosphorylation of these proteins regulates transcription level. The regulation of gene transcription involves phosphorylation of specific transcription factors. Cherry *et al.* (1989) reported that the yeast transcriptional activator *ADR1* is phosphorylated *in vitro* by cAMP-dependent protein kinase and that mutations that enhance the ability of *ADR1* to activate *ADH2* expression decrease *ADR1* phosphorylation. The increase of cAMP-dependent kinase activity inhibited *ADH2* expression *in vivo* in an *ADR1* allele-specific manner. Their data suggest that glucose repression of *ADH2* is in part mediated through a cAMP-dependent protein phosphorylation –inactivation of the *ADR1* regulatory protein. The *CTT1* (catalase T) gene is controlled by oxygen via hem, by nutrients via cAMP, and by heat shock (Belazzi *et al.*, 1991). The transcription of this gene is negatively regulated and mediated by a positive control element. The transcription of *SSA3* gene (HSP70) is also regulated negatively by cAMP (Boorstein and Craig, 1990). The expression of the polyubiquitin gene, *UBI4*, is repressed by cAMP-dependent protein phosphorylation (Tanaka *et al.*, 1988). The expression of the *UBI4* gene is also induced by mild heat shock by a mechanism other than depletion of cAMP. A stress-inducible protein (p110A and B), glycoprotein is induced by stress conditions that lead to cessation of DNA synthesis and cell division and that decrease cAMP arrest growth through a chain of events that include p118 induction (Verma *et al.*, 1988a,b). This induction is negatively regulated by cAMP level, indicating that cAMP regulates gene expression at transcriptional levels through protein phosphorylation, as observed in mammalian cells.

III. Regulation of Metabolism by cAMP Cascade

Strains indicating low cAMP levels have three distinct phenotypes. First, they fail to grow efficiently on nonfermentable carbon sources. Second, they hyperaccumulate the storage carbohydrate glycogen and trehalose. Third, diploid cells sporulate in rich medium. When glucose enters into yeast cells, it becomes available for glycolysis, the rate of which is regulated at the level of phosphofructokinase I, essentially through fructose-2,6-biphosphate. The passage of glucose through the membrane may cause a transient activation of adenylate cyclase. Cyclic AMP is formed in the cell and activates cAMP-dependent protein kinase. The free catalytic subunit can then phosphorylate several enzymes, one of them being phosphofructokinase II, which catalyzes the formation of fructose-2,6-biphosphate from hexose-6-phosphate and thereby activates glycolysis (Rittenhouse *et al.*, 1986). Another substrate for cAMP-dependent protein kinase is trehalase, which when activated by phosphorylation, hydrolyses accumulated trehalose and provides the necessary hexose-6-phosphate for fructose-2,6-biphosphate formation (Uno *et al.*, 1983). Finally, an interconversion of trehalose-6-phosphate synthetase to its less active form by the same cAMP-mediated mechanism (Panek *et al.*, 1987) would guarantee the glycolytic flux by maintaining the levels of glucose-6-phosphate without diverting them into trehalose. However, NAD-dependent glutamate dehydrogenase is negatively regulated by cAMP (Uno *et al.*, 1984). Phosphatidylserine synthetase was phosphorylated by cAMP-dependent protein kinase and the phosphorylation resulted in reduced phosphatidylserine synthetase activity (Kinney and Carman, 1988). Analysis of peptides derived from protease-treated labeled phosphatidylserine synthetase showed only one labeled peptide. Phospho amino acid analysis of labeled phosphatidylserine synthetase showed that the enzyme was phosphorylated at a serine residue.

Thus, cAMP directly regulates activities of several kinds of enzymes through protein phosphorylation, as observed in mammalian cells, and that indicates at least two kinds of regulation systems through gene expression and direct protein phosphorylation.

IV. Role of cAMP in Controlling Cell Division

Regulation of cell division in yeast has been studied by isolating mutants defective in various portions of the cell cycle and by the use of stage-specific inhibitors or nutrient limitation (Pringle and Hartwell, 1981). The cell division cycle *(cdc)* mutations lead to defects in particular stage-specific functions of the cell cycle. If cAMP is essential for growth of cells, it may be possible to isolate cAMP-requiring mutants. In line with this assumption, a number of cAMP-requiring mutants of *S. cerevisiae* were isolated. The growth of these mutants was arrested in the absence of cAMP, and these mutant strains behave as *cdc* mutants (Matsumoto *et al.*, 1983a).

Growth arrest of the *cyr1* cells incubated in the absence of cAMP was characterized by the production of unbudded cells and by causing subsequent delay in the onset of budding upon the shift to the medium that contained cAMP. The length of the delay was positively correlated with the duration of growth arrest in the absence of cAMP (Shin *et al.*, 1987). The characteristics of this type of growth arrest indicate that the *cyr1* cells starved for cAMP entered a resting state equivalent to the GO state of mammalian cells (Baserga *et al.*, 1973). The *cyr1* cells starved for cAMP had an ability to tolerate heat shock (Shin *et al.*, 1988). The *bcy1* cells, which produced cAMP-independent protein kinase, were not arrested in G1 and were sensitive to a heat-shock treatment. Therefore, yeast cells starved for cAMP are arrested in G1 and go into GO. Conversely, cAMP is required for the transition from GO to the growing state. Phosphorylation patterns analyzed by two-dimensional polyacrylamide gel eletrophoresis in the mutants *cyr1-2* and *bcy1* revealed several proteins, the phosphorylation of which was controlled positively or negatively by the cAMP pathways (Tripp *et al.*, 1986). The presence of some of these phosphoproteins was directly associated with the mitotic cell cycle (positive regulation), which was correlated with the cell cycle arrest (negative regulation). Phosphoproteins, the presence of which correlated with cell cycle arrest, were found to be phosphorylated on serine and threonine residues, whereas the major phosphoproteins present predominantly in proliferating cells were phosphorylated only on serine residues. The *cyr1* cells starved for cAMP and the sulfur-starved GO wild-type cells synthesized nine GO proteins, including UBI4 protein, which were synthesized at stimulated rates during G1–GO transition. The *bcy1* or *RAS2*$^{\mathrm{Val19}}$ mutants synthesized no GO proteins even under the sulfur-starved condition. The addition of cAMP to the sulfur-starved cells was effective in repressing the synthesis of these GO proteins. A simple hypothesis explaining these data is that the derepression of synthesis of these proteins results from a failure of the cAMP-dependent phosphorylation of a cellular protein. Phosphorylation of this protein may be required for the transition from GO to G1, but the nature of this protein is not yet known.

All these considerations led us to propose a model in which the cAMP level controlled by the environmental signals should make the choice between the initiation of a division cycle and the entry of the cells into the resting state. However, extensive studies of the cAMP–cascade system in several kinds of eukaryotic cells indicated that cAMP may not be essential for proliferation of cells, suggesting that roles of cAMP on the cell cycle may be variable depending on the species and cell types (for review of the regulation of cell cycle, see Cross *et al.*, 1989).

V. Regulation of Entry into Meiosis

Diploid cells of yeast are able to initiate meiosis and sporulation only under starvation conditions. Diploidy and starvation are both required for meiosis. Diploidy is mediated through the mating type genes, *MAT-α1* and *MAT-α2,* and diploid cells defective in or homozygous for one of these genes are incapable of undergoing meiosis. The *rme1* mutation, when homozygous, suppresses this deficiency (Kassir and Simchen, 1976). The *RME1* gene is transcriptionally regulated by the genes *MAT-α1* and *MAT-α2* (Mitchell and Herskowitz, 1986). *RME1* has no role in the regulatory pathway by starvation because *rme1* mutations do not permit unstarved cells to sporulate (Mitchell and Herskowitz, 1986). Starvation seems to induce meiosis in *S. cerevisiae* through the adenylate cyclase/cAMP-dependent protein kinase (AC/PK) cascade system. Mutations in the adenylate cyclase gene, *CYR1* (Matsumoto *et al.,* 1982, 1983b; Uno *et al.,* 1985a), and in two of its regulators *CDC25 (cyr2)* (Broek *et al.,* 1987) and *RAS2 (CYR3)* (Toda *et al.,* 1985), result in low levels of cellular cAMP. In these mutants meiosis and sporulation take place in rich medium (Shilo *et al.,* 1978; Matsumoto *et al.,* 1983b; Tatchell *et al.,* 1985; Toda *et al.,* 1985; Mitsuzawa *et al.,* 1989). Mutations in the gene *BCY1* making the protein kinase independent of cAMP (Matsumoto *et al.,* 1983a; Toda *et al.,* 1987a,b), when homozygous, result in diploids being meiosis- and sporulation- deficient (Matsumoto *et al.,* 1983a). The *RAS2*^{val19} mutation results in high levels of cellular cAMP and in diploids being meiosis- and sporulation-deficient (Toda *et al.,* 1985). This implies that the reduced activity might be a necessary intermediate step in the regulation of meiosis by starvation.

The gene *IME1* was originally identified by its ability, when present on a multicopy plasmid, to promote sporulation regardless of the constitution of *MAT* (Kassir and Simchen, 1988). Disruption of *IME1* results in a recessive sporulation-deficient phenotype, and diploids homozygous for *ime1* are also meiosis-deficient. These indicate that *IME1* is a positive regulator of meiosis that is normally repressed by *RME1*. Multicopy plasmids carrying *IME1* also enabled sporulation to occur even in rich media (Granot *et al.,* 1989), thus also overriding the requirement for the starvation signal. Furthermore, transcription of *IME1* was shown to be induced by starvation (Kassir and Simchen, 1988). Multicopy *IME1* plasmids overcome the meiotic deficiency of *bcy1* and *RAS2*^{val19} diploids. Double mutants *ime1 cdc25* and *ime1 ras2* are sporulation-deficient. These results suggest that *IME1* comes after the AC/PK cascade. Furthermore, the level of *IME1* transcript is affected by mutation in the AC/PK genes, *CDC25, CYR1,* and *BCY1*. Moreover, the addition of cAMP to a *cyr1-2* diploid suppresses *IME1* transcription. The presence in a *bcy1* diploid of *IME1* multicopy plasmids does not cure failure of *bcy1* cells to arrest as unbudded cells following starvation and to enter the GO state (thermotolerance, synthesis of unique GO proteins). This indicates that the pathway downstream of the AC/PK cascade branches to control meiosis through *IME1* and to control entry into GO and cell cycle initiation independently of *IME1* (Matsuura *et al.*, 1990).

The behavior of multicopy plasmids carrying various segments from the *IME1* gene region suggests that the region upstream of *IME1* contains both positive and negative regulatory sites. Control of *IME1* by the environment and by the MAT pathway acts through negative regulatory sites (Granot *et al.*, 1989). The deduced Ime1 protein is similar to known protein kinases, suggesting that protein phosphorylation is required for the initiation of meiosis.

Cameron *et al.* (1988), however, have isolated mutant *TPK* genes that suppress all of the *bcy1* defects. The mutant *TPK* genes appear to encode functionally attenuated catalytic subunits of cAMP-dependent protein kinase. The *bcy1* yeast strains containing the mutant *TPK* genes respond approximately to nutrient conditions, even in the absence of *CDC25*, both *RAS* genes, or *CYR1*. Together, these genes encode the known components of the cAMP-generating machinery. The results suggest that cAMP-independent mechanisms exist for regulating sporulation. However, this mechanism is not clear.

VI. Phosphatidylinositol Signaling

The metabolism of polyphosphoinositides has been shown to be an important factor in controlling the proliferation of yeast. The responses of mammalian cells to external signals are commonly mediated by intracellular second messengers, among which are the breakdown products of phosphatidylinositol (PI) 4,5-bisphosphate (PIP_2): 1,2-diacylglycerol (DG) and inositol 1,4,5-trisphosphate (IP_3) (Berridge, 1987; Nishizuka, 1984a,b; Rhee *et al.*, 1989). The recent insights in PI signaling was reviewed by Marjerus *et al.* (1990).

In yeast, addition of glucose to starved cells resulted in a rapid increase in the concentrations of ATP, PIP, and PIP_2 (Talwalker and Lester, 1973). Because inositol phosphoceramide exists at a high concentration in yeast, it is difficult to separate from PIP and PIP_2 from inositol phosphoceramide using the method developed for mammalian cells. One must take care with these substances (Smith and Lester, 1974), especially in an experiment using ^{32}P labeling. The monophosphate form of PI has been assumed to be

PI-4-phosphate (PI4P). Recent evidence has established that a PI kinase, which phosphorylates the D3 position of the inositol ring (PI3 kinase), is associated with many activated protein tyrosine kinases and may play an important role in the signaling of cell proliferation in mammalian cells. To determine the evolutionary conservation of this enzymatic activity, Auger *et al.* (1989) investigated its presence in yeast. PI3-kinase activity was present. Preliminary biochemical characterization of the activity suggested that it was different from the mammalian enzyme yet catalyzed the same reaction, that is, phosphorylating the D-3-hydroxyl position of the inositol ring of phosphatidyl-*myo*-inositol. ^{3}H-Inositol labeling of intact yeast cells with the subsequent extraction, deacylation, and high performance liquid chromatography analysis of the lipids demonstrated that PI-3-P was as abundant as the PI-4-P isomer. The conservation of this enzymatic activity from yeast to humans suggests that it has an important functional role in the cell cycle, as observed in mammalian cells. Further, PIP_2 and IP_3 were detected using the Whatman Sax column (I. Uno *et al.*, unpublished observations), and the phospholipase c gene was also isolated.

Probes derived from cDNA-encoding isozymes of rat protein kinase C (PKC) were used to screen the genome of the budding yeast. A single gene *(PKC1)* was isolated that encodes a putative protein kinase closely related to the alpha, beta, and gamma subspecies of mammalian PKC (Levin *et al.*, 1990). Deletion of *PKC1* resulted in recessive lethality. Cells depleted of the *PKC1* gene product displayed a uniform phenotype, a characteristic of cell division cycle mutants, and arrested cell division at a point subsequent to DNA replication but prior to mitosis. Unlike most *cdc* mutants, which continue to grow in the absence of cell division, *PKC1*-depleted cells arrested growth with small buds. *PKC1* may regulate a previously unrecognized checkpoint in the cell cycle. Using staurosporine, an inhibitor of protein kinase C, several kinds of staurosporine- and temperature-sensitive mutants were isolated. One of them, *stt1* DNA complemented *stt1* mutation, was cloned and sequenced. It was identified that *stt1* gene is an allele of the *PKC1* gene (Yoshida *et al.*, 1992). Protein kinase C was detected in the yeast with bovine myelin basic protein as the phosphate receptor (Ogita *et al.*, 1990). The enzyme was purified at least 500-fold by a four-step column chromatographic procedure and acitvated by the simultaneous addition of Ca^{2+}, diacylglycerol, and phosphatidylserine. Free arachidonic acid alone could activate the enzyme to some extent. However, yeast PKC did not respond significantly to tumor-promoting phorbol esters. The yeast enzyme showed substrate specificity distinctly different from that of mammalian PKCs. H1 histone and protamine were poor substrates. With myelin basic protein as a model substrate, yeast PKC preferentially phosphorylated threonyl residues, whereas rat brain PKCs mainly phosphorylates seryl residues. Further, they apparently showed multiple species of yeast PKC, although a proteolytic artifact cannot be ruled out until structural analysis is done. Although the activation signal of yeast PKC is presently unknown, further enzymological and genetic studies may elucidate the physiological role of this protein kinase pathway in growth and cell cycle control.

The relationship between cAMP- and phosphoinositide-mediated signal transduction systems has been studied by examining alteration of inositol phospholipid metabolism in cAMP mutants. Formerly indicated *pim1* and *pim2* genes were not directly related to PI metabolism (Uno *et al.*, 1988), and *pim3* to *pim5* mutations were alleles of *cyr1, ras2,* and *bcy1*, respectively (Uno and Ishikawa, 1990). The incorporation of ^{32}PPi into PIP and PIP_2 was markedly reduced in *cyr1-2* and *ras2,* which produced low levels of cAMP, and increased in *bcy1,* which produced cAMP-independent protein kinase. The addition of cAMP to *cyr1-2* caused a significant increase in ^{32}PPi incorporation into PIP and PIP_2 (Kato *et al.*,1989). Both activities of PI kinase and PIP kinase in membranes were low in *cyr1-2* and *ras2* but high in *bcy1* and *ras1, ras2, bcy1*. The addition of cAMP to *cyr1-2* caused the activation of PI and PIP kinase. These data indicate that cAMP-dependent phosphorylation enhances PIP_2 synthesis through activation of PI kinase and PIP kinase, which may lead to the enhanced production of DG and IP_3. Thus, a close relationship between cAMP- and phosphoinositide-mediated signal pathways has been demonstrated, but mutant analyses suggest that a step that is required for cAMP to forward the cell cycle is different from that for a PIP_2-mediated step. These data indicated the existence of cross-talk between cAMP and PI cascades. A glycoprotein is attached to lipid bilayer through inositol-containing phospholipids (Conzelmann *et al.*, 1988), as observed in mammalian cells, but the biological function of this protein is unknown.

As indicated, the PI cascade system may exist in yeast cells but may not be well known, so that molecular and biochemical analyses were required for understanding.

VII. Ca^{2+}-Related Cascade

In yeast Ca^{2+}-dependent signaling system may exist as indicated next. Calmodulin was purified and characterized (Ohya *et al.*, 1987; Davis and Thorner, 1989). The molecular properties of this calmodulin are essentially similar to those of calmodulin of higher eukaryotes but differ slightly from those of calmodulin in higher eukaryotes. Calmodulin gene was cloned by using synthetic oligonucleotides. Disruption or deletion of the yeast calmodulin gene results in a recessive lethal mutation; thus, calmodulin is essential for growth (Davis *et al.*, 1986). As found in mammalian cells, *Cam*-dependent protein kinase II activity was detected (Lui *et al.*, 1991) and purified from yeast cells (Londesborough, 1989). *Cam*-dependent protein kinase I and II genes were cloned and sequenced (Pausch *et al.*, 1991), and their characters are similar to those of mammalian cells. Further, putative Ca^{2+}-binding proteins were observed; *CDC31* (Baum *et al.*, 1986) and *CDC24* (Miyamoto *et al.*, 1987) gene products may be Ca^{2+}-binding protein, judging from the putative amino acid sequence of these proteins and related with spindle pole body duplication and bud emergence, respectively. Further, Bender and Pringle (1989) screened yeast genomic DNA libraries for heterologous genes that, when overexpressed from a plasmid, can suppress a temperature-sensitive *cdc24* mutation to identify other genes that may be involved in the bud emergence. One of these proved to be *CDC42,* which has previously been shown to be a member of the rho family of genes, and a second is a newly identified ras-related gene, *RSR1,* which is not essential for growth.

These data suggest that the Ca^{2+}-signaling pathway regulates cell proliferation and cell growth in yeast.

VIII. Tyrosine Kinase

In mammalian cells several kinds of receptor-type tyrosine kinase were observed and involved in the signal transduction system. In yeast Schieven *et al.* (1986) demonstrated a soluble activity capable of phosphorylating both exogenous substrates, such as poly(Glu^{80} Tyr^{20}) copolymer, and unknown endogenous yeast proteins on tyrosine. Furthermore, the presence of a kinase activity in yeast plasma membrane capable of phosphorylating casein on tyrosine has also been reported (Castellanos and Mazon, 1985). Dailey *et al.* (1990) found the extracts of yeast contain tyrosine kinase activity that can be detected with a synthetic Glu-Tyr copolymer as substrate. By using this assay in conjunction with ion-exchange and affinity chromatography, a soluble tyrosine kinase activity was purified more than 8000-fold from yeast extracts. The purified activity did not utilize typical substrate for mammalian tyrosine kinases (enolase, caseine, and histones). The level of tyrosine kinase activity at all steps of each preparation correlated with the content of a 40-kDa protein (p40). Upon incubation of the most highly purified fractions with Mn-ATP or Mg-ATP, p40 was the only protein phosphorylated on tyrosine. Immunoblotting of purified p40 of total yeast extracts with antiphosphotyrosine antibodies and phospho amino acid analysis of ^{32}P-labeled yeast proteins fractionated by SDS PAGE indicated that the 40-kDa protein is normally phosphorylated at tyrosine *in vivo*. ^{32}P-labeled p40 immunoprecipitated from extracts of metabolically labeled cells by affinity-purified anti-p40 antibodies contained both phosphoserine and phosphotyrosine. The gene encoding p40 (YPK1) was cloned from a yeast genomic library by using oligonucleotide probes designed on the basis of the sequence of purified peptides. As deduced from the nucleotide sequence of *YPK1*, p40 is homologous to known protein kinases, with features that resemble known protein serine kinase more than known tyrosine kinases. Thus, p40 is a protein kinase that is phosphorylated *in vivo* and *in vitro* at both tyrosine and serine residues; it may be a novel type of autophosphorylating tyrosine kinase, a bifunctional (serine tyrosine-specific) protein kinase, or a serine kinase that is a substrate for an associated tyrosine kinase.

However, the physiological function of tyrosine kinase in yeast is not known but may be involved in the signal transduction system as found in mammalian cells.

IX. Concluding Remarks

The responses of yeast cells to external signals are commonly mediated by intracellular second messengers, such as cAMP, and the breakdown products of phosphoinositides. An important approach is to use mutants defective in the signal transduction systems as the critical controls for the biochemical analyses of the cascade reactions related to the second messenger. The mutant analysis on the cAMP–cascade system revealed that the cAMP-dependent protein kinase is involved in the phosphorylation events of specific proteins that need to regulate growth and differentiation of yeast. In addition to the cAMP-dependent protein kinase, other protein kinases have been found in yeast. Several kinds of protein kinase regulate their activities by phosphorylation. For instance, the protein kinase chain reaction system works as a network that regulates cellular functions.

The complexity of systems regulated by protein phosphorylation depends not only on the number of protein kinase but also on the number of substrates that they can phosphorylate. It is known that the activities of several enzymes are regulated by cAMP-dependent phosphorylation and dephosphorylation mechanisms in yeast (Matsumoto *et al.*, 1985). These enzymes include trehalase, NAD-dependent glutamatedehydrogenase, fructose-1, 6-bisphosphatase, glycogen phosphorylase, glycogen synthetase, and RNA polymerase, but their exact roles on growth and differentiation in yeast have not yet been elucidated. Further work should be focused on the identification of the nature of the proteins that are phosphorylated by cAMP-dependent protein kinase for growth and differentiation in yeast cells.

As mammalian cells PI and tyrosine kinase signal cascades work in yeast. Thus, the signal transduction systems of yeast cells may essentially be the same as those of mammalian cells.

References

Aelst, L. V., Jans, A. W. H., and Thevelein, J. M. (1991). *J. Gen. Microbiol.* **137,** 341–349.

Arndt, K. T., Styles, C. A., and Fink, G. R. (1989). *Cell* **56,** 527–537.

Auger, K. R., Carpenter, C. I., Cantley, L. C., and Varticovski, L. (1989). *J. Biol. Chem.* **264,** 20181–20184.

Baserga, R., Costlow, M., and Rovera, G. (1973). *Fed. Proc.* **32,** 2115–2118.

Baum, P., Furlong, C., and Byers, B. (1986). *Proc. Natl. Acad. Sci. U.S.A.* **83,** 5512–5516.

Belazzi, T., Wagner, A., Wieser, R., Schanz, M., Adam, G., Hartig, A., and Ruis, H. (1991). *EMBO J.* **10,** 585–592.

Bender, A., and Pringle, J. R. (1989). *Proc. Natl. Acad. Sci. U.S.A.* **86,** 9976–9980.

Berridge, M. J. (1987). *Annu. Rev. Biochem.* **56,** 159–193.

Boorstein, W. R., and Craig, E. A. (1990). *EMBO J.* **9,** 2543–2553.

Botsford, J. L. (1981). *Microbiol. Rev.* **45,** 620–642.

Boutelet, F., Petitjean, A., and Hilger, F. (1985). *EMBO J.* **4,** 2635–2641.

Boy-Marcotte, E., Garreau, H., and Jacquet, M. (1987). *Yeast* **3,** 85–93.

Boynton, A. L., and Whitfield, J. F. (1983). *Adv. Cylic Nucleotide Res.* **15,** 193–294.

Broek, D., Toda, T., Mitchell, T., Levin, L., Birchmeler, C., Zoller, M., Powers, S., and Wigler, M. (1987). *Cell* **48,** 789–799.

Cameron, S., Levin, L., Zoller, M., and Wigler, M. (1988). *Cell* **53,** 555–566.

Camonis, J. H., Kalekine, M., Gondre, B., Garreau, H., Boy-Marcotte, E., and Jacquet, M. (1986). *EMBO J.* **5,** 375–380.

Casperson, G. F., Walker, N., and Bourne, H. R. (1985). *Proc. Natl. Acad. Sci. U.S.A.* **82,** 5060–5063.

Castellanos, R. M. P., and Mazon, M. J. (1985). *J. Biol. Chem.* **260,** 8240–8242.

Chardin, P. (1988). *Biochimie* **70,** 865–868.

Cherry, J. R., Jhonston, R. R., Dolland, C., Shuster, J. R., and Denis, C. L. (1989). *Cell* **56,** 409–419.

Cobitz, A. R., Yim, E. H., Brown, W. R., Perou, C. M., and Tamanoi, F. (1989). *Proc. Natl. Acad. Sci. U.S.A.* **86,** 858–862.

Cohen, P. (1989). *Annu. Rev. Biochem.* **58,** 453–508.

Colicelli, J., Field, J., Ballester. R., Chester, N., Young, D., and Wigler, M. (1990). *Mol. Cell. Biol.* **10,** 2539–2543.

Comb, M., Brinberg, N. C., Seasholtz, A., Herbert, E., and Goodman, H. M. (1986). *Nature (London)* **323,** 353–356.

Conzelmann, A., Riezman, H., Desponds, C., and Bron, C. (1988). *EMBO J.* **7,** 2233–2240.

Cross, F., Roberts, J., and Weintraub, H. (1989). *Annu. Rev. Cell Biol.* **5,** 341–395.

Dailey, D., Schlieven, G. I., Lim, M. Y., Marquardt, H., Gilmore, T., Thorner, J., and Martin, G. S. (1990). *Mol. Cell. Biol.* **10,** 6244–6256.

Damak, F., Boy-Marcotte, E., Le-Roscouet, D., Guilbaud, R., and Jacquet, M. (1991). *Mol. Cell. Biol.* **11,** 202–212.

Daniel, J., and Simchen, G. (1986). *Curr. Genet.* **10,** 643–646.

Daniel, J., Becker, J. M., Enari, E., and Levitzki, A. (1987). *Mol. Cell. Biol.* **7,** 3855–3861.
Davis, T., and Thorner, J. (1989). *Proc. Natl. Acad. Sci. U.S.A.* **86,** 7909–7913.
Davis, T., Urdea, M. S., Masiarz, F. R., and Thorner, J. (1986). *Cell* **47,** 423–431.
Defeo-Jones, D., Scolnick, E., Koller, R., and Dhar, R. (1983). *Nature (London)* **306,** 707–709.
Defeo-Jones, D., Tatchell, K., Robinson, L. C., Sigal, I. S., Vass, W. C., Lowy, D. R., and Scolnick, E. M. (1985). *Science* **228,** 179–184.
Engelberg, D., Simchen, G., and Levitzki, A. (1990). *EMBO J.* **9,** 641–651.
Fedor-Chaiken, M., Deschenes, R. J., and Broach, J. D. (1990). *Cell* **61,** 329–340.
Field, J., Broek, D., Kataoka, T., and Wigler, M. (1987). *Mol. Cell. Biol.* **7,** 2128–2133.
Field, J., Nikawa, J., Broek, D., Macdonald, B., Rogers, L., Wilson, I. A., Lerner, R. A., and Wigler, M. (1988). *Mol. Cell. Biol.* **8,** 2159–2165.
Field, J., Vojtek, A., Ballester, R., Bolger, G., Colicelli, J., Ferguson, K., Gerst, J., Kataoka, T., Michaeli, T., Powers, S., Riggs, M., Rodgers, L., Wieland, I., Wheland, B., and Wigler, M. (1990a). *Cell* **61,** 319–327.
Field, J., Xu, H.-P., Michaeli, T., Ballester, R., Sass, P., Wigler, M., and Colicelli, J. (1990b) *Science* **247,** 464–466.
Friedman, D. L., Johnson, R. A., and Zeilig, C. E. (1976). *Adv. Cyclic Nucleotide Res.* **7,** 69–114.
Fujiyama, A., and Tamanoi, F. (1986). *Proc. Natl. Acad. Sci. U.S.A.* **83,** 1266–1270.
Fujiyama, A., Matsumoto, K., and Tamanoi, F. (1987). *EMBO J.* **6,** 223–228.
Garrett, S., and Broach, J. (1989). *Genes Dev.* **3,** 1336–1348.
Garrett, T., Menold, M. M., and Broach, J. A. (1991). *Mol. Cell. Biol.* **11,** 4045–4052.
Gerst, J. E., Ferguson, K., Vojtek, A., Wigler, M., and Field, J. (1991). *Mol. Cell. Biol.* **11,** 1248–1257.
Gibbs, J. B., and Marshall, M. S. (1989). *Microbiol. Rev.* **53,** 1171–1185.
Gilman, A. G. (1984). *Cell* **36,** 577–579.
Gonzalez, G. A., and Montminy, M. R. (1989). *Cell* **59,** 675–680.
Gonzalez, G. A., Yamamoto, K. K., Fischer, W. H., Karr, D., Menzel, P., Biggs, W., III, Vale, W. W., and Montminy, M. R. (1989). *Nature (London)* **337,** 749–752.
Gottesman, M. M. (1980). *Cell* **22,** 329–330.
Granot, D., Margolskee, J. P., and Simchen, G. (1989). *Mol. Gen. Genet.* **218,** 308–314.
Heideman, W., Casperson, G. F., and Bourne, H. R. (1990). *J. Cell. Biochem.* **2,** 229–242.
Hoeffler, J. P., Meyer, T. E., Yun, Y., Jameson, J. L., and Habener, J. F. (1988). *Science* **242,** 1430–1433.
Holland, K. M., Homann, M. J., Belunis, C. J., and Carman, J. M. (1988). *J. Bacteriol.* **170,** 828–833.
Jones, R. H., and Jones, N. C. (1989). *Proc. Natl. Acad. Sci. U.S.A.* **86,** 2176–2180.
Kassir, Y., and Simchen, G. (1976). *Genetics* **82,** 187–206.
Kassir, Y., and Simchen, G. (1988). *Cell* **52,** 853–862.
Katada, T., and Ui, M. (1982). *Proc. Natl. Acad. Sci. U.S.A.* **79,** 3129–3133.
Kataoka, T., Powers, S., McGill, C., Fasano, O., Strathern, J., Broach, J., and Wigler, M. (1984). *Cell* **37,** 437–445.
Kataoka, T., Broek, D., and Wigler, M. (1985a). *Cell* **43,** 493–505.
Kataoka, T., Powers, S., Cameron, S., Fasano, O., Goldfarb, M., Broach, J., and Wigler, M. (1985b). *Cell* **40,** 19–26.
Kato, H., Uno, I., Ishikawa, T., and Takenawa, T. (1989). *J. Biol. Chem.* **264,** 3116–3121.
Kinney, A. J., and Carman, G. M. (1988). *Proc. Natl. Acad. Sci. U.S.A.* **85,** 7962–7966.
Kore-eda, S., Murayama, T., and Uno, I. (1991). *Jpn. J. Genet.* **66,** 317–334.
Krupinski, J., Coussen, F., Bakalyar, H. A., Tang, W.-J., Feinstein, P. G., Orth, K., Slaughter, C., Reed, R. R., and Gilman, A. G. (1989). *Science* **244,** 1558–1564.

Levin, D. E., Fields, F. O., Junisawa, R., Bishop, J. M., and Thorner, J. (1990). *Cell* **62,** 213–224.

Londesborough, J. (1989). *J. Gen. Microbiol.* **135,** 3373–3383.

Londesborough, J., and Suoranta, K. (1983). *J. Biol. Chem.* **258,** 2966–2972.

Lui, Y., Ishii, S., Takai, M., Tsutsumi, H., Ohki, O., Akeda, R., Tanaka, K., Tsuchiya, E., Fukui, S., and Miyakawa, T. (1991). *Mol. Gen. Genet.* **227,** 52–59.

Marjerus, P. W., Ross, T. S., Cunningham, T. W., Caldwell, K. K., Jefferson, A. B., and Bansal, V. S. (1990). *Cell* **63,** 459–465.

Marshall, M. S., Gibbs, J. B., Scolnick, E. M., and Sigal, I. S. (1988). *Mol. Cell. Biol.* **8,** 52–61.

Martegani, E., Baroni, M., and Wanoni, M. (1986). *Exp. Cell Res.* **162,** 544–548.

Masson, P., Jacquemin, J. M., and Culot, M. (1984). *Ann. Microbiol. (Paris)* **135,** 343–351.

Masson, P., Lenzen, G., Jacquemin, J. M., and Dnachin, A. (1986). *Curr. Genet.* **10,** 343–352.

Matsumoto, K., Uno, I., Oshima, Y., and Ishikawa, T. (1982). *Proc. Natl. Acad. Sci. U.S.A.* **79,** 2355–2359.

Matsumoto, K., Uno, I., and Ishikawa, T. (1983a). *Exp. Cell Res.* **146,** 151–161.

Matsumoto, K., Uno, I., and Ishikawa, T. (1983b). *Cell* **32,** 417–423.

Matsumoto, K., Uno, I., and Ishikawa, T. (1984). *J. Bacteriol.* **157,** 277–282.

Matsumoto, K., Uno, I., and Ishikawa, T. (1985). *Yeast* **1,** 15–24.

Matsuura, A., Treinin, M., Mitsuzawa, H., Kassir, Y., Uno, I., and Simchen, G. (1990). *EMBO J.* **9,** 3225–3232.

Mitchell, A. P., and Herskowitz, I. (1986). *Nature (London)* **269,** 738–742.

Mitsuzawa, H., Uno, I., Oshima, T., and Ishikawa, T. (1989). *Genetics* **123,** 739–748.

Mitts, M. R., Grant, D. B., and Heideman, W. (1990). *Mol. Cell. Biol.* **10,** 3873–3883.

Miyamoto, S., Ohya, Y., Ohsumi, Y., and Anraku, Y. (1987). *Gene* **54,** 125–132.

Montminy, M. R., and Bilezikjian, L. M. (1987). *Nature (London)* **328,** 175–178.

Montminy, M. R., Sevarino, K. A., Wagner, J. A., Mandel, G., and Goodman, R. H. (1986). *Proc. Natl. Acad. Sci. U.S.A.* **83,** 6682–6686.

Morishita, T., and Uno, I. (1991). *J. Bacteriol.* **173,** 4533–4536.

Nikawa, J., Cameron, S., Toda, T., Ferguson, K. M., and Wigler, M. (1987a). *Gene Dev.* **1,** 931–937.

Nikawa, J., Sass, P., and Wigler, M. (1987b). *Mol. Cell. Biol.* **7,** 3629–3636.

Nishizuka, Y. (1984a). *Science* **225,** 1365–1370.

Nishizuka, Y. (1984b). *Nature (London)* **308,** 693–698.

Ogita, K., Miyamoto, S., Koide, H., Iwai, T., Oka, M., Ando, K., Kishimoto, A., Ikeda, K., Fukami, Y., and Nishizuka, Y. (1990). *Proc. Natl. Acad. Sci. U.S.A.* **87,** 5011–5015.

Ohkura, H., Kinoshita, N., Miyatani, S., Toda, T., and Yanagida, M. (1989). *Cell* **57,** 997–1007.

Ohya, Y., Uno, I., Ishikawa, T., and Anraku, Y. (1987). *Eur. J. Biochem.* **168,** 13–19.

Panek, A. C., Araujo, P. S. D., Neto, V. M., and Panek, A. D. (1987). *Curr. Genet.* **11,** 459–465.

Papageorge, A. G., Defeo-Jones, D., Robinson, P., Temeles, G., and Scolnick, E. M. (1984). *Mol. Cell. Biol.* **4,** 23–29.

Pausch, M. H., Kaim, D., Kunisawa, R., Admon, A., and Thorner, J. (1991). *EMBO J.* **10,** 1511–1522.

Powers, S., Kataoka, T., Fasano, O., Goldfarb, M., Strathern, J., Broach, J., and Wigler, M. (1984). *Cell* **36,** 607–612.

Powers, S., Michaelis, S., Broek, D., Anna-A, S. S., Field, J., Herskowitz, and Wigler, M. (1986). *Cell* **47,** 413–422.

Powers, S., O'Neil, K., and Wigler, M. (1988). *Mol. Cell. Biol.* **9,** 390–396.

Pringle, R. J., and Hartwell, L. H. (1981). *In* "The Molecular Biology of the Yeast *Saccharomyces*. I: Life Cycle and Inheritance" (J. N. Strathern, E. W. Jones, and J. R. Broach, eds.), Cold Spring Harbor Lab., Cold Spring Harbor, New York, pp. 97–142.

Resnick, R. J., and Racker, E. (1988). *Proc. Natl. Acad. Sci. U.S.A.* **85,** 2474–2478.

Rhee, S. G., Suh, P.-G., Ryu, S.-H., and Lee, S. Y. (1989). *Science* **244,** 546–550.

Rittenhouse, H. I., Haarrsch, P. B., Kim, J. N., and Marcus, F. (1986). *J. Biol. Chem.* **261,** 3939–3943.

Robinson, L. C., Gibbs, J. B., Marshall, M. S., Sigal, I. S., and Tatchell, K. (1987). *Science* **235,** 1218–1221.

Robison, G. A., Butcher, R. W., and Suterland, E. W. (1971). "Cyclic AMP." Academic Press, New York.

Rodbell, M. (1980). *Nature (London)* **84,** 17–22.

Ruggieri, R., Tanaka, K., Nakafuku, M., Kaziro, Y., Toh-e, A., and Matsumoto, K. (1989). *Proc. Natl. Sci. U.S.A.* **86,** 8778–8782.

Sass, P., Field, J., Nikawa, J., Toda, T., and Wigler, M. (1986). *Proc. Natl. Acad. Sci. U.S.A.* **83,** 9303–9307.

Schieven, G., Thorner, J., and Martin, G. S. (1986). *Science* **231,** 390–393.

Shilo, V., Simchen, G., and Shilo, B. (1978). *Exp. Cell Res.* **112,** 241–248.

Shin, D.-Y., Matsumoto, K., Iida, H., Uno, I., and Ishikawa, T. (1987). *Mol. Cell. Biol.* **7,** 244–250.

Shin, D.-Y., Uno, I., and Ishikawa, T. (1988). *Curr. Genet.* **12,** 577–582.

Smith, M. E., Dickinson, J. R., and Wheals, A. E. (1990). *Yeast* **6,** 53–60.

Smith, S. W., and Lester, R. L. (1974). *J. Biol. Chem.* **249,** 3395–3405.

Suoranta, K., and Londesborough, J. (1984). *J. Biol. Chem.* **259,** 6964–6971.

Suzuki, N., Choe, H.-R., Nishida, Y., Yamawaki-Kataoka, Y., Ohnishi, S., Tamaoki, T., and Kataoka, T. (1990). *Proc. Natl. Acad. Sci. U.S.A.* **87,** 8711–8715.

Takai, Y., Yamamura, H., and Nishizuka, Y. (1974). *J. Biol. Chem.* **249,** 530–535.

Talwalkar, R. T., and Lester, R. L. (1973). *Biochim. Biophys. Acta* **306,** 412–421.

Tanaka, K., Matsumoto, K., and Toh-e, A. (1988). *EMBO J.* **7,** 495–502.

Tanaka, K., Matsumoto, K., and Toh-e, A. (1989). *Mol. Cell. Biol.* **9,** 757–768.

Tanaka, K., Nakafuku, M., Satoh, T., Marshall, M. S., Gibbs, J. B., Matsumoto, K., Kaziro, Y., and Toh-e, A. (1990a). *Cell* **60,** 803–807.

Tanaka, K., Nakafuku, M., Tamanoi, F., Kaziro, Y., Matsumoto, K., and Toh-e, A. (1990b). *Mol. Cell. Biol.* **10,** 4303–4313.

Tatchell, K., Robinson, L. C., and Breitenbach, M. (1985). *Proc. Natl. Acad. Sci. U.S.A.* **82,** 3785–3789.

Taylor, S. S., Buechler, J. A., and Yonemoto, W. (1990). *Annu. Rev. Biochem.* **59,** 971–1005.

Thevelein, J. M., Beullens, M., Honshoven, F., Hoebeeck, G., Detremerie, K., Hollander, J. A. D., and Jans, A. W. H. (1987a). *J. Gen. Microbiol.* **133,** 2191–2196.

Thevelein, J. M., Beullens, M., Honshoven, F., Hoebeeck, G., Detremerie, K., Griewel, B., Hollander, J. A. D., and Jans, A. W. H. (1987b). *J. Gen. Microbiol.* **133,** 2197–2205.

Toda, T., Uno, I., Ishikawa, T., Powers, S., Kataoka, T., Broek, D., Cameron, S., Broach, J., Matsumoto, K., and Wigler, M. (1985). *Cell* **40,** 27–36.

Toda, T., Cameron, S., Sass, P., Zoller, M., Scott, J. D., McMullen, B., Hurwitz, M., Krebs, E. G., and Wigler, M. (1987a). *Mol. Cell. Biol.* **7,** 1371–1377.

Toda, T., Cameron, S., Sass, P., Zoller, M., and Wigler, M. (1987b). *Cell* **50,** 277–287.

Toda, T., Cameron, S., Sass, P., and Wigler, M. (1988). *Gene Dev.* **2,** 517–527.

Tripp, M. L., Pinon, R., Meisenhelder, J., and Hunter, T. (1986). *Proc. Natl. Acad. Sci. U.S.A.* **83,** 5973–5977.

Tripp, M. L., Bouchard, R. A., and Pinon, R. (1989). *Mol. Microbiol.* **3,** 1319–1327.

Uno, I. (1988). *Jpn. J. Genet.* **63,** 471–493.

Uno, I., and Ishikawa, T. (1990). *In* "Calcium as a Second Messenger in Eukaryotic Microbes" (D. H. O'Day, ed.). Am. Soc. Microbiol., Washington, D.C.

Uno, I., Matsumoto, K., and Ishikawa, T. (1982). *J. Biol. Chem.* **257,** 14110–14115.

Uno, I., Matsumoto, K., Adachi, K., and Ishikawa, T. (1983). *J. Biol. Chem.* **258,** 10867–10872.

Uno, I., Matsumoto, K., and Ishikawa, T. (1984). *J. Biol. Chem.* **259,** 1288–1293.

Uno, I., Matsumoto, K., Hirata, A., and Ishikawa, T. (1985a). *J. Cell Biol.* **100,** 1854–1862.

Uno, I., Mitsuzawa, H., Matsumoto, K., Tanaka, K., Oshima, T., and Ishikawa, T. (1985b). *Proc. Natl. Acad. Sci. U.S.A.* **82,** 7855–7859.

Uno, I., Mitsuzawa, H., Tanaka, K., Oshima, T., and Ishikawa, T. (1987). *Mol. Gen. Genet.* **210,** 187–194.

Uno, I., Fukami, K., Kato, H., Takenawa, T., and Ishikawa, T. (1988). *Nature* (*London*) **333,** 188–191.

Verma, R., Iida, H., and Pardee, A. B. (1988a). *J. Biol. Chem.* **263,** 8569–8575.

Verma, R., Iida, H., and Pardee, A. B. (1988b). *J. Biol. Chem.* **263,** 8576–8582.

Yamawaki-Kataoka, Y., Tamaoki, T., Choe, H.-R., Tanaka, H., and Kataoka, T. (1989). *Proc. Natl. Acad. Sci. U.S.A.* **86,** 5693–5697.

Yoshida, S., Ikeda, E., Uno, I., and Mitsuzawa, H. (1992). *Mol. Gen. Genet.* In press.

Young, D., Riggs, M., Fieled, J., Vojtek, A., Broek, D., and Wigler, M. (1989). *Proc. Natl. Acad. Sci. U.S.A.* **86,** 7989–7993.

INDEX

D

E

F

G

H

I

K

R

S

W

Y

Z

ISBN 0-12-364542-5